TERRA NOVA

TERRA NOVA

The New World After Oil, Cars, and Suburbs

ERIC W. SANDERSON
Wildlife Conservation Society

Abrams, New York

To my son, Everett

This was the object of the Declaration of Independence. Not to find out new principles, or new arguments, never before thought of, not merely to say things which had never been said before; but to place before mankind the common sense of the subject, in terms so plain and firm as to command their assent, and to justify ourselves in the independent stand we are compelled to take.

Letter from Thomas Jefferson to Henry Lee (May 8, 1825)

List of Illustrations

Part I
The Siren Song

Oil

Cars

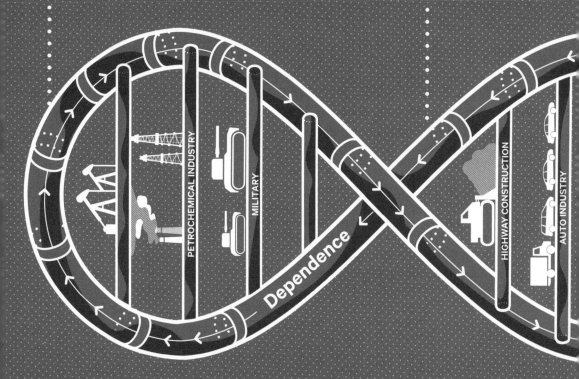

PETROCHEMICAL INDUSTRY

MILITARY

Dependence

HIGHWAY CONSTRUCTION

AUTO INDUSTRY

Figure No. 1
The connections among oil, cars, and suburbs fueled the twentieth-century American economy but now reinforce a lifestyle dependent on long commutes, gas-fed automobiles, and the energy in oil.

Suburbs

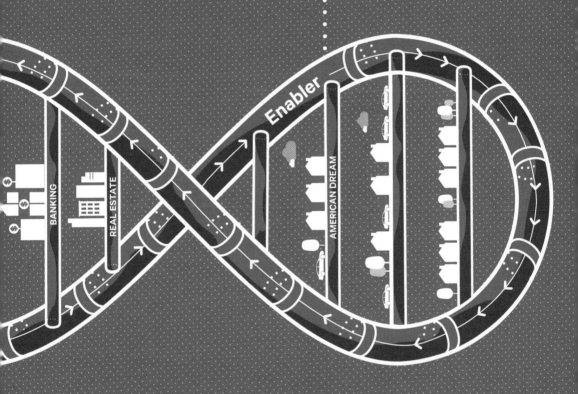

BANKING

REAL ESTATE

Enabler

AMERICAN DREAM

1

How the Sirens Sing

All these things have thus come to an end.

Homer, *The Odyssey*

First of all, you'll run into the Sirens.
They seduce all men who come across them.
Whoever unwittingly goes past them
and hears the Sirens' call never gets back.
His wife and infant children in his home
will never stand beside him full of joy.
No. Instead, the Sirens' clear-toned song
will captivate his heart.

Homer, *The Odyssey*, Book 12

On his long journey home after the Trojan War, the hero Odysseus came to an island where the goddess Circe advised him to avoid the Sirens, beautiful winged monsters whose irresistible song lured mariners to their death. Forewarned but undaunted, Odysseus sailed into peril anyway. His plan: He would listen but not give in. The wily hero packed his men's ears with beeswax and commanded them to tie him to the ship's mast. There he stood as they sailed into treacherous waters; the Sirens called to him, and he heard their song. As Circe had predicted, he longed to go to them, to cast away everything he held dear. He shouted at his men, ordered, then begged them to set him free, but the mast was strong, the rope held fast, and his men couldn't hear his pleas. And so Odysseus did not perish, but emerged on the other side of the Siren song wiser, saner, and prepared to complete his journey home.

Like Odysseus's less fortunate peers, Americans have been hearkening to a Siren call. Monsters have been singing to us for decades, and we have found their music persuasive, beautiful, and often irresistible, even though we know that it beckons us to our own destruction. Despite our best intentions, oil, cars, and suburbs have become the modern American Sirens.

You know the dangers well; they are in the news practically every night, and have been a generation or more. Oil brings us hatred from the people who have the wells; wars in the Middle East to protect the supply; economic shocks and cycles of unemployment, inflation, and foreclosure; poisoned air and waters; and a climate altered by carbon released from its millennial slumber underground. Cars isolate us in metal boxes, discourage us from exercise, expose us to accidents and sudden death, and squander our time on congested freeways, while each year requiring more roads to fragment the landscape, entomb farm fields, cleave neighborhoods, and drain our collective coffers in servitude to lifeless prairies of asphalt. Suburbs, originally conceived as garden cities, are criticized for their monotony, segregation, sprawl, outrageous property taxes, obsequious service to retailers, and aesthetic, social, and cultural barrenness, but in my book the problem with suburbs as currently constituted is that they require automobiles, and automobiles require oil: Suburbs force us to drive.

Oil, cars, and suburbs sing to us constantly and harmoniously. Their song bridges history and landscape to appeal to our identity as a free nation; we hear it on the television and we say to ourselves: We wouldn't be Americans without these things. So the trap is laid. A seduction composed of our own desires, the song is composed of the choices and actions of our parents and grandparents and sustained by our own choices and actions. These things—oil, cars, suburbs—are not monsters on their own, but only monsters as we have made them, and because of what they make us do.

They were never meant to be such a difficulty; they were intended to provide joy, freedom, and wealth. For most of the twentieth century, connecting these three buoyed the American Dream of a better material life for every generation. So successful was the combination that many Americans began to think oil-cars-suburbs *was* the American Dream, confusing means with ends. Yet it seems that these same means have now run their course: For the first time since the founding of the nation, the quality of life in America is in decline; the price, perhaps, of too much of a good thing.

Signs of an era's end are all around us, yet blithely we continue as Americans have done since the time of William Howard Taft and Theodore Roosevelt. The advantages of oil as an energy source are obvious and attractive—especially when it's cheap and abundant. Autos have been a mark of personal liberty for at least a century. A variety of interests represent the detached, single-family home on a cul-de-sac as the epitome of life's ambitions. Oil, cars, and suburbs sing to us all the time.

Not Listening

I have to say I have been caught humming the song, too. I'm a kid from the suburbs, like a hundred million others; gas stations, mini-malls, and sprawling ranch homes defined the landscape of my youth in northern California.* Some of my earliest memories are of squabbling with my brother in the back of the family Ford on the way to the store or the swim club or the innumerable other errands for which driving was the only practical alternative. My family had a modest suburban house near a creek, constructed where an orchard had been. It was two miles from our house to the closest store, a long walk with groceries but just a quick jaunt in the car—whether it was Dad's Rambler, Mom's Ford, Dad's Mercury, Mom's Volvo, Dad's Acura, Mom's Subaru, or Dad's Lexus. I marked my youth by the cars my parents drove.

When I was in high school, my grandparents in rural Colorado offered to help me buy a car. Every young man needed a car, they said. That was especially true where they lived, where it was twenty miles in twenty minutes to the closest town. But I was more interested in my Uncle Larry's dusty and forlorn fifty-four–volume set of the

---

*Here and periodically throughout the book I reflect on incidents in my own life, which in many ways is standard issue for an American born and raised in the late twentieth century. These anecdotes are scarcely interesting in themselves but are included because they illustrate how the Siren song affects us all.

Great Books of the Western World in the den. My grandparents thought I was crazy to spend Friday nights reading John Locke and Adam Smith; my brother and sister suspected it was a ploy to avoid carting them around town.

Later when I went to college at the University of California, Davis, I bought my own car, a beat-up, used blue 1977 Volvo station wagon. My friends and I affectionately named her Brünnhilde, leader of the Valkyrie. She took me home on weekends and let me escape to the Sierra Nevada mountains in a few hours; but around town, I discovered I rather liked to ride my bike. Davis was one of the few places in the country where the weather, terrain, and traffic engineering had conspired to make a bicyclists' paradise. Weeks would pass when the only transportation I needed, I provided by my own pedal power. It was a local freedom, but a sweet one nonetheless.

In the meantime, my parents' marriage ended and, eventually, my dad sold the house. I jokingly offered to take it off his hands in the early 2000s; the combination of his mortgage and his property taxes (kept low by Proposition 13 in California) was less than the rent for my one-bedroom apartment. Dad declined. In the end the house sold for twenty-four times what my parents had bought it for back in 1969, a goodly appreciation of 400 percent even after adjusting for inflation.

About the same time, I emerged from graduate school penury to take a job in New York City, coaxing my decrepit but beloved blue Volvo across the country. I almost left my bike behind, but at the last minute, threw it back in to discover when I reached the Bronx that there were bike lanes from the island where I lived to the green park where I worked, six and a half miles away. Though the climate wasn't as amenable to it, I found my Davis-like wheeled existence could be transplanted to the big city, too; in fact it was facilitated by how close everything was. Riding a bike in the Bronx meant paying more attention to personal safety but it worked better than I thought it would (especially when conjoined with subway trips downtown).

I came to the Bronx as an ecologist to work for the Wildlife Conservation Society (the Bronx Zoo's parent organization), a New York City cultural institution with a century-long dedication to wildlife and wild places around the world. My task was to bring technical aspects of modern geography into its global mission to save tigers, elephants, whales, gorillas, and other charismatic megafauna (the big critters everyone loves and can't imagine a world without). I knew more than a person rightly should about GPS, satellite imagery, geographic information systems, and other techniques spoken of mainly in acronyms. In graduate school, I had learned how to dig deep and patiently into data to see what patterns pertained and ask questions that led to questions that led to other questions. More than most disciplines, ecology thrives on complexity, and ecology in the service of conservation (a subdiscipline called conservation biology) pulls one rapidly into the domains of economics, society, and politics. It was—and is—exciting, heady work.

One of the first projects I undertook for my employer was to make a new kind of map of the world by combining computer-rendered versions of human population density (the number of people living in a place), land use (represented by agricultural fields and urban areas), roads and other transportation networks, and lights detectable by a satellite at night. We found that 83 percent of the earth's land surface was directly influenced by humanity according to one or more of these measures; 98 percent of the places where it's possible to grow rice, wheat, and corn had already been touched by humanity. Rolling off the plotter, the map seemed to blink with digital solemnity: The frontier was gone.

We called our map "the human footprint," but a better name might have been "the human tire track," after the numerous roads crisscrossing Africa, Asia, Europe, Oceania, and the Americas. We electronically painted the map red, black, and purple where there were a lot of people, and forest green where there was the least human influence—in the wild places, the places that we were trying to save. The suburbs were easily identifiable by their pink fleshy color, inflammation around the wine-dark cities, one of many signs of a planet's domestication.

Against the Mast

Still I was lulled by the Siren song. Despite the "tire track," the traffic, the reports of climate change, the speeders on the parkways, the dreadful cost of car insurance, the dead animals on the road shoulder, the digits racing past on the pump—all of these seemed just the price that had to be paid for people to get to work, and since they were tallied on such different accounts, I hardly put them together. I didn't see oil, cars, and suburbs for the interlocking, mutually reinforcing system that they are, touching nearly everything I cared about in both my professional and personal lives.

Looking back, the moment that finally brought me to the mast, the shock that forced me to put the pieces together, was September 11, 2001, the day that planes hijacked by terrorists crashed into the World Trade Center in New York, the Pentagon in Washington, and a farm field in Pennsylvania.

The attacks brought home to me the reality of the American presence in the Middle East and the hostility it had engendered. Our collective and my personal complacency shattered, no one could mistake the vengeance the American republic meted out in return, measured in warplanes and tanks, bombs and missiles, black operations, extraordinary renditions, enhanced interrogations, and drone strikes a world away. We spent trillions of dollars, sacrificed the health and lives of thousands of American soldiers, and killed or saw killed tens of thousands of others, not only in Afghanistan, which launched the terrorists, but also in Iraq, whose link to 9/11 was tenuous at best. Yet strangely, throughout a decade of death and mayhem, we left untouched the origins of Osama Bin Laden's wealth and hatred in Saudi Arabia.

I wondered why it all happened so. One fact stood out above the rest: American dependence on oil fields around the Persian Gulf. Presidents had been speaking of it since the 1970s, from Richard Nixon to Barack Obama, with hardly any effect. We said we would, and we did fight one overt war to protect that oil (the Gulf War of 1991), and we were clearly prepared to fight another, and we did. Oil—and the necessity of protecting it on the other side of the world at all costs—necessitated battling monsters.

But I wanted to know why. Why oil? Why there? With a kind of bookish patriotism, I started reading, and I started calculating. As you will have surmised, questions about oil transmuted into questions about cars and transportation, and those grew into other questions about suburbs and land use, what we value, how and why we value it, and who we are as a people and a country. Did I discover something that had never been said? No. Did I come to realize how much oil-cars-suburbs predicated the shocks and disasters of my time? Yes, I did. What I learned comprises the first part of this book.

In the process, I also learned to listen more carefully to the music of the economy. I heard in the Siren song two themes intertwined: a motif of energy and an anthem of profit, the twin currencies of the natural and human economies, respectively. I discovered that if I ever forgot one or the other in my search, I would quickly become lost in murky waters, drowned in minutiae, distracted by the schools of red herrings that confuse the otherwise ineluctable relationships among oil, cars, and suburbs.

If the first part of this book is an attempt to explain how the Siren song came to be and why it draws us so strongly to a doom we know but can't seem to avoid, the second half is an effort to describe a new way of life, a promise beyond oil, cars, and suburbs, designed to sustain American prosperity, health, and freedom for generations to come. Can you imagine? The daily news often seems so grim and intractable, and the monumentality of our investment in the current model so overwhelming, that many Americans have difficulty conceiving of another way of life. But other ways are possible, even preferable. The goal is to imagine a future without also having to appeal to some miraculous technological fix. Some marvel might be in the offing, but let's not depend on it for our nation's welfare and security. Rather, let us conceive how we can do better with what we already have, by making some rearrangements and banishing some old false presumptions about the way the world works. As we shall see, many of the solutions are already within our grasp; a new form of the American Dream is already being dreamt across a nation rooted in a land of wealth and opportunity.

But first we need to shake off the old nightmares. Think of this book as the mast on Odysseus's ship. Think of yourself as the cunning hero. Ready yourself to hear the Sirens sing. And then relax. My aim is not to make you feel bad for having to drive your car to work or for wanting a house with a garden out of town. Rather, this book is about eluding a trap that we have made together and that, together, we can unmake.

Let us dream of a new world: America after oil, cars, and suburbs.

2

An Ode to Oil

Truth is like the sun.
You can shut it out for a time,
but it ain't goin' away.

Elvis Presley

The Siren song begins with an ode to oil. Oil comes from nature. By *nature* I mean the interactions of soil and rock, air and water, energy and life, that characterize our verdant planet; and by *natural*, I mean the qualities of everyone and everything participating in the great congress of life, including you and me. When we burn oil, the products of combustion are released to nature, where they mix again with air, water, energy, and life. If we want to understand why our culture, politics, and economy are so dependent on oil, we first need to understand where it naturally comes from and where it naturally goes. We need to look up, to the sun and the sky.

The Celestial Campfire

The sun is the ultimate source of all energy on earth, whether it's used by grass in the fields, trees in the forest, or your car on the road. Though poets might prefer a more evocative comparison, astrophysicists liken the sun to a nuclear fusion reactor. Astronomers observe that the sun's diameter is more than one hundred times larger than the earth's, and it is unimaginably hot—nearly 15 million degrees Celsius at its center. Within that heat, the sun packs enormous pressure; the core is forty-three times denser than a diamond. Under these extreme conditions four protons slamming together make one helium atom through nuclear fusion. When that happens, about 0.7 percent of the mass of the protons is turned into energy ($E = mc^2$), and about 0.000000045 percent of that energy eventually comes flying in our direction in the form of sunlight. That doesn't sound like a lot, but it's enough to power all life on earth, and more. In fact, the energy in sunlight arriving on earth contains about twelve thousand times more energy than humanity uses in a year.

Sunlight is made of photons, each of which carries a small packet of power. Although each photonic pedestrian literally travels at the speed of light, inside the sun it travels no more than a centimeter on average before bumping into something else—think of a Fifth Avenue sidewalk on a sunny spring day, but with a hundred million more occupants. With each bump, each photon loses a small quantum of energy. Depending on the number of bumps, the photon originating at the sun's core will take anywhere from 10 to 170 thousand years to leave its crowded precincts. Once a photon escapes, it takes only eight minutes to travel the 93 million miles to earth.

The sun has been the campfire around which earth and the other planets have warmed since the solar system formed from a gassy stellar nursery 4.6 billion years ago. After the sun's first billion years of brightening, enough light was reaching the earth that life had a chance to get started. Fast forward 2.5 or 3 billion years, and multicellular life finally managed to find a way to live on the planet. Initially life lived mainly in the oceans, where it developed some neat tricks.

Some of the ancestral organisms came up with a way to convert the sunlight into food. That trick is called photosynthesis. Photosynthesis takes energy from the sun, combines it with molecules of carbon dioxide (CO_2) from the atmosphere and water (H_2O) from the earth, and uses it to form carbohydrates (like sugar and starch).

Over time, other metabolic processes developed, which turned those carbohydrates into proteins, fats, DNA, and the other molecules of life; in other words, and eventually, into you and me. Our bodies—and the bodies of nearly all living things, from trees to elephants to plankton—are built from the energy of the sun trapped through photosynthesis and carried by carbon-based molecules. The process isn't terribly efficient—only about 0.3 percent of the sunlight that hits the earth is captured by living things, even today. But because the sun is generous with its energy, and sunlight has been coming for a very long time, life has evolved into the enormous range of forms that scientists call biodiversity. The sun, our celestial campfire, is not only the giver of life: It is what gives life form, agency, and action.

How Nature Makes Oil

Catching a fraction of the sun's energy is the most abundant kind of life on earth, one that we almost never see (a red tide or two notwithstanding): plankton. Plankton are microscopic creatures that float in the water, tossed and cradled by the waves. Examined under a microscope, they appear as delicate, crystalline structures, sometimes opaque, sometimes fully transparent. They include representatives of all major kinds of life from bacteria to archaea (like diatoms and protists), from plants to animals; some ocean giants, like tuna and cod, begin life in planktonic form. The name *plankton* comes from Greek, and means "errant" or "wanderer"—which is appropriate since plankton can't swim, but instead are swept along with the currents. Estimates vary, but somewhere between two-thirds and four-fifths of all biomass on earth is plankton; they are the broad platform of the food pyramid on which all other forms of aquatic life depend—the grass of the ocean plains. Many live for only a few days or weeks, reproduce, and then die, and when they die, they are either eaten by other organisms or sink to the bottom of the sea.

Some of the plankton that fall into the deep are covered in sediments before they decompose. Their silent burial is more likely to happen in the springtime, after winter storms have stirred nutrients from the depths up into the water column, and when swollen rivers carry sand and silt from land into large freshwater lakes or shallow seas, like the ones that once covered part of what we now call Pennsylvania, Texas, and Saudi Arabia. If the bodies of the wanderers are buried quickly enough, or if an oceanographic feature called a thermocline (a layer of warm water trapped above and below by cold water) forms, oxygen is excluded, decomposition prevented, and the plankton bodies begin yet another journey in the deep.

Buried under the ocean floor, geologic processes take over. Over millions of years, the sands and silts entrapping the plankton bodies are folded into the earth's rocky

Oil Begins with Plankton

A sample of the immense variety of the most abundant kind of life on earth: plankton. These tiny organisms floating in the ocean over 100 million years ago provided the natural resource (i.e., their dead bodies) from which today's oil was formed.

Source: Redrawn from Horsman (1985).

Meroplankton

Acanthochiasma spp.

Peridinium spp.

Oligotrich

Chain Diatom

Globigerina spp.

Halosphaera spp.

0   100μ (0.1 mm)   1 mm

How Oil Forms

The process of converting plankton to oil takes around 100 million years and involves six unlikely events. Plankton from the age of dinosaurs die in ancient seas and are then covered by sediments from ancient rivers. The plankton-enriched sediments slowly become buried in the crust of the earth, subducted to depths of 7,500–10,000 feet, where temperature and pressure cook them into crude oil. That oil, less dense than the surrounding rocks, migrates upward toward the surface, where some is captured in reser- voir rock surmounted by an impermeable cap, held in place for some lucky human to find.

Drawn from descriptions in Deffeyes (2006) and Head et al. (2003).

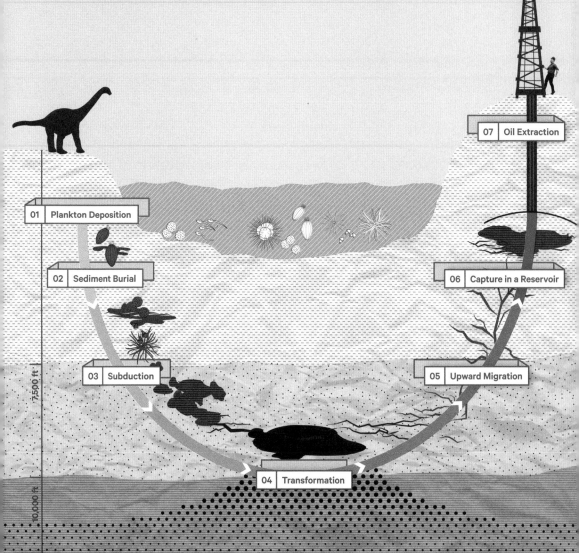

01 | Plankton Deposition

02 | Sediment Burial

03 | Subduction

04 | Transformation

05 | Upward Migration

06 | Capture in a Reservoir

07 | Oil Extraction

7,500 ft

10,000 ft

Geothermal Heat

100+ million years

*Note: Not drawn to scale.

mantle, deeper into the planet's crust. It is hot in that rocky tomb, being closer to the center of the earth, and the pressures are immense. When depths exceed 7,500 feet and temperatures rise in the range of 120–320 degrees Fahrenheit, the molecules that once made up the plankton cells begin to change. They break and simplify; they loop; they re-form, link up, and complicate—and if things go just right they eventually become the mixture we call crude oil. The carbohydrates and other molecules formerly known as plankton have been converted to hydrocarbons.

The entire process is remarkably inefficient, with energy given away at every step. Not only is life poor at capturing the rain of photons from the sun, but only a minuscule portion of all the plankton that have ever lived has actually been trapped in sediments. And then the cooking has to be just right. If the depth is too great and the heat too high, instead of oil, dead plankton becomes natural gas, a mixture of simple, volatile hydrocarbons that caused the gushers of Texan lore; if the rocks don't get cooked enough, we get thick tars (also called "oil sands"), like the ones currently being extracted in Alberta.

If oil does form, it begins to ooze upward—now being liquid and less dense than the rocks around it—sometimes all the way to the surface, in which case most is lost to evaporation, oxidized, or eaten by bacteria—some 90 percent of all the oil ever formed has been lost that way. The other 10 percent is caught in a trap somewhere near the surface, in what's known as reservoir rock, geological formations pierced with billions of tiny interconnected pores, and capped off with denser, impermeable rocks above.

All this explains why oil and natural gas are so rare—at least six unlikely events playing out over eons are needed to create these fossil fuels: plankton deposition, sediment burial, subduction (the burying of organic materials in the earth's crust), transformation at the right temperature and pressure, upward migration, and capture in a reservoir. And then humans need to discover the reservoir and safely extract the oil. Because any one of those events is unlikely, the combination of them is much more unlikely. By one estimate, less than 0.1 percent of land and continental shelves of our enormous planet holds oil, and the distribution of those places is highly irregular. Oil is always an unexpected treasure.

The End of Oil?

Given its strange and unlikely origins, oil—and other fossil fuels, made through similar processes from different organic materials—cannot be renewed, at least in our lifetimes, indeed in the lifetime of our species. While the human species has been on earth for only the last million years or so, the energy supply you're burning as you move your car down the highway formed at least 100 million years ago, perhaps longer. It will be millions of years again before oil is made from the plankton of today.

What's more, because of how it is made by nature, oil is radically and unfairly distributed around the world. Geology does not care a whit for the factors of human history—the personalities, politics, and conflicts that shape the boundaries of nations on the surface. As a result some countries sit atop enormous reservoirs of oil, while other nations have none.

These factors make it very difficult to say how much oil there is, which has led to endless debates about whether we are running out of oil or not. Estimates of "proved reserves" are essentially informed guesses—often reflecting political facts as much as geological ones. For example, in Saudi Arabia, proved reserves jumped by 85 billion barrels in the 1989–90 run-up to the Gulf War (no new fields were announced), and have remained strangely flat ever since, even though the Saudis pump over three billion barrels per year. Although we might not know how much oil is left, what we can say with some certainty is how much we have already used, because the people who do the discovering and the recovering keep records.

One such record keeper was the outspoken petroleum geologist M. King Hubbert, of Shell Oil Company. He knew that people had been looking for oil in earnest for nearly a century when he took up the search in the 1950s and that oil's economic value ensured that each generation had brought its best technology to bear on the problem. Hubbert also knew that for all practical purposes, no new oil was being created, so he wondered: What if the rate of discovery of oil is related to the amount of oil left undiscovered? In other words, he reasoned, if the amount of oil is finite, then for each oil field discovered, there would be one less field left unfound.

Imagine clamming on the seashore. The first people to start digging find clams quickly and easily, but those who come later have to expend more time and effort to find the same number, because fewer remain to be found. The rate of discovery declines regardless of the number of clams under the beach to begin with, and if you track that decline until the rate of discovery reaches zero, you can estimate how many clams the beach once held. (The analogy isn't perfect, because clams grow back if left alone, faster than oil can form, but I hope you get the idea.)

Using some mathematical reasoning based on this concept, Hubbert predicted in 1956 that oil production would peak between 1965 and 1970 in the United States for the simple reason that by then most of the oil out there to be found would have already been found, given the observed rate of discovery. In fact oil production in the United States did peak in 1970. In subsequent work and with some additional data, Hubbert predicted that global discoveries would peak around the turn of the millennium, and he was not far off—global production has, in fact, begun to level off. (Hubbert's model did not include the effect of economic recessions that have slowed down the rate of oil exploitation.)

Hubbert drew a graph of oil discovery—past, present, and future. He rather famously stretched out the timeline for thousands of years before and after the year 2000, to emphasize that the oil dependence of modern society was and will be necessarily

Hubbert's Insight

Because it takes nature so long to produce oil from plankton, it is of necessity a finite resource, at least on human time scales. Shell Oil petroleum geologist M. King Hubbert tried to make this point through math, scientific reports, conference presentations, and eventually this graph, which shows the rate of commercial extraction over the last 2,000 years. He stretched the x-axis another 2,000 years to suggest that a time might come when we could no longer depend on oil.

———————

Source: US Energy Information Administration (2012a), Etemand & Luciani (1991), US Energy Information Administration (2012c). Values prior to 1860 and for 2012 are estimated.

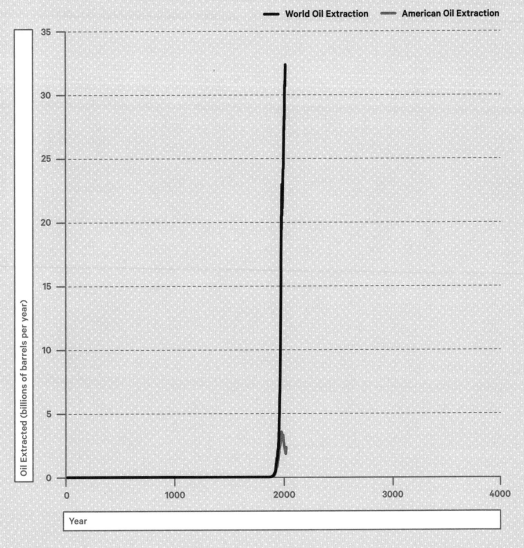

The Powerful Country:
US Oil and Natural Gas Fields

After 150 years of looking, we have a fair idea where the oil and gas is (or was). The geography of oil distribution has no regard for politics, but does follow closely the shorelines of the ancient seas that once submerged the middle part of the country.

Source: Redrawn from Biewick (2008).

Active Oil and
Natural Gas Fields

Dead Holes
Gas
Oil
Oil + Gas

of relatively short duration—likely only 150–200 years or so in the long history of humanity, which itself is only a brief blip in the long history of the earth. His graph looked like a steep mountain, and has since become known as Hubbert's Peak. It's led to the expression "peak oil," referring to the point in time when we cross from the upward side to the downward side of the graph.

Hubbert's work does not say for certain that there is not some enormous oil field yet to be found in the world; it just says that after 150 years of searching, it's highly unlikely. You wouldn't bet your house on winning the lottery; similarly Hubbert and other scientists suggest it is also ill-advised to bet the fate of modern industrial civilization on perpetual oil consumption.

Decline of course is not the same as absence. Being near the top of the peak means we still have a lot of oil to burn. In fact, having surmounted Hubbert's Peak implies we have a bit less than halfway still to go, the oil executives must surely note, rubbing their hands together. In the next period of oil sales, prices will continue to turn in favor of those who have the oil and against those who do not, with ever-greater profits to be made. Peak prices are the significance of peak oil, not the end of oil per se.

Past-peak pricing is ubiquitous in America today. For example it is the most important, but least reported, factor in the Deepwater Horizon disaster in 2010. The Deepwater Horizon platform was drilling 13,000 feet below the ocean bottom in mile-deep waters in the Gulf of Mexico, on the edge (or "horizon") of what is technically possible, because the oil on land and in shallower waters had already been tapped. Profit, however, outweighed the hazard, until the accident occurred. Similarly the current monumental efforts to extract the oil sands in Canada and exploit shale gases in the United States—and the emerging international competition over the Arctic Ocean depths, conveniently unveiled by melting ice—are all parables of the easy oil being gone. These "newly found" resources are exploitable, at great cost, because the price of oil has risen to a point where the markets deem that the risk is worth taking.

Black Gold

Given our current appetite for oil, it might seem strange that it has not always been the fountain of profit that it is today. In fact, for most of history, oil was just another of the strange and curious gifts of nature, like peacocks or lodestones, interesting but hardly worth the fuss. As Daniel Yergin, the energy consultant and historian, writes in his Pulitzer-winning encyclopedic history of oil, *The Prize: The Epic Quest for Oil, Money and Power,* in ancient times people scooped oil from natural seeps in the Middle East to plug things up. The pharaohs imported bitumen, a naturally occurring form of asphalt, from the Dead Sea to seal mummies in their tombs. Noah's ark and Moses's basket were probably caulked with tar to make them waterproof. Later the Byzantines mixed crude oil with lime to make Greek fire, a flammable mixture, to fling at terrified Crusaders from Europe. Pliny, the Roman doctor, in 79 AD prescribed oil to heal wounds, treat cataracts, and cure aching teeth.

The man who made oil a modern economic resource was the nineteenth-century American polymath George Bissell. Over a long and rambling career, Bissell was a teacher of Latin and Greek at Dartmouth College, a journalist in Washington D.C., a superintendent of schools in New Orleans, and an attorney in New York City. In 1853, traveling north to New Hampshire through remote western Pennsylvania, he observed a motley group of people collecting oil by soaking blankets in seeps along a river named Oil Creek. They called the goo "rock oil" or sometimes "Seneca oil," after the local Indians, and used it primarily as medicine. Seeing a vial of oil when he returned to Dartmouth, and knowing it was flammable, Bissell had an idea that would change the world: He wondered if rock oil could be used to make light.

He was not the only one who wondered. In the mid-nineteenth century most artificial lighting came from burning whale oil and beeswax candles. With just a little bit of pressure from the evolving industrial economy of the time, the worldwide population of whales was being rapidly depleted, causing a run on prices and a search for new sources. As scientists on both sides of the Atlantic were experimenting with obtaining oil from coal, tar, and other sources, Bissell hired Benjamin Silliman, Jr., a chemistry professor at Yale University, to run some experiments using a novel analytical technique called fractional distillation. Silliman showed that boiled rock oil could be distilled into kerosene, a substance only recently described, which when burned in the right kind of lamp, created a bright, golden light that competed successfully with oil boiled from whale's blubber. The whales were saved from extinction and a new commodity was born.

There was still the problem of getting the oil out of the ground in sufficient quantities to make it profitable; collecting the drippings from oil-soaked blankets was not going to do it. While pausing in the shade one hot day in New York City, Bissell had another flash of genius. Looking into the window of a pharmacy, he noticed the label of a patent medicine that showed a derrick used to drill for salt water (a common practice in China, recently imported to the United States). Bissell's mind leapt forward to a ludicrous thought—what if you could drill for oil, too?

To put his plan into motion, he formed a company in New Haven, Connecticut, and enlisted investors. One of those investors happened to know a slightly larger-than-life, recently unemployed railroad conductor named Edwin Drake. Drake knew nothing about oil or mining, but he did have a free railroad pass; Bissell and Co. offered him a job. To ensure his good reception, the company sent several letters addressed to "Colonel" Edwin Drake before his arrival. When he finally got to town after a difficult journey by rail and mail wagon, the newly esteemed "Colonel" was welcomed into the struggling metropolis of Titusville, Pennsylvania, population 125.

Drake leased an oozing black spot and convinced a blacksmith and his two sons to drill, offering a dollar per foot penetrated. Not a bargain, as the work was difficult and slow and no one knew if it would work. Drake started running out of money, and back east Bissell and the other investors were running out of hope. In August 1859 the company pulled the plug and wrote Drake an order to stop. But before the letter arrived, Drake and his men hit oil at sixty-nine feet down.

Drake's initial well supplied a stately fifteen barrels of oil per day, but that was enough; the rush was on. The Pennsylvania oil rush and subsequent crash set a pattern for oil that has subsequently been repeated many times in many places. Prices sky-rocketed as speculators and others rushed in to take advantage of the new discovery; derricks sprung up overnight; production overwhelmed the market; prices collapsed; and investments evaporated, all within a frantic eighteen months. Oil brought other dangers, too: The first gusher, a well where oil flowed freely, caught fire in April 1861, immolating nineteen people and blazing out of control for three days. Oil also brought the crowds: Despite the Civil War, which opened in South Carolina the same week as the inferno, Americans wanting a chance for riches streamed along the muddy roads to create instant towns in an area newly anointed Oildom, the Great Oildorado, Petrolia, or more simply, the Oil Regions.

Booms and busts were inevitable because coordination and conservation were entirely absent. Each driller rushed to achieve as much production as he individually could, which meant that the productivity of a field was less overall. Though Drake et al. didn't know it, oil is typically found under pressure from natural gas, a consequence of its formation. Inserting a well releases the pressure, drawing the oil to the surface. If everyone takes as much as possible as quickly as possible, this valuable pressure is exhausted, reducing the total amount of oil that can be obtained. Even today with greater knowledge and modern techniques, about half the oil in a commercial field remains underground after conventional extraction. Modern "unconventional" technologies try to force more oil out with combinations of chemical solvents, water, or other materials, which must be pumped in the millions of gallons, sometimes mixing with natural groundwater supplies. (Hydraulic fracturing, fueling an early-twenty-first-century oil and natural gas boom in the United States, is a related application, using pressures great enough to break rock to spring fuels from stingy reservoirs.)

The law of the time (and today) also encouraged producers to pump quickly. Oil discoveries were governed by the "rule of capture," developed in medieval England to resolve hunting disputes. It stated that if a deer or a bird moved from one estate to another, the latter estate's owner could kill the animal with no recompense to the former, for no one can say how and why deer or birds move; they are part of the commons shared by all. Similarly, landowners had the right to draw whatever wealth lay beneath it, even if drilling sucked their neighbor's property dry. As one English judge wrote, the rule of capture applied because no one really understood what happened in the "hidden veins of the earth." In effect, the rule of capture meant take as much as you can as fast as you can, before your neighbor does the same to you. Conservationists and economists know this problem as the "tragedy of the commons." It is a tragedy founded in ignorance.

Once the oil was aboveground, stored in old wine casks, and safely labeled as someone's private property, the next challenge was getting it to the closest refinery or railroad crossing. Wagons were slow, and Oil Creek didn't flow with enough force to transport the barrels efficiently. Someone suggested damming the creek, then breaking the dam to let the water go in a rush, as was sometimes done to move timber. Because containers were in short supply, oil workers also filled open wooden skiffs, "bathtubs of tar," as one historian describes them. In the flood, the barrels and skiffs would be carried downstream with much spillage, danger, and excitement; crowds would line the route to feel the cool breeze caused by rushing waters and to watch the fun. In the busiest weeks of the boom, two manufactured floods, or freshets, washed down Oil Creek per week. Today we still measure oil by the barrel, equivalent to forty-two gallons.

Eventually prices stabilized, pipelines replaced the floods, and refineries sprung up to distill the kerosene from the crude, in the oil regions and elsewhere. Alas for Drake! He lost his money in fruitless speculation and died in poverty, supported by a small pension from the state of Pennsylvania. George Bissell did better, moving to Oil City, at the confluence of Oil Creek and the Allegheny River. He turned his oil and land investments into a bank and a hotel, becoming a wealthy man and living out his later years comfortably in Manhattan.

The person who made real money was a Clevelander named Rockefeller. John D. Rockefeller's talent was to consolidate, integrate, and ruthlessly out-compete everyone else. Between 1870 and 1890 his Standard Oil Company captured 90 percent of the US market for oil, from wells to pumps, an early example of what later business analysts would call advantageous vertical integration. Theodore Roosevelt had another name for it: monopoly. When the Standard Oil trust was finally broken up in 1911, Rockefeller made even more money from the stock in the daughter companies he was forced to create. Rockefeller's stake made him the wealthiest man in the history of the world.

None of this could have happened until Yale's Professor Silliman distilled rock oil into kerosene and showed that it could be burned in a lantern, generating light again from oil, effectively reversing the process that began millions of years before in the sea. Silliman also discovered that other "waste" products were created when refining oil. Some of these compounds were too volatile or smoky to be used in lamps and would come to have other uses. One of those volatile compounds, if left alone, evaporated into a noxious gas, and so people started calling it gasoline.

3

Flexible Power

There is no doubt about our absolute and complete dependence upon oil. We have passed from the stone age, to bronze, to iron, to the industrial age, and now to an age of oil.

Harold Ickes, *Saturday Evening Post* (February 16, 1935)

Oil's utility, and its attraction, derives not only from its tremendous, pent-up energy, but also from what you can make from it, which is almost anything. As if realized from a long-lost formulary of medieval alchemy, much of modern industry is founded on the gifts that biology and geology provide via crude. How and why that is takes some explaining, including a brief lesson in—here I will apologize in advance— organic chemistry.

A Brief Chemistry Lesson

Because oil is created from formerly living things, just like those living things, it contains thousands of different organic chemicals based mainly on hydrogen and carbon (hence "hydrocarbons"). These different chemicals have elaborate, even elegant, structures, reflecting the many ways that carbon can be combined with itself and other elements. That's a good thing; it meant when the evolutionary process stumbled upon what we humans later called organic chemistry, life had a pliable putty to experiment with. Arrange the carbon and hydrogen (and a little bit of nitrogen, oxygen, and phosphorus) in the right way and you get DNA, which allows your father's good looks and your mother's intelligence to be passed on to you. Arrange the atoms a different way and you get the proteins that build your muscles and the carbohydrates that give you energy. Organic chemistry is what allows your eyes to see these words and your brain to process what they mean.

In fact, the strange arrangements of a very few kinds of atoms are exactly what make organic chemistry painful for undergraduates and fascinating for their professors. Subtle distinctions in the way a chain of carbons branches, whether the bond between two atoms is single or double, or even if the molecule twists toward the right or to the left make large differences in how that molecule will react with others. Put certain hydrocarbon compounds together and you get bubble bath; other combinations make a plastic garden chair, a candle, an aspirin pill, an insecticide, a road surface, or a bomb.

Producing these chemical arrangements in the laboratory is energetically expensive and difficult to achieve from scratch, which is why finding a ready-made source of complex organics in oil has been of critical importance to modern industry. It is the natural resource on top of which many different profit centers can be built. The constituents of oil each have their own distinctive chemical properties, not the least of which is the ability to hold energy. However, to use these different components, they need to be teased apart: Crude requires refinement.

Crude Refinement

If you remember back to chemistry class or watching Hawkeye make gin in his tent on the TV show *M*A*S*H*, one way to decompose a mixture is to slowly cook it over a flame. Different molecules boil at different temperatures, allowing them to be captured through distillation. (Hence a "still" is an apparatus for separation.) Hawkeye boiled a mixture of fermented grain, which yielded alcohol. The same process of separation can be applied to oil, but rather than the 200-plus degrees Fahrenheit required for gin, crude oil needs to be heated to over 600 degrees, yielding the refined mixtures called gasoline, kerosene, diesel fuel, and asphalt.

Not all crude is the same, of course. It varies in composition and appearance depending on age, origin, and geological circumstances. Some crude oils are as thin as water, and others are as thick as maple syrup; some are golden with traces of green, while others are black as the night. Trade names reflect provenance, like wine: West Texas Intermediate, Brent Blend, Dubai Crude, Bonny Light, Malaysian Tapis, or Midway Sunset Heavy connote to engineers sweetness, texture, and price.

Arriving by ship, truck, or pipeline, oil is pumped into the numerous oil refineries that dot the American shoreline from the Gulf of Mexico to the coast of California to the side of the New Jersey Turnpike. The bright lights, massed conduits, and smoking stacks of the oil refinery all have a singular purpose: to get as much value out of every barrel of oil as possible. Crude enters enormous distillation towers and is slowly brought to boil, heated by burning a stream of oil diverted for the purpose. Inside the tower are horizontal plates pierced with small holes. The rising gases pass through holes in the plates. The gases cool and condense into liquids as they drop below their boiling points, each at its own level; the dripping fractions are channeled off the tower into separate tanks. Smaller, simpler molecules with lower boiling points go right to the top: propane and butane, which fuel barbeque ovens and power plants. Middle fractions of middle-length molecules are drawn off to form motor fuels, including jet fuel, gasoline, and diesel, and the precursors of plastics and explosives. Heating oil and fuel oil for ships (bunker) comes off next, leaving at the bottom thick, heavy residuals, which are waxes for candles, lubricants for moving parts, and tar and asphalt for roads and roofs.

Refineries not only separate oil into fractions, they also "crack" the fractions—breaking long-chain hydrocarbons into shorter ones—and combine constituents to generate longer molecules, sometimes manufacturing new compounds that nature inconveniently fails to provide. In addition they remove unwanted residues, particularly the small portions of sulfur, nitrogen, and metals that promote air pollution, and dump them in landfills; refiners also add various emulsifiers, demulsifiers, detergents, dispersants, and anti-icing, anti-corrosion, anti-oxidation, anti-static, and anti-etc. agents to create the final products, which then leave the refinery by ship, truck, or pipeline, headed for the local gas station or chemical factory.

In the process, something magical happens: Even before the additives, more volume comes out of the distilling column than went in. For every forty-two-gallon

Figure No. 6 — **Organic Oil**

A barrel of crude, derived from plankton, contains thousands of elegant, complex, useful organic chemicals. Some of these compounds provide the precursors for soap, plastics, pesticides, pharmaceuticals, and bombs; others provide gasoline, jet fuel, and diesel.

Source: Redrawn from Downey (2009).

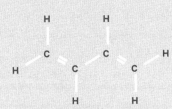

Butadiene (C4H6)

Toluene (C7H8)

Ethylene (C2H4)

Octane (C8H18)

Benzene (C6H6)

Methane (CH4)

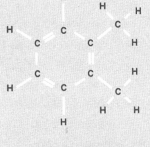

ortho-Xylene (C8H10)

Acetylene (C2H2)

barrel of oil processed in the United States, approximately forty-four gallons of refined products are produced. That's called refinery gain, and it happens because the products of refining have lower densities than crude oil.

In America most refining aims for the sweet spot, where the highest value lies. An average forty-two-gallon barrel of crude oil processed in a Gulf Coast refinery yields approximately twenty gallons of gasoline, ten gallons of diesel fuel, four gallons of jet fuel, and ten gallons of other components. (Remember the refinery gain.) Depending on the time of year, engineers steer the process to promote some fractions over others: Before and during the summer driving season, for example, refining works in "max gasoline mode," expecting that we will drive more; for most of the rest of the year, refiners focus on jet fuel and heating oil, to provide flights home and warm winter holidays. Meanwhile the "other components" are riding on the back of the market for transportation fuels. Ever wondered why plastic bags are so cheap they can be given away for free? Thank ExxonMobil.

Unintended Consequences

Of course free isn't always good, especially when it comes to the wonders of chemistry. The same organic chemistry that makes these remarkable products possible also means that these products can react with our similarly constituted organic bodies, sometimes in unintended ways; that's exactly why so many of them are irritating, toxic, and/or carcinogenic. DDT (dichlorodiphenyltrichloroethane) provides one of the most famous examples. Heralded as the greatest pesticide of its time in the early twentieth century, this product of the oil age killed insects without having (it was thought then) any known effects on people, even when children were doused with the stuff; soon, however, the effects of DDT were being felt downstream. Birds—particularly certain large birds—were disappearing all over the country. Slowly biologists discovered that DDT broke down in water into compounds that, when eaten by fish and then by birds that ate the fish, created a chemical that prevented the shells of bird eggs from thickening; in a tragedy worthy of myth, the eggs in the nest literally cracked under their mother's weight.

Who knew? In fact no one knew and no one would discover it today except by chance—how could you ever devise a test to predict such an indirect outcome? By the 1950s, bald eagles, which once flew in the hundreds of thousands, were down to fewer than five hundred breeding pairs in the lower forty-eight states. Many other species, including peregrine falcons, the fastest animal in the world when diving on prey, became critically endangered. After Rachel Carson's manifesto, *Silent Spring*, was published and Congress heard testimony about its effects, DDT was finally banned in 1972—but my point is not about what kind of chemical testing was or was not appropriate to protect us (or the birds) or what kind of policy should or should not have

How Crude Is Refined

It takes energy to refine crude into gasoline. Raw oil is boiled at over 600° F to separate all the fractions that fuel the industrial economy. While the main stream of oil enters the distillation tower, a small amount is diverted to feed the flames. The process of distilling gasoline in a typical US refinery is about 86 percent energy efficient.

Source: Redrawn from US Energy Information Administration (2010a); refinery efficiency from Wang et al. (2004).

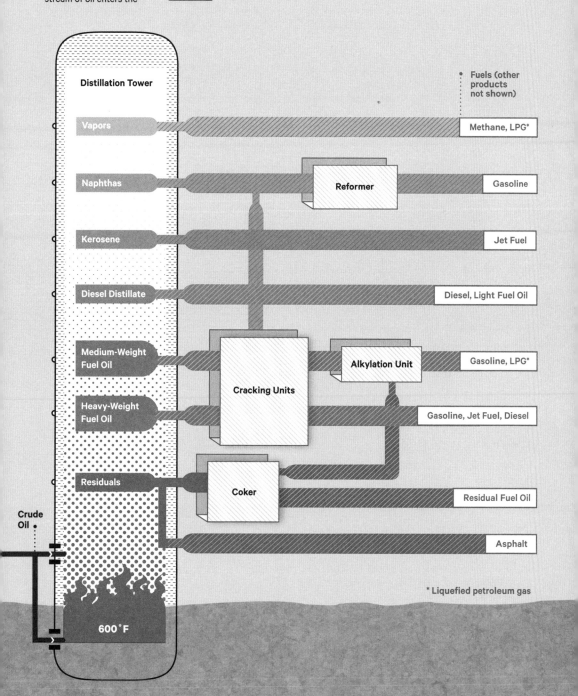

Distillation Tower

Fuels (other products not shown)

Vapors — Methane, LPG*

Naphthas — Reformer — Gasoline

Kerosene — Jet Fuel

Diesel Distillate — Diesel, Light Fuel Oil

Medium-Weight Fuel Oil — Cracking Units — Alkylation Unit — Gasoline, LPG*

Heavy-Weight Fuel Oil — Gasoline, Jet Fuel, Diesel

Residuals — Coker — Residual Fuel Oil

Asphalt

Crude Oil

600°F

* Liquefied petroleum gas

What Flows from a Barrel of Oil

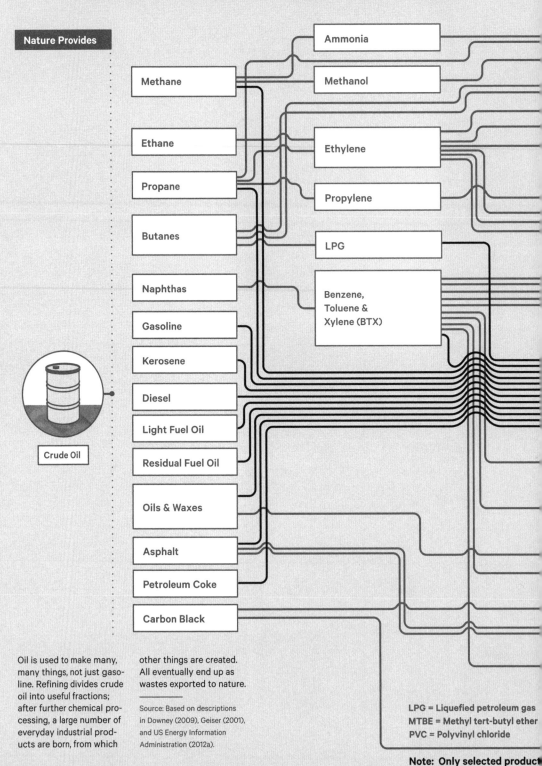

Nature Provides

Methane

Ethane

Propane

Butanes

Naphthas

Gasoline

Kerosene

Diesel

Light Fuel Oil

Residual Fuel Oil

Oils & Waxes

Asphalt

Petroleum Coke

Carbon Black

Crude Oil

Ammonia

Methanol

Ethylene

Propylene

LPG

Benzene, Toluene & Xylene (BTX)

Oil is used to make many, many things, not just gasoline. Refining divides crude oil into useful fractions; after further chemical processing, a large number of everyday industrial products are born, from which other things are created. All eventually end up as wastes exported to nature.

Source: Based on descriptions in Downey (2009), Geiser (2001), and US Energy Information Administration (2012a).

LPG = Liquefied petroleum gas
MTBE = Methyl tert-butyl ether
PVC = Polyvinyl chloride

Note: Only selected products and connections shown

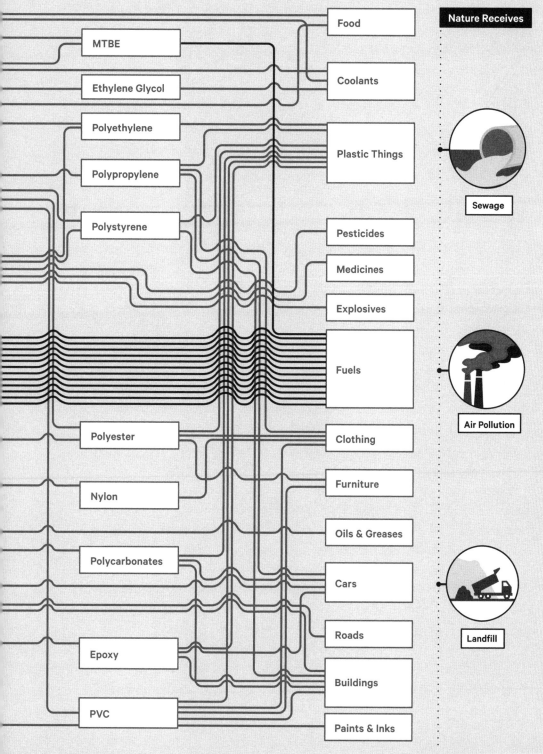

Oil for Fuel

Oil for Other Things

Food

MTBE

Coolants

Ethylene Glycol

Polyethylene

Plastic Things

Polypropylene

Polystyrene

Pesticides

Medicines

Explosives

Fuels

Polyester

Clothing

Nylon

Furniture

Oils & Greases

Polycarbonates

Cars

Roads

Epoxy

Buildings

PVC

Paints & Inks

Nature Receives

Sewage

Air Pollution

Landfill

been in place. Rather, the reason DDT affected bird populations in such a peculiar and unpredictable way is that DDT was ubiquitous—and that's because oil was ubiquitous. DDT would not have been so cheap and therefore so abundant if its precursors, acetaldehyde and benzene, were not readily available by-products of refineries aiming to make gasoline. Indeed many other chemicals, and many objects made from those chemicals, are so abundant and, relatively speaking, so cheap precisely because of the vast river of oil flowing through our economy, with consequences for the health and welfare of more than eagles.

Which makes it all the more ironic that oil, an alchemist's dream, more valuable than Harry Potter's sorcerer's stone, more magical than Merlin's staff, is hardly valued for the elaborate, amazing, flexible structures it contains. Rather, oil is valued for what is held by those structures: energy.

Joules, Watts, and MacKays

When oil is burned, it releases a lot of energy (and carbon dioxide and water, the reversal of photosynthesis)—in fact, the long hydrocarbon chains of oil are among the most energy-dense natural fuels known. Refining further concentrates the most energetic fractions to make that terrifically energy-demanding activity called driving possible. But what do we mean by "energy-dense" or "energy demanding"? How do we measure energy anyway? And what is it exactly?

Thus we embark on one of the most difficult passages of our journey together: the nonintuitive, practically mystical, nature of energy itself. The standard textbook definition describes energy as "that which changes the physical state of a system." Take the car in your driveway as a concrete example. The car is a system. Start the engine, accelerate down the street, and you have changed the physical state of the system from rest to motion. That change required energy—in this case energy from the sun, trapped in dead plankton, converted to oil, refined into gasoline, and activated by a spark in your car's engine. Your body is also a system; the energy in food allows it to move, grow, and think, changing your body's physical state. (Yes, thinking changes your physical state, as electrochemical impulses race along your neurons.)

In enabling these changes, energy is always conserved, meaning it is never created and never destroyed. Rather, energy changes form and increases its "disorder" with each change, which might seem odd, but it's the law. In fact this is one of the best demonstrated laws of the universe; the Second Law of Thermodynamics has never been violated as far as we know. In your car, the energy in gasoline changes from a chemical energy form to a kinetic energy form (i.e., motion), and also heat and noise; during breakfast, your cereal changes from toasted oats into you running for the train. Seventeenth-century scientists, just getting a handle on energy, called it the *vis viva*, the living force.

Although energy is never created or destroyed, it is not always used productively (as can be readily observed among a group of teenagers). When a driver guns a car's engine, some of the chemical energy of the gasoline is transformed into motion, but most is spent producing the roar of the engine and the heat of the exhaust. The heat is energy disordered, that is, less organized relative to the chemical energy that went in, which satisfies the thermodynamic principles, but no one else.

The ratio of the energy you productively get out to the energy you put in is called efficiency. There is no such thing as a perfectly efficient motor, which would have an efficiency of 100 percent. Most modern internal combustion engines are 18–25 percent efficient, meaning for every five units of energy put in, only one is actually used to get your car down the road. Although that doesn't sound great, it's better than most biological systems can do. Ecologists estimate as a rule of thumb that only about 10 percent of the energy in each exchange in a food chain is productively transmitted. This is why there are a lot of plants, some herbivores, and only a few predators in any given ecosystem; there just isn't enough energy to support many predators. It is also why omnivorous humans put more pressure on the rest of nature when we choose to eat meat over vegetables, because each pound of meat requires ten times more energy to produce than each pound of veggies. In fact one of the secrets of maximizing energy efficiency is to reduce the number of transformations energy makes from where it comes from to what you want to use it for: beans instead of chicken for dinner; south-facing windows rather than fossil fuels for heat; walking instead of driving; and so on.

Not surprisingly, it took many years to figure out how to measure "that which changes the physical state of a system." In the end a professional brewer and amateur physicist in Britain named James Prescott Joule came up with a solution, described in a paper from 1845. He conceived of a unit called the foot-pound, which was defined as the amount of energy required to lift a one-pound object one foot high, working against the gravitational pull of the earth. Joule showed that foot-pounds could also be used to measure the energy used in heating water or compressing a gas or discharging a battery—in effect showing that energy is the same, no matter what form it is in. Modern physicists recognized Joule's accomplishments by naming the basic unit of energy after him. A joule is defined as the application of a force of one newton (the international unit of force, named after Sir Isaac Newton) through a distance of one meter, equivalent to about three-quarters of a foot-pound.*

The energy-related unit one more commonly sees is named after the Scotsman James Watt. Watt developed the first reasonably efficient modern steam engine in 1776, after being given the task of repairing one of Thomas Newcomen's engines at the University of Glasgow. Newcomen's apparatus, developed initially in 1712, managed

---

*I have provided a table of unit conversions on page 334.

Energy Density of Various Fuels

Fuels vary dramatically in how much energy they contain. (Energy per weight is known as energy density; energy is "that which changes the physical state of a system.") Fuels derived from crude oil are notable for their high energy density—less than nuclear fuel rods and hydrogen, but little else. "Minutes microwaving" is an invented unit of energy that refers to how much energy a kitchen microwave uses running at full power for one minute.

Source: Compiled from Argonne National Laboratory (2012), Chevron Corporation (2006), McDonalds.com (2012), Shukla et al. (2001), and Smil (2007). Fossil fuel energy densities are based on higher heating values that include the energy committed to vaporization of water during combustion.

How Many Minutes Microwaving Does One Pound of Fuel Produce?

26,126,952
Enriched Uranium

1,072
Liquid Hydrogen

380
Propane

363
Jet Fuel/Kerosene

362
Conventional Gasoline

304
Biodiesel

226
Ethanol

181
Coal

156
Wood

81
Big Mac Sandwich

4
Lithium-Polymer Battery

3
Lithium-Ion Battery

1
Lead-Acid Battery

0.01
Water at 100m Dam Height

Energy Use for Various Tasks

Driving is remarkably energy intensive. Getting in your car, accelerating to 60 mph, and driving one mile uses approximately the same amount of energy as running the AC for a day and night. Because driving requires so much energy, trading time for space the way we do now requires an energy-dense fuel like gasoline.

Source: Google searches for typical consumer devices; see Chapter 11 for calculations of transportation energy.

Task	Power Consumption (watts)	Duration (mins)	Energy Requirement (mins microwaving*)
Household Tasks			
Heat a cup of water in the microwave for one minute	1,000	1	1.0
Use a tablet computer for ten hours	3	600	1.5
Turn on a 40 W electric light bulb for one hour	40	60	2.4
Listen to baseball game on radio for three hours	15	180	2.7
Work on a desktop computer with LCD monitor for one hour	115	60	6.9
Operate a vacuum cleaner for ten minutes	1,600	10	16.0
Watch three episodes of *Law & Order* on a big-screen TV	150	180	27.0
Bake batch of cookies in electric oven	3,000	20	60.0
Play game console on TV for four hours	290	240	69.6
Dry clothes in electrical tumble dryer	2,500	40	100.0
Keep food cool in refrigerator for one day	600	1,440	864.0
Run air-conditioner for 24 hours on a summer day	1,200	1,440	1,728.0
Driving			
Accelerate a passenger car from 0 to 60 mph	1,243,188	0.5	621.6
Drive a passenger car one mile at 10 mph	2,850	6	17.1
Drive a passenger car one mile at 30 mph	123,760	2	247.5
Drive a passenger car one mile at 60 mph	958,300	1	958.3

* One minute microwaving equals the energy used by a 1000 W microwave oven running for one minute.

only about one percent efficiency; Watt saw a way to improve it to 10 percent. That difference launched the Industrial Revolution. It finally allowed human beings to quasi-efficiently transform the energy contained in fossil fuels (initially coal, later oil) into useful motion. Watt deployed his device to pump water out of mines in England so that more coal could be dug out, and later his engine was adapted to other purposes, such as the railroad locomotive, which enabled the coal and people to get around. For his accomplishment, the unit of power, defined as one joule per second, is called a watt. When you put a 100-watt bulb in a lamp and turn it on, you know that the bulb will transform energy at the rate of 100 joules per second into light (and some heat) for you to use while reading.

The joule and the watt are thus complementary—the joule measures the amount of energy, and the watt measures the flow of that energy in time, whether during generation (as in a windmill converting wind energy into electricity) or consumption (a motor converting electricity into motion). If energy is like a stream, the joule measures the amount of water in the stream and the watt measures how fast that water is flowing by.

The problem with the joule as a unit of measurement is that it is impractically small, equivalent to the energy transferred from an apple's physical state to a physicist's head after falling one meter. Tracking such a small dollop of energy is not so useful when we are measuring car engines, windmills, or the energy expended by the US economy in a year. One could, and people do, use units like megajoules, gigajoules, terajoules, even exajoules, but more often, scientists come up with other units that more naturally fit what they are trying to measure, such as the kilocalorie (equivalent to 4,184 joules, good for food), the British Thermal Unit or BTU (equivalent to 1,055 joules, good for heating), and the quad, which is one quadrillion BTUs, or about 1 exajoule (10^{15} joules, good for economies).

For everyday uses of energy, a more practical unit is the kilowatt-hour (kWh), defined as the amount of energy that a one-kilowatt device (such as a 1000-watt microwave) would use in one hour. The kilowatt-hour is the standard unit of electrical power bills, equivalent to 3,600 joules, or the amount of energy used by your television displaying ten hours of programming, your laser printer producing 600 pages of text, or your car moving about one quarter of a mile. However, because microwaving is more familiar than "kilowatt-houring," I'll use the idea of "minutes microwaving" as our standard unit of energy from here on. Just remember that an hour heating in a small microwave is the same as a kilowatt-hour. Gasoline contains about 34 kWh of energy per gallon, which means that the energy in one gallon of gas could run your microwave for thirty-four hours straight! Gasoline is potent stuff.

To measure the flow of energy in time, I like the suggestion of David MacKay to use kilowatt-hours per day (1 kWh per day = 60 minutes of microwaving per day). MacKay, a physics professor at Cambridge University and advisor to the British government on energy and climate change policy, contributed a book of singular importance to those of us living in the twenty-first century: *Sustainable Energy— Without the Hot Air*, published in 2009. In it MacKay shows in a straightforward, no-nonsense way the physics of different forms of energy generation and consumption, testing various sustainable-energy scenarios for the United Kingdom. (It has a fascinating appendix on how much easier things would be if he lived in the United States.) He is simultaneously a realist and an idealist, and his book is an excellent complement to all that follows from here.

As MacKay writes, one kilowatt-hour per day is "a nice human-sized unit," since most personal household devices use energy at that scale. For example, one 40-watt bulb left on for twenty-four hours would use almost one kilowatt-hour per day (40 W × 24 hours = 960 Wh ≈ 1 kWh); your 1000-watt microwave left running continuously day and night would use, by definition, 24 kWh per day. One kWh per day is also roughly equivalent to the amount of work you or a human servant can do in a day.

MacKay's book is so clear and his contributions are so important that I propose we name a new unit of energy after him: the MacKay, equivalent to one kilowatt-hour per day.

The Energy that Drives America

However we measure energy, having ready supplies of "that which changes the physical state of the system" has transformed nearly everything about our lives. It is power in its most elemental form. Our economy, our culture, our use of space and time have all dramatically changed since the days of Drake in Titusville, largely because of oil and other age-old natural resources gotten from the ground. The Industrial Revolution of Joule and Watt's time was a revolution precisely because the new sources of energy made possible entirely new ways of being and acting human; the changes continue today with each new electronic or mechanical device.

Nowadays, because abundant energy is so commonplace, we forget how powerful we have become. Gone are the days when human labor and animal muscle had to do most of the work. Now we fly in steel birds and whisk down the highway behind engines pulling with the power of two hundred horses without thinking twice. Statistics from the US Energy Information Administration (EIA) show that in 2011[*] for the 309 million people who lived in the United States, the economy deployed 85 billion MacKays, or about *274 energy servants per day for each American.* You may not feel like it, but you are more powerful than kings of yore, at least in terms of the energy that you use.

---

[*]Keeping up with the statistics describing the Siren song is complicated by the schedules of the reporting agencies. I've attempted to use the most up-to-date statistics as of this writing (in mid-2012), except in cases where the economic downturn made them unrepresentative. Source notes can be found at the end of the book.

Energy Use in America, 2011

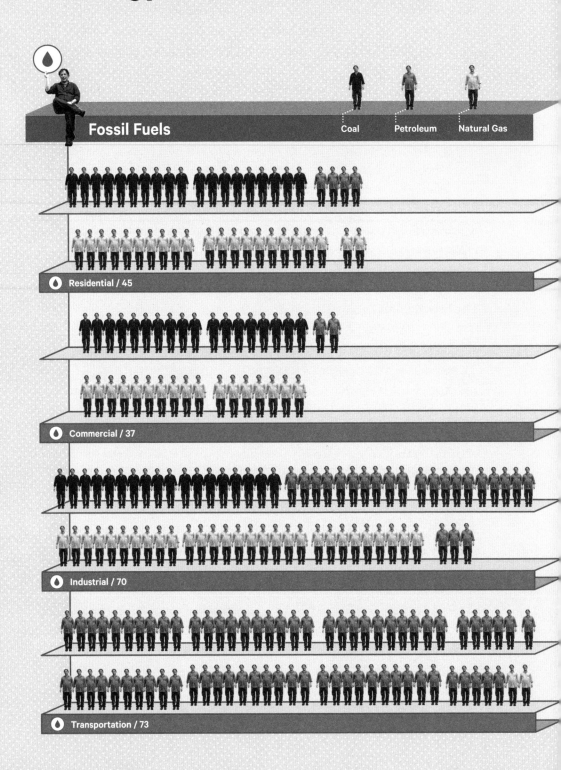

Fossil Fuels

Coal   Petroleum   Natural Gas

Residential / 45

Commercial / 37

Industrial / 70

Transportation / 73

We don't always appreciate how powerful we've become. If energy were recast in terms of human effort, then in 2011 every American was aided by 274 energy servants (measured in MacKays per person, where a MacKay is a kilowatt-hour per day). Those servants warmed our homes, lit our offices, fueled our factories, and moved us around. The servants come from different sources. For the residential, commercial, and industrial sectors of our economy, obedient, diligent, tireless MacKays come from a combination of fossil fuels, nuclear energy, and renewables (like wind, solar, and geothermal). In contrast, the transportation sector is supplied almost entirely by one fuel: Cars depend on oil, and so America does as well.

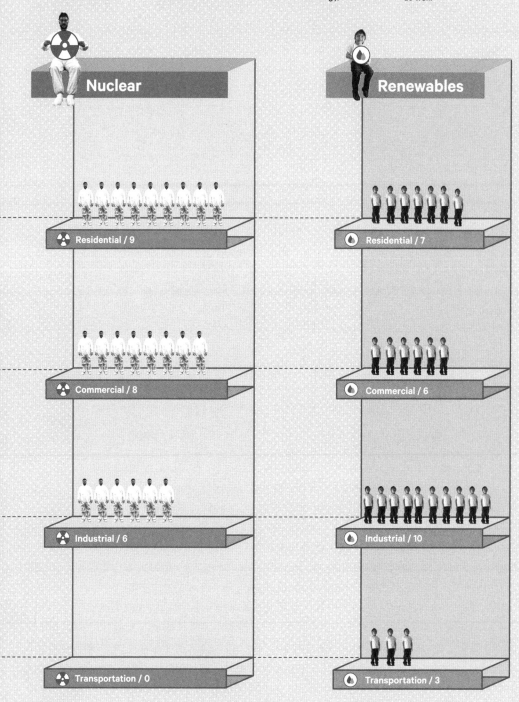

Nuclear

☢ Residential / 9

☢ Commercial / 8

☢ Industrial / 6

☢ Transportation / 0

Renewables

⬤ Residential / 7

⬤ Commercial / 6

⬤ Industrial / 10

⬤ Transportation / 3

Not all of those MacKays come from oil, of course—we derive our energy from a variety of sources. If you imagine your 274 MacKays arranged in your backyard, then only 100 of them would be black and oily, representing crude oil and imported petroleum (36 percent of the 274 energy servants); 55 would be dark and shiny like coal (20 percent); and another 70 would be wispy phantoms, supplied from natural gas (26 percent). In other words, 225 of the 274 MacKays supporting you on an average day—82 percent of the American energy supply—are obtained from fossil energy. The remaining MacKays milling around in the background are 23 glowing yellow servants, electricity from nuclear fission reactors (9 percent), and 26 smiling green ones, from renewable sources like windmills, solar panels, geothermal plants, and hydroelectric power (9 percent).

We are not all equally in charge of these MacKays—some of them are our direct servants, available via switch or dial, but many others are indirect, embedded in the goods and services we consume. At the macroeconomic scale, analysts divide energy use into four major sectors: commercial, residential, industrial, and transportation. (Sometimes electricity generation is also broken out, but I have added it back into the other four sectors for the moment; electricity complicates things, but in an interesting way, which we will return to in the chapters to come.) Commercial and residential uses are largely about heating and cooling and running all our gadgets. Industrial uses, like oil refining or car manufacturing, create goods for sale. Transportation moves us and our stuff across the country and around the world.

Reported separately are government uses of energy, which include the military. During the Iraq War, in 2008, the US government deployed 888 million MacKays on our collective behalf, 714 million of which marched in our armed forces. That's 646 MacKays per active-duty military person per day to move the tanks, fly the planes, and get the job done. During the Iraq War the US military consumed nearly 1.7 million gallons of fuel per day. (Ironically, as the insurgency in Iraq against American forces developed, convoys transporting fuel became a major target, which required heavier armor on the Humvees guarding the convoys, which increased fuel usage and therefore required longer convoys, which required more guards, who were endangered by more attacks, and so on—the very definition of a vicious circle.) During World War II, War Department logistics assumed that 50 percent of the tonnage moving to the battlefield was ammunition, 20 percent food, water, and supplies, and the remaining 30 percent fuel; today the fuel component is closer to 70 percent by weight.

The skies are swept clear by more energy-demanding military equipment. The Air Force alone uses 9 percent of all the energy consumed by the federal government. The F-22 Raptor, an "air superiority and ground support" fighter, reputedly gets 0.4 miles per gallon, depending on speed and conditions, but can make Mach 2 when it needs to; the C-5 Galaxy, a strategic airlifter, gets just 0.07 miles per gallon, flies slower, but can carry 270,000 pounds at 35,000 feet above the ground. The C-130 Hercules, a smaller kind of cargo plane, toted fuel refined in the United States to Iraq during the war. A round trip might use 19,200 gallons of fuel to transport 7,000 gallons to the battlefield. In total, those millions of silent, serviceable MacKays mean that the US military is the largest institutional user of crude oil in the world, deployed primarily to protect the largest national user of crude oil in the world, the United States.

Let's return to the MacKays in your backyard and ask them to line up in rows by sector, so that we can see how different energy sources supply different kinds of consumption. In the residential and commercial queues you will see similar mixtures of MacKays—45 and 37 black, 9 and 8 yellow, and 7 and 6 are green, representing the predominant use of fossil fuels (mainly coal and natural gas for electricity, and some heating oil) and electricity from nuclear and renewables. The residential line is slightly longer than the commercial queue (61 vs. 51 MacKays), representing 22 percent and 19 percent, respectively, of total energy consumption. The industrial line is even longer, with 70 black, 6 yellow, and 10 green MacKays. (Let's remember the MacKays diverted to boil the oil during refining.) We hand out a bright-red letter "A" (for Automobile) to 19 of the 86 industrial MacKays, honorary members of the transportation queue, recognized for the energy consumed to refine oil.

Turning to transportation itself, we see a phenomenon unique in our economy— a long, sinuous stretch of uninterrupted black fossil fuel MacKays, 73 of them, with just 3 green servants and a partial very faint yellow one shoving for a place at the back. If you get closer, 71 of those blackened figures are dripping with crude. Those are the petroleum MacKays, dutifully waiting to push your car down the expressway. Though we can use oil for many things, our cars are the only ones that depend on it.

4

The Cheap Oil Window

The cow was milked too hard. Moreover, she was not milked intelligently.

Captain Anthony F. Lucas, Beaumont, Texas (1904)

At this point in my investigations, I was starting to see some of the connections hidden in plain sight by the Siren song. Because of how nature makes oil, we know only how much we pull from the ground, not how much is left. Having gotten it, we found it to be so powerful and flexible that entire economies could be built upon it, yet our dependence today is driven by a single factor: the massive amounts of petroleum we pour into transportation.

Because that dependency entails wars, political strife, and hatred, the obvious question is: Why can't we supply the oil from here in America? Why rely so heavily on the Persian Gulf, on the other side of the world? Ours is a big country after all, and we know there is oil beside the north shore of Alaska, under the Gulf of Mexico, off the coast of California, and who knows where else. What about Canada and Mexico? It must be those damned environmentalists. "Drill baby drill," some have been known to chant, catching the deep, percussive rhythm of the Siren song.

Where Oil Used to Come From

The sad truth is once upon a time the rocks of the United States were the world's greatest oil depot—in fact, for most of the twentieth century, American oil dominated global supply. We still don't do badly, all things considered, with over 360,000 active wells pumping today, more than all other countries combined. The tremendous geological variety of America and our continental scope meant that after Drake and crew demonstrated the potential of drilling in 1859, many corners were left where oil might be hiding.

After Pennsylvania, the next great American oil discoveries were in Ohio and Indiana (1884), Kansas (1892), California (1894), and Texas (1901). In January 1901 wildcatters brought in a gusher on a hundred-foot hill called Spindletop in East Texas, which produced oil at an unheard-of rate of seventy-five thousand barrels per day. The well spewed for nine days, creating a lake of oil, and as before, caused an oil rush, a boom in production, and a collapse in prices. Within months, 214 wells crowded Spindletop's denuded crown, and the price of oil had dropped from over a dollar a barrel to three cents. (Meanwhile in nearby Beaumont, Texas, water was selling for five cents a cup.) The engineer who brought in the Spindletop gusher, Captain Anthony Lucas, was invited back in 1904 to see what his discovery had wrought. Everyone had gone home, the wells had dried up, and water was cheap again. He pronounced: "The cow was milked too hard. Moreover, she was not milked intelligently."

Milking intelligently was going to be hard to accomplish with so much money at stake. The pattern of strike, boom, and bust would be repeated over and over again during the early twentieth century, in different places by different players— in Louisiana and Oklahoma, along the Caspian Sea in Russia, in Romania, then in the Ottoman Empire (modern-day Iraq) and in Persia (modern-day Iran). By 1908, the year Henry Ford's first Model T rolled off the assembly line, the United States was producing over 178,000 barrels of oil per year; the rest of the world put together 108,000. Demand was growing, but not in proportion to supply, leading to one of the ironies of oil history: the supply problem. For most of the oil industry's history, the problem has been too much oil, not too little. And the problem with short demand and long supply, when governed by the rule of capture, was that it led to low prices, low margins, and strong incentives to pump and sell oil as quickly as possible.

Eventually institutions developed to help stabilize the price by controlling the amount of oil coming out of the ground. Not surprisingly, the first to do so were the oil companies themselves, or at least one company: Rockefeller's Standard Oil. Rockefeller stabilized prices for kerosene for lighting, and then for gasoline, through the simple expedient of monopoly. He offered to buy out competitors, and if they didn't agree to reasonable terms, he undercut them, providing a simple choice—sell or go under. By 1890 Standard controlled 90 percent of the refined oil product in the United States and had ended, temporarily, the effects of what he called "ruinous competition." The price of oil was what Mr. Rockefeller said it would be.

Amid the corrupt years of America's Gilded Age, when politicians were routinely bought and sold, 1890 shines out, like sunlight piercing a bank of clouds, for in 1890, against all odds, Congress passed the Sherman Antitrust Act. A trust is a business arrangement created for the purposes of curtailing competition and controlling prices; the Sherman Antitrust Act made those practices illegal within the United States. The act was intended as a swipe at the practices of the railroads, but fifteen years later, it had its greatest effects in oil. Egged on by the muckraking journalist Ida Tarbell, who had seen firsthand Rockefeller's work in the Oil Regions of Pennsylvania, President Theodore Roosevelt ordered his Justice Department to sue to break up the Standard Oil trust. In 1911 the Supreme Court agreed; Standard Oil was a trust and it must be broken into thirty-four pieces. These new companies, which were subsequently renamed again and again, still dominate the oil industry. Some of the most important "Standard Oil" descendents were:

Standard Oil of New Jersey, the largest of the Standard daughter companies, was renamed Esso in some territories, Exxon in others, and later merged with Standard Oil of New York to form the largest oil company of them all, ExxonMobil;

Standard Oil of New York became Socony, then Socony-Vacuum, then Mobil, and now ExxonMobil;

Standard Oil of Ohio, then Sohio, was bought out by BP;

Standard Oil of Indiana, then called American Oil Company (Amoco), was also slurped up into BP; and

Standard Oil of California, or SoCal, was renamed Chevron, then after merger, ChevronTexaco, and is now known as just Chevron again.

These companies, plus Texaco, Gulf Oil, Anglo-Persian (then British Petroleum, now BP) and Royal Dutch Shell (now Shell), comprised "the majors" that came to control the global oil supply for most of the twentieth century.

While the dissolution of Standard Oil ensured competition would continue in some form within the United States, elsewhere the major producers were under no such constraints, and moreover were faced with a glut of oil in the 1920s coming out of the newly Communist Soviet Union, which was desperate for hard cash. Several presidents of Standard Oil companies met with their counterparts at Anglo-Persian and Royal Dutch Shell and concluded a secret agreement while grouse hunting in the Scottish Highlands in 1928. The executives agreed to leave things "as is," meaning they agreed to produce only in proportion to total demand and to maintain the relative shares as they existed at the time of the agreement. So if Anglo-Persian supplied 40 percent of the European market in 1928, it was agreed that Anglo-Persian would continue to supply 40 percent of that market into the future, even as European demand grew. Same with other parts of the world. Pleased with themselves, in 1929 the companies generously enlarged the pact to include the Soviets.

It was not to last; there was always someone who wouldn't play along, especially among those who were not part of the game. More annoying, oil kept being found in places where the majors didn't expect it, including Oklahoma (1905), Wyoming (1908), California (1911), and Texas again (1911). Huge new fields were found in Mexico (1901) and Venezuela (1914). Each time independents with more to gain and less to lose undercut the big companies' prices. A power even greater than the oil companies themselves was needed to manage the flow of oil into the economy, and the oil companies found what they were looking for in the government.

Texas Steps In

Slowly countries around the world were realizing that they had an interest in oil beyond ensuring domestic prosperity, because oil had become essential to wrecking the domestic prosperity of others. World War I demonstrated how much the modern military depended on oil, as nineteenth-century cavalry and infantry charges met the horror of twentieth-century tanks and flamethrowers on the fields of Flanders. The Austro-Hungarian Empire and Germany launched the war with huge stockpiles of iron and coal, but it was energy-dense oil that proved decisive for the new strategies of "total war" based on mechanized land movement, air power, and oil-fueled navies. The new First Lord of the Admiralty, Winston Churchill, made his mark by committing the British Navy to oil in 1912, and ordered the attack on Gallipoli in 1915 to keep open the flow of oil shipped through the Dardanelles from Caspian fields. In the Middle East, after the war, British bureaucrats used their victory—secured at last with American soldiers, machines, and oil in France—to redraw the political boundaries of the newly

The History of Oil, 1859–2012

Oil for light, and then for fuel, has driven a global obsession with finding and extracting crude. In the US oil extraction grew steadily from 1859 until 1970, when it peaked at 3.5 billion barrels per year (see below and, closer up, at right). Consumption, unfortunately, has not relented, except when the economy stum-bles, folks lose work and drive less. The shortfall between production and consumption is the amount we trade for, 3.2 billion barrels in 2012. Meanwhile global extraction continues to increase, even after 150 years of exploitation, though it has slowed dramatically since the 1970s. (Global pro-duction equals the rate of consumption and now exceeds 32 billion barrels per year.) As prices go up, tar sands, oil shales, and deep ocean basins become the next targets. The ques-tion isn't: When will we run out of oil? The question is: When will we run out of money to pay for it?

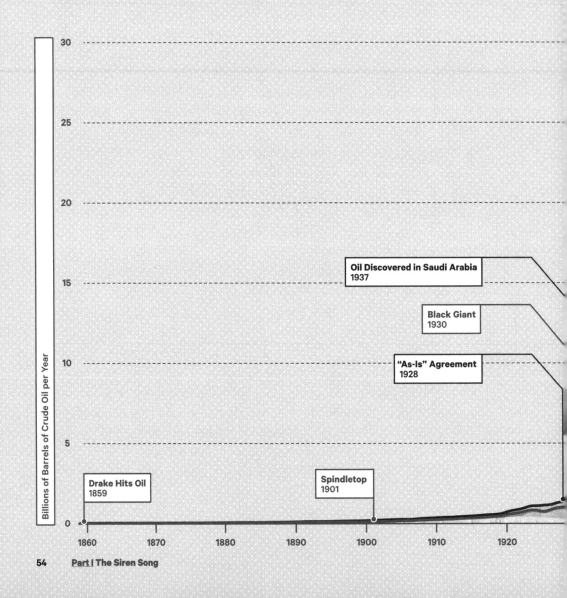

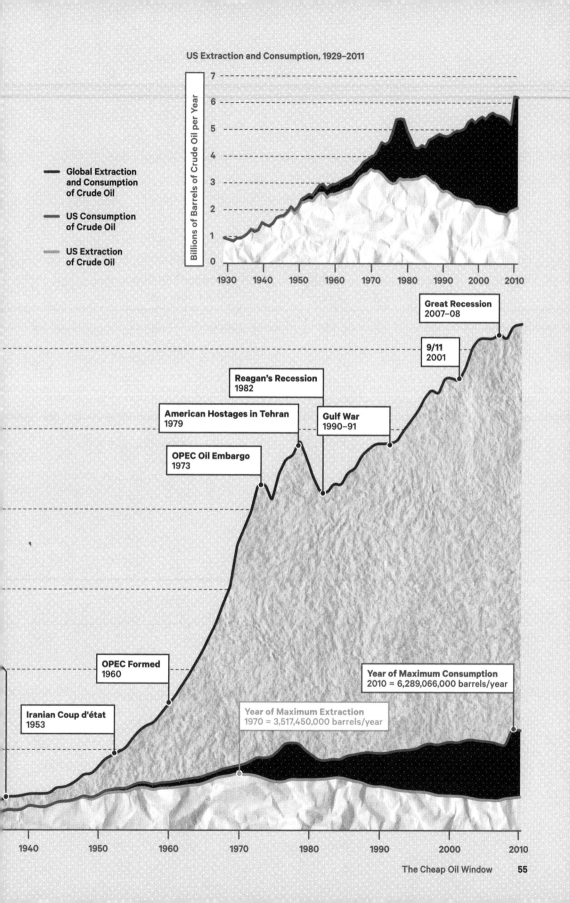

US Extraction and Consumption, 1929–2011

Billions of Barrels of Crude Oil per Year

- **Global Extraction and Consumption of Crude Oil**
- **US Consumption of Crude Oil**
- **US Extraction of Crude Oil**

Great Recession
2007–08

9/11
2001

Reagan's Recession
1982

American Hostages in Tehran
1979

Gulf War
1990–91

OPEC Oil Embargo
1973

OPEC Formed
1960

Year of Maximum Consumption
2010 = 6,289,066,000 barrels/year

Iranian Coup d'état
1953

Year of Maximum Extraction
1970 = 3,517,450,000 barrels/year

created client states of Iraq, Jordan, and Palestine, wrested from the defeated Ottoman Empire.

Cartographers at the time had no idea how influential their straight lines across the desert would become. The oil companies thought they knew, however; a consortium of companies (including Standard Oil of New Jersey and Socony-Vacuum) drew their own "red line" around the new protectorates to keep Middle Eastern oil for themselves and away from their competitors. Within the area they agreed to share among themselves evenly. Kuwait, Persia (Iran), Bahrain, and the still undiscovered fields of the Arabian Peninsula lay just outside.

Back in North America, the US government took newly discovered oil fields on public land in Wyoming and California out of production and made them naval petroleum reserves, but the temptation to capitalize on them became too much to resist. After the war officials leased the fields back to oil companies in return for bribes and kickbacks, generating the tempest known as the Teapot Dome Scandal (named after a domestically shaped geological formation in Wyoming). Seeking to distance himself from the trouble before the election of 1924, President Calvin Coolidge appointed the Federal Petroleum Conservation Board to advise on production for national security, justifying the new extension of federal power with what has become a tenet of US foreign policy ever since: "The supremacy of nations may be determined by the possession of available petroleum and its products."

Until then, the care and management of that petroleum and its products in the United States had been a responsibility of individual states. For example, since 1915 the Oklahoma Corporation Commission had been entrusted with regulating production of oil to avoid "economic waste," by matching supply to demand through a process called proportional rationing, or prorationing. Prorationing meant the state decided how much oil would be produced and then distributed quotas to various extractors proportionally, depending on capacity. Established oil extractors, like the big, integrated oil companies, liked the quota system because it solved the problem of boom and bust production and ensured a steady flow of oil, keeping things, more or less, "as is." It also bought them political cover—was prorationing about stabilizing prices or conserving oil supplies for the future? It wasn't clear, and the oil companies worked hard to keep it that way. Independent oil producers, in contrast, hated the policy, because it kept them from getting a bigger slice of the pie, or in some cases, any pie at all. They circumvented prorationing by selling "hot oil" to dealers in other states, until the federal Connally Hot Oil Act of 1935 (named after Senator Thomas Connally of Texas) banned the interstate commerce of oil produced without quotas. Thus, government and industry made common cause to meter the oil, establish a fair price and a "decent profit," and conserve as a matter of national security.

Prorationing was put to the test in 1930, in the midst of the Great Depression, when Columbus "Dad" Joiner found the Black Giant field. Just as with Spindletop thirty years before, no one had thought that there was oil in the pinelands of East Texas; like Lucas, Dad Joiner proved "them" wrong on October 30, 1930 by bringing in the largest field discovered up to that time. Independents scrambled to get a piece of the action. Predictably, the price of oil plummeted, from a dollar a barrel in 1930 to less than ten cents by the summer of 1931, as production from the Black Giant swelled to over a million barrels per day. Soon more than half of all US demand was being supplied from one massive deposit.

The Texas Railroad Commission (TRC) managed oil in Texas, but unlike its counterpart in Oklahoma, the TRC could not by law regulate production for economic reasons, only to avoid "physical waste." The commission tried shutting down production anyway, arguing that overproduction from the Black Giant was wasting the long-term potential of the field (which it was), but courts repeatedly found that the commissioners had overstepped their authority. Thus in August 1931, Texas Governor Ross Sterling, former president of Humble Oil, declared that East Texas was in "a state of insurrection and open rebellion," and ordered National Guardsmen and Texas Rangers to shut down the wells. (The Rangers rode in on horses because torrential summer rains had made the roads impassable to motor vehicles.) The clampdown worked, at least temporarily, to the relief of the oil industry, though consumers struggling through the early years of the Great Depression may have disagreed.

In November 1931 the governor forced a law through the Texas legislature authorizing the TRC to avoid "economic waste," and thus prices slowly ascended to the dollar-a-barrel level that the oil majors and government felt balanced economic profit and national security. Other states agreed to coordinate their production with the Texas price. The TRC, with support of the federal government and the major oil companies, would, more or less, control the price of oil produced in the United States for the next forty years.

"A War of Engines and Octanes"

As if it needed repeating, World War II again demonstrated the value of oil in war some seven billion times—that's the number of barrels of Allied oil consumed to win the conflict; six billion of them were extracted from American wells. The Axis powers—Germany, Italy, and Japan—entered the war oil-impoverished, and so early in the conflict moved rapidly to secure supplies in Romania, North Africa, and the Dutch East Indies (today's Indonesia). Germany's strategy of blitzkrieg, or "lightning war," depended on oil's energy, and German jerry cans to carry fuel (reverse engineered by the United States) became ubiquitous behind the lines on both sides. The Desert Fox, German Field Marshal Erwin Rommel, felt the pinch first when he outdrove his supply lines while attacking Egypt from Libya in 1942. The trucks supplying him used more fuel coming and going to the front line than they could carry; eventually his tanks ground to a halt and were defeated by better supplied British and American forces. That same summer, the German Sixth Army was sprinting toward Russian oil fields beyond the Caucasus Mountains but became bogged down by stiff resistance at Stalingrad. In the darkest hours of the war, Soviet Premier Joseph Stalin proposed a grim toast at dinner with British Prime Minister Churchill: "This is a war of engines and octanes. I drink to the American auto industry and the American oil industry."

WWII also provided new demonstrations of the remarkable chemical flexibility of oil as feedstock for industry. In 1933 the New Jersey–based Standard Oil Research Development Company announced a new laboratory method to make toluene, the key ingredient of TNT, from oil; from 1940 to 1945, US companies produced 484 million gallons of toluene for explosives. When the Japanese invaded the Dutch East Indies in 1940, they took control of 90 percent of the world's supply of natural rubber, obtained from tree sap. To compensate, organic chemists developed an improved method to manufacture synthetic rubber from oil; by the war's end, the production of synthetic rubber for tires, shoes, guns, and other war goods was more than double what trees could supply.

Meanwhile in Germany, hobbled again by the lack of domestic oil, scientists learned how to squeeze oil from a lump of coal. IG Farben and other German chemical companies had developed methods to make synthetic gasoline (syn-gas) and jet fuel from coal during the 1920s. The process was inefficient—requiring five parts of energy for every two parts fuel energy returned—but it worked. Adolf Hitler's plans for victory depended on synthetic gas until the German army could overrun oil fields in Romania and Russia. Later syn-gas plants were built strategically close to the concentration camps, so that slave labor could help create the fuel for the German tanks, planes, and U-boats. It all came to a dramatic end during a bombing raid in 1944. The architect of Hitler's wartime economy, Albert Speer, said: "I shall never forget the date May 12 [1944]. On that day the technological war was decided." It was three weeks before D-Day.

Back home in America, people grumbled, but managed, under gas rationing. "After all, there is a war on," they told themselves, counting their tickets for tires and fuel, while waiting for the streetcar to come by; meanwhile, their leaders were already worrying about where the oil would come from for the next war. President Franklin Delano Roosevelt's Secretary of the Interior, Harold Ickes, was characteristically pessimistic, writing in 1943: "If there should be a World War III it would have to be fought with someone else's petroleum, because the United States wouldn't have it. America's crown, symbolizing supremacy as the oil empire of the world, is sliding down over one eye." The drive to secure oil supplies outside the United States would rule American foreign policy into the next century.

The end of World War II unleashed an enormous pent-up demand to get back on the road again in the United States, supplemented, as we shall see, by economic policies that built the roads, fueled construction, and encouraged suburbanization. Though transportation led the way, markets were also rapidly developing for the other by-products of refined crude oil: plastics, pesticides, drugs, home heating oil, and asphalt. Taken together, US demand for oil was rising by leaps and bounds, which challenged the industry's ability to supply it and maintain a level price. The search was on for new supplies. Although war had delayed development, postwar America could now focus on a discovery made by a petroleum geologist from Standard Oil of California (SoCal) back in 1938: Oil lay beneath the sands of Arabia.

The Oil of Arabia

Though oil had been known in what are now Iran and Iraq since antiquity, the new "elephant" oil fields found in the Arabian Peninsula were unprecedented in size and quality. Filled with sweet, light oil, they were also remarkable for the ease by which crude could be lifted from the earth. The first and overriding US interest was to make sure that American companies, not our erstwhile friends in Europe and the Soviet Union, secured the supply. So in February 1945 after the Yalta Conference with old allies Churchill and Stalin, FDR took a detour to meet America's new ally, King Abdul Aziz of Saudi Arabia, known as Ibn Saud, on a navy ship moored in the Suez Canal. Ibn Saud's tribe had unified most of the Arabian Peninsula in 1932, motivated by a deep and conservative form of the Islamic faith and a distrust of British interests that traced back to World War I. We don't know exactly what Roosevelt said to Ibn Saud, but we do know that an important commitment was made in the wake of their discussions.

The new arrangement had two parts. Roosevelt's successor, Harry Truman, affirmed the first part in a letter to Ibn Saud in 1950: "I wish to renew to Your Majesty the assurances which have been made to you several times in the past, that the United States is interested in the preservation of the independence and territorial integrity of Saudi Arabia. No threat to your Kingdom could occur which would not be a matter of immediate concern to the United States." In other words, a pledge of security was given to the Saudi king that the American republic has ever since upheld.

The second part was an exclusive long-term lease for a consortium of US companies to extract Saudi oil. SoCal and Texaco had an agreement covering extraction, but they needed other partners with bigger markets to absorb the massive surge of oil coming down the line: Standard Oil of New Jersey and Socony-Vacuum (aka Exxon and Mobil) in the eastern United States fit the bill. The British, French, and Soviets were all cut out of Saudi oil. In exchange, the American consortium (Aramco) assumed all the risks and invested all the money required to find, pump, distribute, and sell the oil. Aramco sent a royalty to the king for the privilege, four shillings per ton of crude oil, payable in gold.

Figure No. 13

How Prices Have Changed, 1775–2012

Prices change over time, not only in response to supply and demand, but also as a function of how many dollars are in the world. Economists correct for these changes using the consumer price index, which measures a standardized basket of goods and services that everyone buys (e.g., food, housing, gas). When prices go up, we call it inflation, measured by the slope of this line. Inflation occurred during the Revolutionary War, the War of 1812, the Civil War, and both World Wars, but nothing compares with the mountain we've climbed since 1971.

Source: Drawn from data compiled by Sahr (2012), indexed so the price level in 2009=1. The 2012 index value is an estimate. Also see US Bureau of Labor Statistics (2012a).

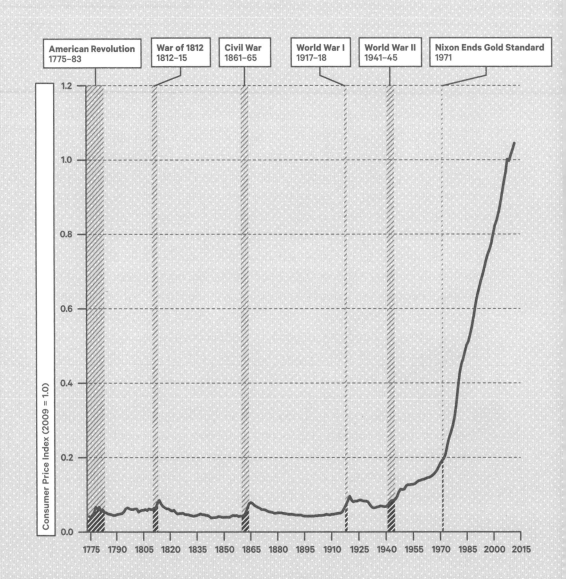

American Revolution 1775–83 | War of 1812 1812–15 | Civil War 1861–65 | World War I 1917–18 | World War II 1941–45 | Nixon Ends Gold Standard 1971

Consumer Price Index (2009 = 1.0)

The deal was soon tested. As usual the independents were coming up with all kinds of unconventional notions. In 1947 a consortium of American independents (the American Independent Oil Company, or Aminoil) won the concession for the Kuwaiti portion of the "Neutral Zone," a boxy area of desert along the coast shared by the Saudi king and Kuwaiti emir, originally created by thoughtful British cartographers for nomadic Bedouins after WWI. The Aminoil consortium offered the emir $7.5 million and a million-dollar yacht up front and guaranteed at least $625,000 per year in royalties going forward, whether or not oil was found. Not to be bested, the next year J. Paul Getty offered the king $9.5 million, and a minimum of $1 million per year in royalties for the Saudi portion of the Neutral Zone. Getty and Aminoil were forced to work together, to no one's pleasure. But six years and $30 million dollars later, they finally struck oil. By 1957 Getty was the richest man in America.

50–50

Meanwhile, a new approach to natural resources was being explored in Venezuela, where American companies had been extracting oil on behalf of the Venezuelan people since 1914. During World War II, Caracas passed a new law that insisted that petroleum profits be split 50–50 between the government and the companies. With Washington focused on war on two fronts and the implicit threat of nationalization in the air, FDR refused to pick a new fight in South America, and the companies grudgingly accepted the deal. Soon the new 50–50 principle would sweep the world.

After the Getty deal and Venezuela, it was clear to the Saudis that the American consortium, Aramco, could afford to pay more too. In 1950 the Saudis demanded 50–50. Aramco wanted no part of a renegotiation, considering it deeply unfair, a betrayal of the remarkably profitable deal they had struck only a few years before. Eventually the American government stepped in with an accounting solution that kept both the powerful companies and the Saudi king happy.

The new deal took advantage of an interesting feature of the US tax system—the foreign tax credit. This credit was originally instituted in 1918 to encourage US investment abroad; it allows US companies to deduct the taxes they pay to foreign governments from their federal taxes. Under the new proposal, the Saudis and the companies would transmute their royalty payments, which were treated as a business expense and therefore not deductible, to foreign taxes, which would be deductible; the result was a transfer of millions of dollars from the US treasury to the king's vault, with little difference to the companies' bottom line. In effect, the American taxpayer paid off the Saudis to ensure access to oil.

The effects were dramatic for all concerned: In 1949 Aramco paid $43 million in US taxes and $39 million in royalties; in 1951 Aramco paid $6 million in taxes to Washington and $110 million to Riyadh. Kuwait, Iraq, and Iran were soon insisting on similar deals.

The sword that foreign countries held over the necks of the American companies was nationalization, which meant that countries, not oil companies, would vertically integrate and sell their oil on their own, without assistance from outside. Mexico had done it, kicking out American and European interests in 1938, which taught the industry a painful lesson, especially when FDR refused to back the industrialists during the run-up to World War II. (He decided a neutral Mexico was in our best interests.) Fifteen years later, in 1951, the new democratically elected Iranian government voted to nationalize Iranian oil supplies after the Anglo-Iranian Company (aka BP) turned down a 50–50 offer; British and American intelligence responded by sponsoring a coup to topple the government and reinstall the Shah of Iran in 1953. The Shah gratefully accepted a 50–50 split with a new consortium of British, French, and American companies, and then drove a ruthless bargain with the Italians, taking three parts profit for every four parts extracted in a deal signed in 1957. As Getty had also discovered, you pay a premium for coming late to the party.

Another place where the 50–50 deal would not apply was back in the United States. At home, the key to oil wealth had always been to obtain property rights above the oil. In most states, subsurface rights followed surface rights, and having obtained the right to insert a straw, the rule of capture applied to what could be sucked out. Before prorationing became the norm, everyone sucked as hard as possible. On private lands, royalties were worked out between the owner of the land and the company through leases. On public lands, the General Mining Law of 1872 stipulated that any person could lay a claim to "locatable" minerals—gold, silver, cinnabar, copper, or "other valuable deposit"—by marking the location, filling out a form, and paying $2.50 or $5.00 per acre, depending on the kind of deposit (vein, lode, or placer). After establishing a claim, whatever you pulled from the ground was yours, subject only to income taxes (not established until 1913) and selected excise taxes on products made from those deposits, such as a 10-cent-per-gallon tax on kerosene illuminants.

Not surprisingly, lenient rules sparked an oil rush on public lands in the early decades of the twentieth century. After the Teapot Dome scandal, Congress passed the Mineral Leasing Act in 1920, which established new procedures for public petroleum concessions. Auctions for oil and natural gas prospecting rights would be held for ten-year leases. Companies would offer a bonus (a onetime, up-front fee to the government, on which the bid would be decided), a fixed rental of $2 per acre per year, and a royalty of 12.5 percent of the gross value of the oil or gas produced. The Outer Continental Shelf Act of 1953 extended analogous provisions offshore. These rules are still in place today. In other words, America sold (and continues to sell) its oil cheap, receiving profit on its hundred-million-year-old natural resource at a rate of one part in eight on land (or one part in six at sea), rather than the one part in two that most other governments insist upon.

The Cheap Oil Window

In the middle part of the twentieth century, oil and therefore gasoline were dependably cheap. On an inflation-adjusted basis, the price of a gallon of gas dropped by a third between 1931 and 1971. The cheap oil window

lasted four decades, long enough for the American economy to become addicted to the energy density of petroleum. In recent years the price of oil has climbed to heights not seen since the nineteenth century.

Source: US Energy Information Administration (2011b, 2012b) lists the domestic crude oil first purchase price and US EIA (2011c) gives annual average motor gasoline regular retail price, supplemented by Pogue (1921) for 1913–18. Inflation adjustment from Sahr (2012) to constant

2009 dollars. The 2012 oil price is the average value, Jan.–Oct. 2012.

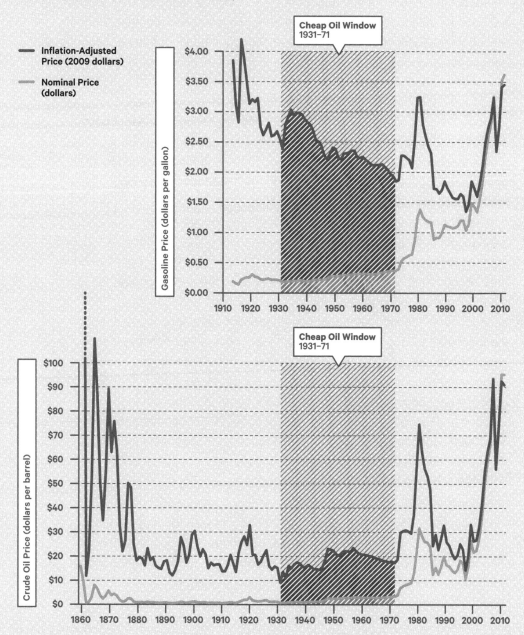

Inflation-Adjusted Price (2009 dollars)

Nominal Price (dollars)

Cheap Oil Window 1931–71

Gasoline Price (dollars per gallon)

Cheap Oil Window 1931–71

Crude Oil Price (dollars per barrel)

Thus, through combinations of negotiations, diplomacy, interventions, tax write-offs, and supply manipulation, the price of foreign oil remained under the control of the major companies; in America compliant state officials controlled the price of oil. Producers and governments shared the same goal: to keep oil cheap enough to encourage demand and expensive enough to generate a profit. Everyone wanted to avoid the "demand destruction price," when consumers might actually change their behavior so they didn't use as much. The result was the cheap oil window, forty years from 1931, when Texan oil was brought under economic prorationing, to 1971, when the TRC ended quotas. Compare prices over time to see how well the system worked: In 1950 the nominal (non–inflation-adjusted) price of gasoline was 27 cents per gallon; in 1960, it was 31 cents; and in 1970, 36 cents. Adjusting for inflation, the real price of gas in America dropped 40 cents per gallon between 1950 and 1970.

The Cheap Oil Window

The price for anything, including oil, is a negotiation, implicit or otherwise. I go to work at the zoo and the zoo gives me money; I go to the gas station, insert my credit card, and fill my tank. It is extraordinary, in a completely ordinary kind of way, that a person can work at a job, even write a book, and then transmute that effort into dinner, or a new pair of jeans, or a gallon of gas. Money is an economic sorcerer's stone. Like energy, which changes the physical state of a system, money converts one kind of human activity into another, all the while adjusting the flows of different goods and services through prices that send signals about the things we want, in the proportions that we collectively want them. And no one is forced to buy or sell anything unless they want to, at least in theory.

Interestingly, the prices we pay are a function of not only how much money we have in our pocketbook, but also of how much money there is in the world. For money, different from energy, is not conserved; the amount is defined by governments that either tie the currency to a physical substance like gold, or just declare how much there will be (a fiat currency). If the amount of money were fixed, then prices would reflect only supply and demand. If the amount of money increases, because the government prints more bills or adds more to its electronic accounts, then the value of money decreases and prices increase to compensate—that is called inflation. The Federal Reserve Bank (or the "Fed"), which had taken charge of the money supply in 1913 and arguably botched the job during the Great Depression, was extremely cautious through the 1960s, especially after World War II, when other currencies were tied to the American dollar, which was tied to gold. The amount of gold was ultimately limited by what could be found underground and on a day-to-day basis by what was held in Fort Knox and other government vaults. In the years after World War II, American money was the golden bridge to prosperity.

Or was it oil? Addiction to the energy in oil deepened dramatically during the cheap oil window. In 1949 Americans used 5.8 million barrels of crude oil per day; by 1970 we were using two-and-a-half times that much—14.7 million barrels per day. America's economy was running on gasoline, warming with heating oil, burning crude for power, and selling a bounty of new industrial goods: plastics, fertilizers, pesticides, and pharmaceuticals. Back in 1931 James Truslow Adams, a writer and historian, wrote an essay that defined the American Dream as "life should be better and richer and fuller for everyone, with opportunity for each according to ability or achievement." Although Adams qualified, "It is not a dream of motor cars and high wages merely," by the 1950s, a private house, a car, and all the modern conveniences were paving the way for higher living standards for millions of Americans. They all depended on oil.

There were dissenters of course. Rachel Carson's book came out in 1962, proclaiming the silent spring that had settled over America, in part due to by-products of oil. Although a definitive link between chemicals in the environment and cancer in bodies remained elusive, cancer rates were ticking up. Scientific conclusions were (and remain) difficult to draw because a control population (i.e., an American population not exposed to the organic chemicals derived from oil) was missing; everyone lived (and lives) in a world saturated by the by-products of oil.

The Federal Trade Commission continued to rumble about price fixing, discreditable practices during the war, and other irregular agreements. Congress investigated these charges in 1952, but the intensifying Cold War meant that Harry S. Truman's White House put national security concerns ahead of worries about economic collusion. After all, oil was cheap, so why worry about additional competition that would make it cheaper still? In testimony and in public statements, the oil companies claimed with good reason that they were securing American interests and a way of life that they were in no small part creating with their products. Although they evinced sharp elbows at home, selling neatly differentiated fuels in tidy gas stations across the land, overseas the companies were increasingly "cooperating" to protect their supply.

Sometimes the companies' worst enemies were themselves. To fight a glut of oil pouring into the world economy from the cash-strapped Soviet Union, Exxon announced in August 1960 it had decided to lower the posted price it would pay for oil. Since foreign taxes were calculated on the basis of this price—like the Saudi deal— Exxon had in effect unilaterally cut the budgets of oil-producing countries around the world. Those governments were not happy.

The same year, powerful Congressional Democrats from Texas, Representative Sam Rayburn and Senator Lyndon Johnson, pushed through a higher tariff on foreign oil imported into the United States. The tariff was introduced at the behest of domestic producers being undercut by cheap oil from overseas. In retaliation Venezuela, Iran,

Iraq, Saudi Arabia, and Kuwait decided to form a cartel to coordinate production—later expanded to include other oil-producing nations including Nigeria, Gabon, Ecuador, Indonesia, Qatar, Algeria, the United Arab Emirates, and Libya, but pointedly excluding the United States. The Organization of Petroleum Exporting Countries, or OPEC, was born.

The Window Slams Shut

In the early 1970s the system of production controls established in 1931 in the wake of the Black Giant went off the rails for several reasons. First, global demand finally exceeded global supply, as consumption rose to within one percent of production. With such a slim margin, any small upset in the long chains of distribution that wrapped the world could, would, and eventually did, cause large swings in price.

Second, in 1970 US domestic production peaked and began its slow decline, even with the discoveries of new oil fields on the north slope of Alaska and under the waters of the Gulf of Mexico. The age of American oil domination was over, spent to fight World War I, World War II, and the conflicts in Korea and Vietnam, and to build a new oil-dependent American landscape of cars and suburbs. In acknowledgment of the inevitable, the Texas Railroad Commission formally ended the system of prorationing in 1971. Texan producers, and by extension everyone else, could now pump as much oil as they desired, to squeeze as much as remained out of their depleted fields as quickly as possible.

Third, as we have seen, the exporting nations had bargained hard and organized to protect their own interests, learning lessons from what American and European interests had been doing for nearly a century. The OPEC cartel, with the support of our allies in Saudi Arabia and virulent anti-Americanists like Muammar Gaddafi, was the new power in petroleum. Nationalization was creeping across the oil fields of the earth, and from now on it would be the countries, not the companies, that would drive the hard bargains.

The Yom Kippur War of 1973 precipitated a new use for oil in war—withholding it. On October 6, 1973, Egypt and Syria launched simultaneous attacks on Israel. Six days later US and Western allies moved to resupply the Israeli military with fuel, and ten days after that, OPEC ministers agreed to deploy the "oil weapon," immediately raising the price of a barrel of crude by 70 percent, to the unheard-of price of $5.11 a barrel. A few days later, they cut off the oil supply entirely, causing prices to soar. Neither Saudi Arabia nor any of the other Arab states would ship oil to the United States. The war in the eastern Mediterranean ended before Halloween, but the embargo continued into the following spring, causing high prices, long lines, gasoline shortages, and outrage on the part of politicians and the public alike. How dare they cut off our oil? When it was finally over in March 1974, the point had been made; from now on OPEC would set the price of oil according to its own system of priorities.

Now inflation began to soar. The Federal Reserve Bank had been slowly weakening the American dollar's exchange against gold, and in 1971 President Nixon closed the gold window forever. Foreign investors and American citizens could no longer turn to the government to redeem their dollars in shiny metal; instead the dollar became a fiat currency, backed by the full faith and promise of the US government. The value of the dollar plummeted. Moreover, under the advice of economists like the famously vociferous Martin Feldstein, the Fed adopted a moderate, pro-inflationary policy to encourage borrowing and business investment.

The "oil shock" of 1973 didn't help. (An oil shock is a sudden, unexpected increase in the price of oil.) Economists still argue about why changes in oil prices ripple through the economy to such an extraordinary degree since across a broad set of commodities, oil amounts to only about 5 percent of the total national expenses (particularly when it's cheap). But many economists agree that sudden surges in the price of oil create broad secondary waves in prices across the board. Demand for oil is often described as inelastic, meaning that it is less sensitive to changes in price than other commodities; a person can go without a vacation, but forgoing the drive to work is harder. Oil and the general level of prices are linked because increased costs to make and ship commodities are passed along to consumers, and because consumers, with less money in their pockets from higher gas prices, spend less, slowing activity in other aspects of the economy. In such circumstances it is tempting for the Federal Reserve Bank to stimulate the stumbling economy through fiscal stimulus, by lowering interest rates and increasing the money supply. Over the long term, more money in circulation drives inflation.

If 1973 didn't make the point, the oil shocks of 1978 and 1979 provided further learning opportunities. Nixon had instituted price controls in the wake of the oil embargo, but they were of no help when the pro-Western Shah of Iran was deposed and replaced by a theocratic state led by Islamic mullahs hostile to the American nation, led by the Ayatollah Ruhollah Khomeini. Islamist students and militants stormed the US Embassy in Tehran, Americans were taken hostage, Iranian oil production dropped, and President Jimmy Carter banned Iranian imports to retaliate. (We too had an oil weapon; only it was pointed at our own heads.) The price of oil skyrocketed again—in inflation-adjusted terms, it rose over 70 percent in less than six months in 1979—not only because of Iranian supplies and OPEC price manipulations, but also because American dependence on foreign oil was growing out of control.

Even with the completion of the trans-Alaska oil pipeline in 1977, American production lagged sorely behind consumption. The deficit was 1.3 million barrels per day in 1970, 4.1 million barrels per day in 1975, and 6.3 million barrels per day in 1979, before the recession. Conservation measures were tried, and for once, they seemed to take hold. As much as people laughed at Carter for wearing his sweater around the White House, policy incentives were slowly easing the American people out of heating our homes and powering our dynamos with oil.

Instead we started depending on oil's fossil fuel cousin, natural gas, and nuclear fission for electricity. We even built some windmills and began to experiment with solar panels; new tax credits for all kinds of energy paved the way for new energy investments. But overall the trend was still ominous: In 1970, 12 percent of the oil consumed in America was imported; by 1980, 39 percent came from abroad.

Not coincidentally, we continued to assert our bellicose stance toward Persian Gulf oil. In 1980, the president stood before the nation after the Soviets invaded Afghanistan to declare the Carter Doctrine. In his Southern drawl he warned: "Let our position be absolutely clear: An attempt by any outside force to gain control of the Persian Gulf region will be regarded as an assault on the vital interests of the United States of America, and such an assault will be repelled by any means necessary, including military force." Under this doctrine, the CIA began supporting the mujahideen, Islamic militants fighting in Afghanistan against the Soviet army, including an earnest, wealthy Saudi scion named Osama Bin Laden. Economic and military aid flowed to Iraq fighting Iran in a bloody war, 1980–88. Then in 1990–91, President George Herbert Walker Bush rallied the world community to expel Saddam Hussein's Iraqi forces from Kuwait, before they could enter Saudi Arabia.

Over the last forty years the US economy has had but one answer to oil shocks: recession. The recessions of 1973, 1980, 1981, 1990, and 2008 have all been associated with major run-ups in the price of oil. The recession of the early 1980s was particularly deep because of the Federal Reserve Bank's drastic attempts to bring inflation— unleashed during the terms of three prior presidents—into check, by throttling back the money supply and raising short-term interest rates. Fed Chairman Paul Volker and President Ronald Reagan were credited with saving the economy, despite the high unemployment rates and economic dislocation that their policies engendered; less well remembered is that simultaneously OPEC members had a major falling-out, which led to a glut of oil on world markets, causing oil prices to plunge and oil companies to absorb massive losses, and undoubtedly helping the US economy as a whole revive.

In the latest instance, the average price of a barrel of oil rose from $66.52 in 2007 to $94.04 in 2008, and this time no major political events in the Middle East were the cause. Rather the oil shock of 2008 seems to have been driven by speculation in the markets. Another Reagan-era policy shift was to deregulate the price of oil; now, no one was controlling the price—not the companies, not the government, not even OPEC; prices were decided on the free and open market, where producers, consumers, and speculators all took their chances. We learned that speculation (and not just on oil) could bring an economy down.

Nine Percent

Remarkably despite a century of gushers, overproduction, underproduction, price fixing, trust busting, threats, wars, deals made, deals broken, and continuously mounting consumption, there still is something called US oil production—in fact, in 2009, a down year because of the effects of the Great Recession, US suppliers still managed to supply 44 percent of the US market and exported a small amount of oil abroad. By mid-2012 domestic production exceeded 50 percent of domestic demand, spurred on by high gasoline prices and falling sales because of the economic recession. Our largest crude oil import partner is not from the Middle East at all, but rather Canada, with a 12 percent share; then Mexico at 7 percent. Together the OPEC nations supply 27 percent of the US market—including contributions of crude from Saudi Arabia (6 percent), Venezuela (6 percent), Nigeria (3 percent), and even Iraq (3 percent), whose valuable oil fields are finally back on the world market in an appreciable way (released from sanctions and partially rebuilt after war). Nowadays the Persian Gulf subset of nations (Iran, Iraq, Kuwait, Saudi Arabia, Bahrain, Qatar, and UAE) supply in total only 9 percent of American demand, about 647 million barrels, for which we paid about $3.53 per gallon at the pump and another $170 billion to fight the wars in Iraq and Afghanistan (approximately $550 per person in the United States in 2011).

When I found out that such a slim proportion of our contemporary oil consumption comes from the Middle East, I was surprised, even shocked. After all, we fight wars and lavish garrisons of troops, carrier battle groups, and other weapons of war to protect that small percentage. It's fair to ask why.

There are two reasons. Frankly, we can't stomach the thought of anyone else having access to those oil fields without our say-so. And our requirement is so high that we need every drop we can get. Since we haven't succeeded in diminishing our demand in the four decades since American oil production peaked in 1970—or actually in any decade since oil was discovered—we have to secure every slice of supply, even if it lies in a troubled, repressive part of the world that's none too keen to have American soldiers barracked there. I imagine the Canadians would feel the same way about American troops stationed outside Calgary. (Not to mention Canadians posted outside Houston.)

Meanwhile, after decades of tacitly supporting government price controls on their industry, the oil companies have found tremendous success in a world where they no longer call the shots but ride the market to ruin or riches. Prior to 1970, the petroleum and gas industry usually reported after-tax profits of $5–20 billion per year (in inflation-adjusted 2009 dollars), the exception being the Black Giant years of 1931–33, when they posted a series of small losses before the TRC brought supply under control.

Since 1970 profits have gone places undreamt-of by past generations of oil executives. In 1974 the oil industry pulled in $40 billion (2009 dollars). In 1980 the American oil industry as a whole posted a $64 billion profit, despite the recession and events in Iran and Afghanistan. The companies, to their deep dismay, also discovered that great gains can be followed by great losses in this strange new world; in 1986 oil companies toted up an ugly $21 billion one-year loss, dragged down by OPEC's inability to control production (and price). In Reagan's America oil was too cheap to make

Oil Trading, 1859–2012

Importer or Exporter?
The trade balance of a commodity is defined as exports minus imports; a positive trade balance shows that exports exceed imports in a given year. From the Civil War to World War I, America was a net exporter of oil, then again during the Great Depression and World War II, when American wells supplied six billion of the seven billion barrels combusted to defeat the Axis powers. Since 1949 we have been oil debtors.

A Deepening Dependency
From 1929 to 2012 (note the change in y-axis magnitude) American dependency on foreign oil grew dangerously, requiring a long series of extreme measures to ensure that petroleum supplies continue to flow.

Where the Oil Comes From
Despite the abundance of high-quality, easily extracted oil found near the Persian Gulf, countries in that region have never provided more than a fraction of the oil that moves America. The single most important exporter to the US today is Canada, but it is only one among many; in the first half of 2012, the US received imports of 500,000 barrels or more from fifty-eight countries.

Source: Calculated from US Energy Information Administration (2012b). The first two graphs based on crude oil imports and exports only; the third reports trade in crude oil plus refined petroleum products. 2012 values are estimated.

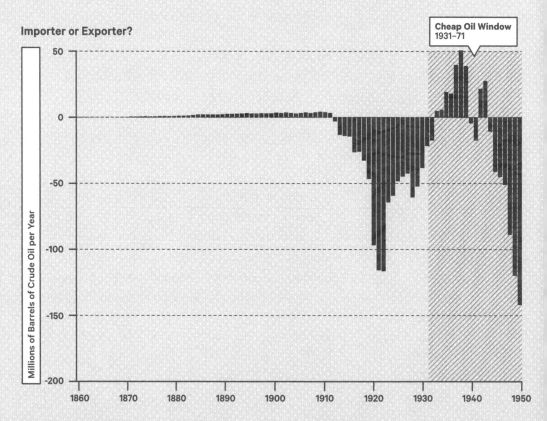

Importer or Exporter?

Cheap Oil Window 1931–71

Millions of Barrels of Crude Oil per Year

Where the Oil Comes From

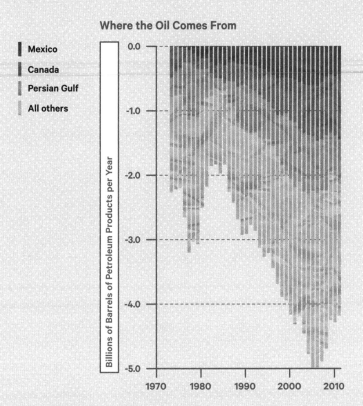

Mexico
Canada
Persian Gulf
All others

Billions of Barrels of Petroleum Products per Year

0.0
-1.0
-2.0
-3.0
-4.0
-5.0

1970   1980   1990   2000   2010

A Deepening Dependency

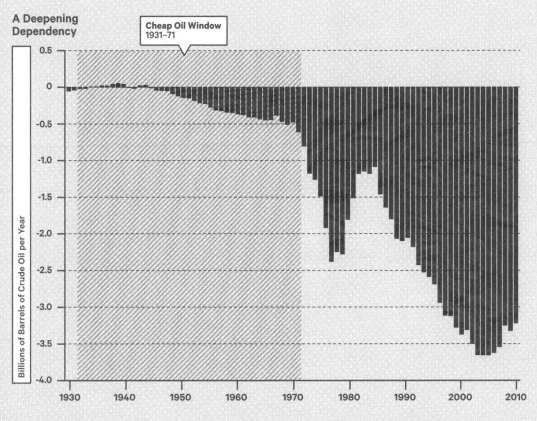

Cheap Oil Window
1931–71

Billions of Barrels of Crude Oil per Year

0.5
0
-0.5
-1.0
-1.5
-2.0
-2.5
-3.0
-3.5
-4.0

1930   1940   1950   1960   1970   1980   1990   2000   2010

Oil Shocks and Recession, 1971–2012

An oil shock is a sudden, unexpected increase in the price of oil. Since 1973, the American economy has experienced five major oil shocks, each immediately followed by a recession, defined as a significant decline in economic activity. Whereas the recessions in 1974, 1979, 1991, and 2001 were all associated with conflicts in or originating from the Persian Gulf region, the oil shock of 2008 appears to have been a result of market speculation. The 1981–1982 recession resulted from efforts by the Federal Reserve Bank to tame the inflation unleashed by increases in the money supply during the 1970s.

Source: Redrawn from Federal Reserve Bank of St. Louis (2012a). Oil prices are the posted price for West Texas Intermediate prior to 1982 and the spot price afterward. Recessionary periods are defined by National Bureau of Economic Research (2012).

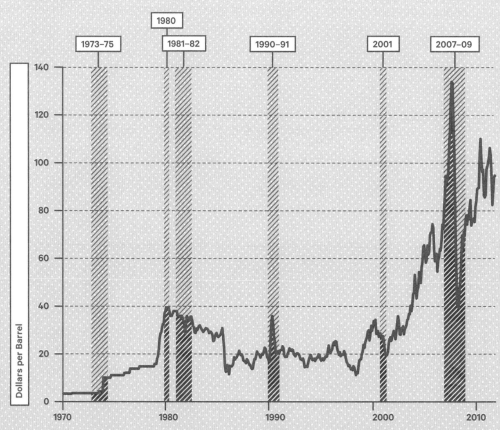

— **Price of West Texas Intermediate Crude Oil ($, not adjusted for inflation)**

//// **Recessions**

1980

1973–75   1981–82   1990–91   2001   2007–09

Dollars per Barrel

140

120

100

80

60

40

20

0

1970   1980   1990   2000   2010

money on, and though the public cheered, the industry lost its shirt. Once OPEC got back on its feet again in the 1990s, and after the practically painless (unless you were a soldier) Gulf War of 1991 and with oil prices shooting upward, the first decade of the twenty-first century was extraordinarily profitable for the oil and gas industry: $38 billion in 2001, $53 billion in 2004, and $75 billion in 2008. All told, between 1929 and 2009, including the years of losses, the industry cleared more than $1.7 trillion in profit, one-quarter of which has come since 2000.

It's easy to see why—they know how to play the game. The cheap oil window got us hooked on the ride up Hubbert's Peak, and now like a heroine captured by a mustached villain, we are tied to the tracks as an oil-dependent economic locomotive barrels down on us. The cruel irony of Hubbert's Peak is that long before the oil runs out, we will have run out of money to pay for it. Even in bad economic times our thirst is prodigious. In 2009 as the United States economy wallowed in the doldrums of the Great Recession, we still drank in over 6.9 billion barrels of oil, an amount equivalent to the full production of the United States for the sixty-odd years between 1859 and 1922—everything that John D. Rockefeller ever refined, plus some. Or to see it another way, 6.9 billion barrels is more oil than the United States has ever produced in any single year—more than the year after Spindletop blew (89 million barrels, 1902), more than the year the Texas Rangers rode into town to control production from the Black Giant (851 million barrels, 1931), and more than when American oil production peaked in 1970 at over 3.5 billion barrels of oil.

Which means you have to expect a miracle to believe there is a way to drill ourselves out of the mess we have gotten ourselves in. Though the high price of oil means that more expensive sources of oil have become more economically feasible—the tar sands of Canada, oil shales in North Dakota, the deep waters in the Gulf of Mexico, and whatever lies beneath the Arctic Ocean—it also means we are more dependably desperate. The demand curve continues to climb upward while the supply is tipping down, as predicted by an oil company geologist fifty years ago. Increased demand and diminished supply, economists somberly tell us, pushes the price predictably up, beyond the effects of general inflation, and because the shocks can come from any direction— the Middle East, West Africa, South America, Russia, Wall Street—the price of oil is less certain, more volatile, harder to plan for. No, we are not running out of oil, the Sirens sing, sharp teeth glinting in the fading light, but you can expect that as long as you and everyone you know have to have all those oily MacKays for your motorcar, the American way of life is going to be a lot more costly and undependable.

5
Time for Space

I will build a car for the great multitude.

Henry Ford

So it happened that when I drove my car across the continent—from Davis, California, to the Bronx, New York—to take a new job in 1998, the price on the pump was $1.05[9] per gallon, almost the lowest price of the decade before or the decade since.

I made the trip in April across Interstate 10 to avoid any lingering snow farther north, crossing southern California, then tilting through the vast arid landscapes of Arizona, New Mexico, and Texas, hitting various national parks along the way. In the piney woods of East Texas, I turned northeast to meander through the fields, forests, and swamps of Mississippi and Tennessee, pulling over at various Civil War sites, then drove up into the old Appalachian mountain chain to take the view from on top of the Great Smokies. I descended to Asheville, North Carolina, and caught the famed winding Blue Ridge Parkway as the mountain rhododendrons and dogwoods broke into exaltations of early spring flower. I especially wanted to see Thomas Jefferson's house and gardens at Monticello in Virginia, after which, refreshed and inspired, I set myself for the long, last, trafficky leg up the New Jersey Turnpike and across the George Washington Bridge into New York. (After two weeks of glorious weather, as I pulled into the toll plaza, it began to rain.)

Across eleven states in sixteen days, I drove 4,052 miles. 'Hilde, my Swedish chariot, built in 1977 out of one and a half tons of steel, made about 18 miles to the gallon, meaning my trip was enabled by some 225 gallons of gas. That fuel contained enough energy to have kept my microwave humming continuously for nearly a year, but instead it was used to transmit one person, one car, and a load of stuff across the country. Those gallons of gasoline were refined from a bit less than thirteen barrels of oil (adjusting for refinery gain and efficiency), and assuming a uniform nationwide sampling in 1998, were drawn 45 percent from US wells and 55 percent from abroad. I could have sold my car in California and taken an airplane instead. Flying direct would have consumed less energy and oil—only the equivalent of two months of microwaving, and eleven barrels of crude for my share of the jet fuel. More stuff would have had to go on the truck that carried the bulk of my earthly belongings across the country in a container via Los Angeles to the City that Never Sleeps, expending a prorated 678 hours of microwaving equivalent energy, that is, diesel from approximately two barrels of crude, for my share.

A drop in the bucket of our nationwide consumption, my trip hardly registered in the prodigious travel logs of millennial Americans. A 2001 government survey of our traveling habits found that as a people we collectively traversed four trillion passenger-miles that year. (A passenger-mile is one mile traveled by one person; a car with three passengers driving one mile would generate three passenger-miles.) A 2009 update to the National Household Travel Survey showed that in the midst of the Great Recession, travel dropped by several percentage points, to a mere 3.7 trillion passenger-miles nationally. That's the equivalent of every American making four laps between New York and San Francisco—on average 12,159 miles per person per year—and that includes babies and the housebound. Because we sometimes ride together, those distances result in a total of 2.9 trillion *vehicle* miles traveled in 2009 ("vehicle miles traveled" is universally abbreviated VMT in the transportation literature), 98 percent of which were made in cars, trucks, motorcycles, and other personal motor vehicles. And those 2.9 trillion VMT are the main reason why we use so much petroleum, are forced to import oil from other countries, become upset when gas prices spike, and lose our shirts when the economy sputters and wanes after an oil shock.

Fair enough—we have seen oil's advantages for transportation: It's portable, energy dense, and for a long time (and again briefly in 1998), very cheap. But why do we drive so much? Where are we going? Looking at the television commercials, one would think that cars are mainly for family trips to the countryside and pick-up trucks for manly men to splash through trout streams, but in fact, most travel is for hum-drum reasons to everyday places: Commuters need to get to work, kids need to get to school, and we all need to go shopping, eat a meal out, visit friends and family, attend services, and take care of all the other tasks and events crammed into our busy lives. Meanwhile, behind the scenes, our stuff is on the move—more than sixteen billion tons of freight in 2009, traveling some 3.6 trillion ton-miles, trading time for space by truck, train, plane, barge, and pipe.

Trading Time for Space

Time and space are fundamental dimensions of the universe. It is a minor miracle of the modern world that we can get out of bed in the morning, take a car or bus to the airport, and be on the other side of the continent in a few hours. In doing so, we have traded a small amount of time (half a day) for a huge amount of space—some three thousand miles. And in those new spaces we can meet new people and do new things that weren't possible in the spot we started.

For most of humanity's history, our inability to trade time for space efficiently limited the circle of our interactions. Before the Industrial Revolution, you could move only as fast as muscle power could get you there—on either your own legs or those of an animal, at least on land. Over the water, given a sail, you might get some help from the sun, wind, and tide, but otherwise transportation was ultimately limited by how much energy could be gotten from sunlight (via food) into muscle, and thus into motion. For most folks, that meant living within walking distance of where they worked, which was often either in an adjacent room or in the fields nearby. Riding a beast or a coach enabled people to cover greater distances, but for most of human history, travel was lengthy, arduous, and expensive, the exclusive province of those with wealth, leisure, and time to burn.

The Industrial Revolution gave us the ability to release the energy trapped in fossil fuels like coal, and eventually oil, and use it to further human ends, like getting around. Within a few decades, Watt's improved steam engine combined with Henry Bessemer's new method for making steel led to the development of the steam locomotive pulling cars with steel wheels running along steel rails. Steel wheels running on steel rails, which generate about 400 percent less friction than rubber tires over pavement, allowed greater speeds for less energy. (That is, they were more energy efficient.) The locomotive outpaced all other forms of transportation, bridging North America by 1869, dropping the travel time between New York and San Francisco from four months by ship to six days by train. Growth was explosive, especially considering the kinds of technology available at the time to build the rails, mainly mules and men. In 1830 the total length of track in the United States was 23 miles. It grew to 2,808 miles by 1840—more than in all of Europe, where the first trains had been designed. By 1850 the United States had 9,021 miles of track—more than half the mileage in the world. By 1860 American rails extended to 30,026 miles, mainly in the north; by 1880, 93,267 miles; by 1900, 193,346 miles. In some parts of the country, wagon roads fell into disrepair and were abandoned altogether as railroads became the only way to go.

Initially, the railroad's expansion was driven by simple economics; new connections made new economies that redounded to the benefit of the railroad companies and the communities they connected. However after 1850, the government accelerated extension of the railroads into as-yet unsettled areas by giving away massive quantities of land. In the midst of the Civil War, President Abraham Lincoln signed the Pacific Railway Act of 1862, which enabled him and his successors to give away 129 million acres of land in the western parts of the country—about 10 percent of the land area of the United States—in less than nine years. At the beginning, land grants were given to the states, which then redistributed them to promote the railroads, an arrangement ripe for corruption. Later federal land grants were made directly to railroad companies. The railways then did what they pleased with the land, harvesting what natural resources they could, setting up company-owned stores supplied exclusively by the company-owned railways, and selling what land they didn't want to others. Other acts gave away land to timber, mining, agricultural, even educational interests, as in "land-grant" universities, like the one in Davis. And for the private citizen, there was the Homestead Act, which provided 160 acres to anyone willing to improve the land and live on it for five years.

For the railroads and the economy, farms connected to towns, towns connected to cities, and small cities connected to bigger ones, enabling economies of scale and new ways of living to develop. As the economy grew, the material quality of life improved for millions of Americans, many of them new immigrants, who were often bewildered by the rate of change. The memoirist and historian Henry Adams, speaking of his own

The Railroad Acts

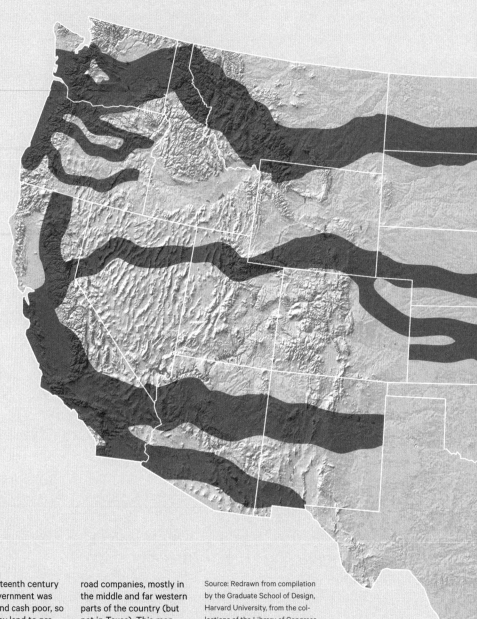

In the nineteenth century the US government was land rich and cash poor, so it gave away land to promote transportation and development. In all, we deeded about 10 percent of American land—over 129 million acres—to the railroad companies, mostly in the middle and far western parts of the country (but not in Texas). This map shows the approximate extent of the grants, which were given out in checkerboard fashion within the shaded areas.

Source: Redrawn from compilation by the Graduate School of Design, Harvard University, from the collections of the Library of Congress (digital id: mhsalad 120033).

childhood, claimed that "the American boy of 1854 stood nearer the year 1 than to the year 1900." The Industrial Revolution transformed what it meant to be American.

Economically the great problem of the 1870s was deflation, not inflation. The Industrial Revolution achieved dramatic increases in production, aided by massive influxes of cheap natural resources made available through expropriation, genocide, and expansion of the railroads. Prices began to drop like a stone. Recall one of Rockefeller's justifications for the Standard Oil trust in that era was that cooperation was necessary in industry to keep business afloat; otherwise everyone would suffer as a result of low prices, the effect of "ruinous competition."

Although you would think that low prices would provide for the general welfare, in fact, they caused trouble for people who had commodities to sell, like farmers. In a deflationary environment, a farmer taking a loan for spring planting would have to pay back that loan after the harvest with money worth more in real terms than the money originally borrowed, cutting into profits. Meanwhile selling the commodity—say, wheat—netted fewer dollars because of decreasing prices. Thus, deflation worked to the interests of bankers and against the interests of borrowers. (Inflation has the reverse effect, encouraging us all to be debtors.)

Prices, as we have seen, reflect more than the relative balance of supply and demand for a commodity; they also reflect the balance between the supply of money and the total amount of transactions people want to make. If the currency is fixed to a standard like gold, there can only be as much paper money value as there is gold in the vault (or some ratio of it); if more products chase less money, then prices drop, causing deflation. Thus, nineteenth-century farmers and populists like William Jennings Bryant wanted the monetary standard to be extended to silver, allowing more money to be minted and thus to circulate, allowing prices to rise (to "inflate"); bankers and others holding money advocated for keeping to the gold standard, which limited the amount of currency in circulation, because they stood to benefit from lower prices.

And benefit they did. The railroad barons, landlords of new estates and purveyors of monopolies at the railhead, gathered unto themselves enormous sums of money, the likes of which had not been seen before. But the real winners were the men who supplied the railroad companies and the expanding economy the raw materials that enterprise required. In turn-of-the-twentieth-century America, John D. Rockefeller became America's first billionaire and eventually the richest person who has ever lived selling oil; Andrew Carnegie, steel magnate, amassed a fortune worth $400 million; and J. P. Morgan, the financier, had $119 million to his name. These men translated their wealth into political power, and then used that power to push for more, both against the government, which ceded the basis for their wealth, and against the laborers who helped create it. The Gilded Age, 1870–1900, saw huge production gains and extraordinary inequity, driven by the incitements of practically unlimited profit.

In America, the story was told, anyone could be a millionaire; it didn't matter how you were born, only what you could do and who you knew. But having so much power in the hands of so few threatened the democratic experiment. How could there really be equal voices and equal votes when some Americans could "pocket" politicians through election contributions? How could a person get ahead when the fix was in for corrupt companies and their government minders? Corruption and fraud were rampant and for the most part unrepentant. The American psychologist William James diagnosed it sharply in 1907: "The moral flabbiness born of the exclusive worship of the bitch-goddess SUCCESS. That—with the squalid cash interpretation upon the word 'success'—is our national disease." It's an illness we've never quite shaken.

New Ways to Move

In the gilded cauldron of growing population, new technologies, expanded economies, filling lands, and growing inequity, another new technology was born that would revolutionize the relationship between energy, motion, and money: the internal combustion engine.

Combustion is just another name for fire. It reverses the process of photosynthesis, releasing energy in the form of light and heat, water, and carbon dioxide from organic materials in the presence of oxygen. The spark that starts the fire adds enough energy— the activation energy—to the organic materials to get an energy-producing reaction going. Then the combustion of materials itself spreads the reaction until the fuel is exhausted and has a lower, less orderly state, a plume of smoke in the wind.

Steam engines use combustion to generate motion. In locomotives, an engine burned coal to boil water to make steam to turn a wheel to propel the train. Internal combustion engines, in contrast, skipped the energetically wasteful intermediate steps of making and capturing steam by instead setting off a series of small, carefully controlled explosions inside an airtight cylinder. The force of the gases released from the explosion pushes a piston from one end of the cylinder to the other, which turns a crank, which turns the wheel. The spent gases—carbon dioxide, water, and other side products—are then released to the atmosphere when the moving piston reveals a small valve; the pressure gone, the piston falls back in place (or is driven back by a sidearm), and another explosion starts the process over again. (Diesel engines invented by Rudolf Diesel in 1892 work similarly, but because diesel fuel is so energy-dense the compression of the cylinder is sufficient in itself to set off the explosion. No spark plug required.)

The first internal combustion engines were European inventions. Two brothers, Nicéphore and Claude Niépce, working in Napoleon's France, were the first to link combustion to motion, building a boat they called ostentatiously *le Pyréolophore* (roughly translated: "the fire-wind producing machine"). Powered by a combination

Railroads and Streetcars Over Time

Railroads

Engines running on rails swept across the American landscape from 1830 to 1916, with over 250,000 miles of track laid during that time. Since World War I, there's been a long period of slow senescence, most markedly after 1970, though in recent times, there has been a slight uptick in railroad mileage. The United States still retains a world-class freight rail network, with over 162,393 miles of track in 2011, traversed by 24,250 locomotives pulling over 380,000 freight cars.

Source: Compiled from American Association of Railroads (2009, 2012), US Bureau of the Census (1975), and US Census Bureau and Social Science Research Council (1949).

Streetcars

After they were invented in the 1880s, electric street railways grew rapidly, fueling the first suburban housing boom. Then the cheap oil window and bad deals did them in; by the mid-1980s, we were left with only 384 miles of track nationwide, down from a high of 32,548 miles in 1917. Since the Reagan era, light rail/streetcar systems have grown by 285 percent to 1,477 miles of track providing over 466 million passenger trips in 2009.

Source: Compiled from American Transit Association (1971), US Bureau of Transportation Statistics (2012), US Census Bureau and Social Science Research Council (1949), and US Federal Transit Authority (2011).

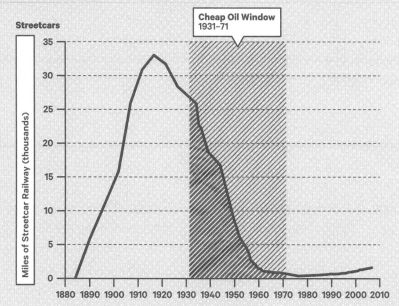

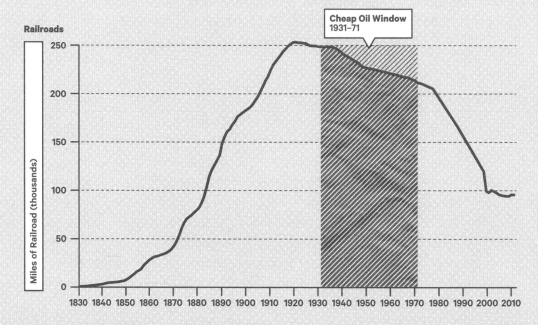

of *Lycopodium* moss, coal dust, and resin, it managed to propel a small craft against the current of the River Saône, near Lyon, in 1807. The emperor awarded the brothers a patent for their engine, and Nicéphore Niépce went on to make important contributions to photography. Almost simultaneously, François Isaac de Rivaz, a French-Swiss politician in retirement, created an engine that used hydrogen and oxygen gas—an early kind of fuel cell—to create a controlled explosion and move a self-propelled charrette (cart) on land; he too received an imperial patent.

But fuels less explosive than hydrogen gas and more energy-dense than moss were the way of the future. Rockefeller's kerosene and by-products like gasoline and diesel fuel fit the bill. Karl Benz, eventually of Mercedes-Benz, patented the four-stroke gasoline engine, driving his three-wheeled Motorwagen in 1886. Initial versions had problems, especially with steering, but rapid improvements led to the first commercial sale of an automobile in 1888. That August Benz's wife, Bertha, demonstrated the Motorwagen's long-range potential by taking their two young sons for a sixty-four-mile ride through the Bavarian countryside, scaring the startled country folk and providing a hint of things to come.

Others had also glommed onto the idea that the products of oil might be a better fuel. In 1872 an American engineer, George Brayton, patented a two-stroke kerosene engine design he called the Ready Motor. Brayton showed his Ready Motor at the Centennial Exposition in 1876 in Philadelphia, where George Selden, an inventor by avocation, a patent lawyer by occupation, admired it. He took Brayton's idea and designed a concept for a similar engine attached to a wheeled carriage, applying for his own patent in 1879. Though he seems never to have built a working prototype, Selden did manage to keep his patent application alive through a long series of carefully orchestrated legal maneuvers, incorporating new design features, some developed by others, into his concept along the way. The patent was finally awarded in 1895. In patent law, the prize goes to the first, not necessarily the best or the most beautiful. This meant that Selden's patent required all manufacturers of internal combustion gasoline road-engines in the United States to pay him a royalty, reducing profits for more skilled engineers such as Henry Leland (of Cadillac fame), Ransom Olds (of Oldsmobile), and Henry Ford (of the Ford Motor Company).

A Spark

Meanwhile, elsewhere in the Industrial Revolution, other scientists were hard at work divining the strange and mysterious qualities of electricity. Electricity is not energy itself but a carrier of energy; it is like gasoline, but more transient. The ancients knew how to generate charge by rubbing a piece of amber with cat's fur, and Benjamin Franklin, experimenting in the 1750s, showed that those electrostatic charges were remarkably similar to the sudden, silver bolt of lightning. What the storm and the rubbed amber generate, we know now, are disassociated charged particles, electrons, which individually weigh only an infinitesimal amount (about 2.0×10^{-30} pound), but en masse can turn rotary mixers, enliven computer chips, or propel freight trains. Electrons move along conductive materials as if they were water in a stream; when they move, they can carry energy via an "electric" current. When a kid sticks his finger in the socket, he becomes part of that stream, and he feels the current.

Michael Faraday, an English inventor and arguably the greatest experimental scientist of the Industrial Age, showed in 1821 that rotating an electrical conductor, like a spindle of copper wire, through a magnetic field generates an electric current; the magnet excites the electrons in the copper to move. Faraday then demonstrated that running an electric current back into the generator will rotate the conductor in the other direction; in other words, an electric motor is just a generator running in reverse. Reversibility meant that if you generate electricity with a dynamo (which uses water or steam to turn the conductor) at one end of a conductive wire, then at the other end that current from the wire can turn an electric motor. Electricity via a wire carries energy from one place to another.

Light was exactly what Thomas Alva Edison let out in 1873 when he invented the light bulb. (In his case, the light came from electricity heating up a resistor until it glowed merrily.) Less than a decade later, Edison installed the first electric transmission system in New York City, powered by six coal-fed dynamos in a basement on Pearl Street. On September 4, 1882, he switched on the lights in lower Manhattan, just a few blocks from the Standard Oil offices, producing enough light to outshine, and eventually outsell, Rockefeller's kerosene-fueled gaslights. (Kerosene lights flickered, smoked, and were dimmer than electric ones. Plus they had shown an occasional propensity to explode.)

Bottling the electricity to store it until you wanted to use it turned out to be electricity's biggest challenge, and it remains the problem for electric transportation today. Chemistry provides some options. Nineteenth-century chemists discovered that some compounds, especially some metals, turn out to be generous with their electrons; if you place them in solution, the electrons will float away and certain other metals that are greedy for charge will soak them up. The movement of electrons in solution causes a reverse movement of electrons in a wire connecting the generous and the greedy terminals (more commonly referred to as positive and negative, or more accurately, donor and acceptor), and creates what we know today as a battery. The term "battery" we owe to the ingenious Franklin again; he coined the term after connecting several early charge-holding jars together—and called them a "battery" because, like a group of cannons, they could be massed for effect.

While Edison was working to invent electric lights in New York, a French physicist, Gaston Planté, showed that running a current back into a battery could reverse the

Figure No. 19
Activation Energy

Activation energy—a small amount of input energy that starts a combustion reaction—is useful as both data and metaphor. In this case, the combustion of iso-octane, one of the constituents of gasoline,

is started by an electrical discharge from a spark plug, releasing the large amount of stored energy held by the fuel. Other kinds of changes also require enough energy to get things started.

Source: Based on data from the National Institute of Standards and Technology (2012).

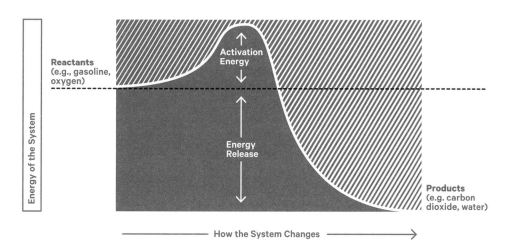

chemistry and store the charge until it is wanted, thus creating the first commercial, rechargeable batteries (the lead-acid battery, the same as in most cars today). Those batteries, connected to an electric motor, could turn a set of wheels or do other jobs until the charge ran out. Recharge them using another fuel source—an idea that Faraday had demonstrated decades earlier—and you were good to go again. Edison was not impressed; he thought rechargeable batteries were just European bunk, declaring in 1883, "The storage battery is, in my opinion ... a mechanism for swindling the public by stock companies. The storage battery is one of those peculiar things which appeal to the imagination ... [but] just as soon as a man gets working on the secondary battery [i.e., a rechargeable one], it brings out his latent capacity for lying." It turned out later that Edison had some latencies of his own.

The First American Cars

For decades, nineteenth-century city dwellers had ridden horse-drawn passenger railways—also known as trams, tramways, trolleys, trolleycars, or streetcars—along regular routes marked out by steel railways in the roadbed. They picked up and dropped off passengers when flagged or at designated stops, and, unlike a private taxi carriage, didn't need to be pre-hired. John Stephenson, an Irish-American entrepreneur, got them rolling first down the Bowery in New York City in 1832. They were successful, but problems rapidly developed with the power source. Horses have to be fed, which required massive supplies of hay and oats to be brought into the city, and once fed, horses created wastes that fouled the streets and made walking an unpleasant, unhealthy, and malodorous business, especially during fly season. Horses also die; a dead horse in a Manhattan street had to be hauled to the river and dumped in. Sanitation records from New York in 1880 showed fifteen thousand horses per year removed from city right-of-ways.

Electric streetcars were a blessing in contrast. Adapting Edison's new electric cables and dynamos to transportation was easy and remarkably popular. Propelled by overhead lines or underground connections and the occasional battery in rural areas, streetcars moved at a rate of 8–15 miles per hour, providing brief, affordable, pollution-free commutes. In 1889 America had fifty lines (or "roads") running over a hundred miles of electrified streetcar track; a year later there was twelve times more track than the year before. By 1902 there were 21,908 miles of streetcar lines, and fifteen years later, 44,835 miles—comparable in length to the modern Interstate Highway System. In 1912 streetcars provided over twelve billion passenger trips nationwide for ninety-five million Americans, an average of 126 trips for every man, woman, and child in the country. People rode them in every big city, most small ones, and many little towns, even in the countryside. Because streetcars required electricity, they drove the first electrification of many outlying areas. During World War II, soldiers, sweethearts, and war materials moved by streetcar. They were how everyone got around.

Gas vs. Electric

Alas! Streetcars did not last, despite their many advantages, a story we will return to later. Rather the future of American transportation at the turn of the century focused on a technological battle royale over personal transport—and at first it seemed that electric vehicles might carry the day.

Electric car pioneer Pedro Salom described the contrast between gas and electric vehicles this way, in 1896. "Of course there is absolutely no odor connected with an electrical vehicle, while all gasoline motors we have seen belch forth from their exhaust pipe a continuous stream of partially unconsumed hydrocarbon in the form of a thin smoke with a highly noxious odor. . . . Imagine thousands of such vehicles on the streets, each offering up its column of smoke as a sacrifice for having displaced the superannuated horse, and consider whether such a system [a system of gasoline vehicles, that is] has general utility or adaptability."

Salom had a better idea. "The electrical vehicle is almost noiseless. It does not begin to make the noise that even a horse makes on a dirt road, while a gasoline vehicle must, perforce, make a continuous puffing noise, due to the exhaust from the motor, and this is not only disagreeable, but to some people alarming."

Streetcars Everywhere

In 1902 US streetcars provided Americans 4.7 billion trips (up from just two billion in 1890). Each mile of track supported over 200,000 trips per year at a time when the population was one-quarter its current size. Even small cities and towns had streetcars, as shown in this tabulation of ridership in communities with fewer than 25,000 residents that year.

Source: Steuart (1905).

Town or City	Population	Streetcar Tracks (miles)	Streetcar Passengers	Trips per Person
Fort Smith, AR	11,587	8.9	731,553	63.1
Riverside, CA	7,973	9.5	547,051	68.6
San Diego, CA	17,700	16.6	2,220,000	125.4
Santa Barbara, CA	6,587	8.5	814,400	123.6
New London, CT	17,548	8.5	1,320,791	75.3
Stamford/Greenwich, CT	18,417	12.7	1,327,617	72.1
Pensacola, FL	17,747	9.0	998,290	56.2
Athens, GA	10,245	6.4	356,969	31.8
Alton, IL	17,487	12.3	1,497,130	83.6
Cairo, IL	12,566	9.7	870,838	69.3
Vincennes, IN	10,249	8.0	450,000	43.9
Burlington, IA	23,201	14.5	1,600,000	69.0
Ottumwa, IA	18,197	10.0	1,211,028	66.6
Wichita, KS	24,671	18.5	1,400,000	56.7
Shreveport, LA	16,013	8.8	1,450,000	90.6
Biddeford/Saco, ME	22,267	8.2	728,909	32.7
Benton Harbor/St. Joseph, MI	11,717	10.5	1,198,826	102.3
Menominee, MI	12,818	6.7	529,764	41.3
Vicksburg, MS	14,834	8.8	1,188,289	80.1
Springfield, MO	23,267	19.1	1,700,715	73.1
Great Falls, MT	14,930	11.9	939,436	62.9
Concord, NH	19,632	12.7	1,510,856	77.0
Laconia, NH	8,042	8.9	436,171	54.2
Long Branch/Belmar, NJ	16,148	23.7	3,737,541	231.4
Perth Amboy/Metuchen, NJ	19,485	9.1	880,128	45.2
Dunkirk/Fredonia, NY	15,743	7.0	681,770	43.3
Kingston, NY	24,535	9.2	2,217,334	90.4
Ashtabula, OH	12,949	5.8	999,857	77.2
Zanesville, OH	23,538	10.0	1,800,000	76.5
Greenville, SC	11,860	7.0	537,603	45.3
Austin, TX	22,258	13.9	1,213,703	54.5
Waco, TX	20,686	16.3	1,605,525	77.6
Ogden, UT	16,313	11.0	861,910	52.8
Burlington/Winooski, VT	22,423	11.2	1,270,136	56.6
Everett, WA	7,838	9.7	971,650	124.0
Janesville, WI	13,185	7.4	304,398	23.1

Salom was talking about his preference for the Electrobat, a lead-acid battery–propelled car he had designed with partner Henry Morris, which first trundled down Broad Street in lower Manhattan in 1894. Policemen marched in front to clear horses and pedestrians out of the way. The lead batteries were heavy: The first Electrobats weighed 4,200 pounds, about half as much as a modern Hummer, but within a year, the intrepid engineers had come out with the Electrobat II, weighing only 1,650 pounds, a little less than a 1960s Volkswagen Beetle. Soon a small fleet of Electrobat taxis was competing with streetcars for passengers on New York streets at startling velocities of nearly ten miles per hour. (A horse and carriage typically travels at three to five miles per hour.)

Of course electric-car batteries needed recharging—the energy has to come from somewhere. Salom and Morris had an idea for that too. They designed special garages where a mechanic could swap in a new battery in less than seventy-five seconds using a system of ropes, pulleys, and winches; the depleted battery could then be recharged from generators overnight in the shop for the next customer to use. They envisioned battery-recharging stations—like modern service stations but with batteries instead of gas—sweeping across American cities, fueling vast fleets of electric cars.

Unfortunately Salom and Morris were engineers first, and businesspeople second. They placed too much trust in their primary creditor, the Electric Storage Battery Company (ESB), which soon swallowed their company whole, leaving the hapless engineers on the street. ESB was absorbed into EVC, the Electrical Vehicle Company, and the EVC tried, as was standard business practice at the time, to establish a monopoly. The "Lead Cab Trust" wanted to control all ground transportation, including batteries, cars, recharging stations, even the roads. Eventually they acquired rights to the Selden patent for gasoline combustion engines in 1901, but the cartel failed, both as a monopoly and as a transportation company, in the face of the new developments on the internal combustion side.

Ford's Gift

Edison's industrial empire had many branches, including one in Detroit, where in 1896 a young engineer named Henry Ford designed and built his own self-propelled internal-combustion vehicle, which he called the Ford Quadricycle. Edison met young Ford and generously encouraged his non-electric, gasoline-fueled pursuits, but Ford had an independent streak; he quit to go off on his own. Like Salom and Morris, Ford also needed to learn business skills, bankrupting two companies in less than three years. But in 1903, the Ford Motor Company was born and the following January gave an exhibition of a new auto design on an icy lake in Michigan; Ford himself drove the test car one mile in 39.4 seconds, setting a new land-speed record of 91 miles per hour.

Ford was well aware of the competition on all sides, from electric cars, streetcars, the well-established railroads, and competing gas-powered car makers like Oldsmobile, which already was selling the Curved Dash, the first mass-produced, inexpensive internal-combustion American car. (The Dash cost $650 in 1901.) Also competing was Henry Leland, a nationally famous machinist renowned for his precision work on Samuel Colt's guns and rifles, who brought high-tolerance machining to the new Cadillac brand in 1902. But as the market was still young and relatively open, there was plenty of room for Ford's Model A in 1903, which turned a pleasant profit with cars priced at $750, saving Ford from another failure. Soon Ford was churning through an alphabet of new models. He tried luxury cars with the Model K of 1906, a six-cylinder model advertised as the "gentleman's roadster," but it didn't sell, so Ford redoubled efforts to build a better, cheaper car for everyman.

It really was every *man,* because most women of the time preferred electric vehicles. Electric carriages started reliably at the flip of a switch, didn't have gears, used a tiller to steer, and unlike gasoline engines, didn't require hand-cranking to start; cranks displayed an unladylike propensity to break arms and heads. Electric carriages weren't as fast or as powerful as the gasoline motors, everyone admitted, but for in-town driving, they were more pleasant, even genteel. They didn't make loud noises, and, as Salom noted, they didn't stink. Unfortunately they also weren't cheap, often running $2,000 or more prior to World War I, which meant that electric vehicles acquired a whiff of privilege that grated against leveling tendencies in Theodore Roosevelt's Progressive Era. Early adopters included Edison's wife, Mina, and Clara Ford, Henry Ford's wife, who categorically refused to ride in what she dubbed "explosion cars." Rockefeller drove both gas and electric vehicles, but was known to favor the bicycle.

Some see the defeat of electric vehicles by Ford's gas-powered, internal-combustion engine models as inevitable, especially after the Model T came out in 1908. The Model T was simple to drive, easy to repair, faster and more powerful than electric cars, and cheap: at first selling for just $825, reduced to $590 in 1912, and again to $345 in 1916. It was no car for rich old women. Ford had found the success he was looking for, and he pursued it doggedly on all fronts, breaking the legal stranglehold of the Selden patent in 1911, introducing moving assembly-line techniques on the factory floor in 1913, and in the spirit of the times, unveiling a profit-sharing plan with employees in 1914. With wages offered at an unheard-of $5 per day and a regulated forty-hour workweek, an employee on the Model T assembly line could own his own Model T with less than a year of savings. And that car cost less than a penny per mile to operate, fuel and maintenance included.

Ford's intentions were clear: "I will build a car for the great multitude. It will be large enough for the family, but small enough for the individual to run and care for. It will be constructed of the best materials, by the best men to be hired, after the simplest designs that modern engineering can devise. But it will be so low in price that no man making a good salary will be unable to own one—and enjoy with his family

The First Cars

At the turn of the twentieth century, it wasn't clear whether Americans preferred quieter, less polluting, more genteel electric cars like the Electrobat, or noisier, smokier, faster cars like the Curved Dash, powered by internal combustion. In the end, "explosion cars" won the day, riding a wave of cheap oil from American wells. The Model T was the greater charmer: simple to drive, easy to repair, and more affordable than other cars.

Source: Electrobat redrawn from *Outing Magazine* (1908); Curved Dash and Model T redrawn from photographs in the Library of Congress (e.g. digital id: cph 3c18747 and txuruny 04974, respectively).

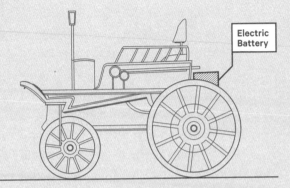

Electric Battery

Electrobat, 1896

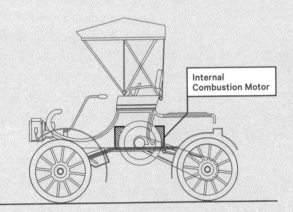

Internal Combustion Motor

Curved Dash, 1901

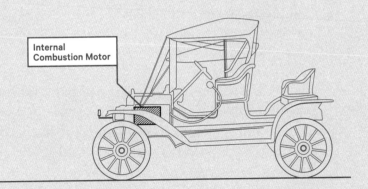

Internal Combustion Motor

Model T, 1908

the blessing of hours of pleasure in God's great open spaces." Americans loved it. In 1910 Ford factories churned out twelve thousand Model T gasoline-powered cars; by 1925 they were building nine thousand *per day.*

The advent of urban explorers motoring around the countryside was not always appreciated as an advance. Some of the horse-and-buggy set worried that the car would supplant the horse, which it eventually did, while others saw cars as tangible symbols of inequity, which they were, particularly before the Model T. "Our millionaires, and especially their idle and degenerate children, have been flaunting their money in the faces of the poor as if actually wishing to provoke them," an anonymous essayist warned in 1906. Woodrow Wilson, later president of the United States, gave a speech warning that "nothing has spread socialistic feeling in this country more than use of the automobile." Others just didn't like amateur drivers on their private property, with their accidents, breakdowns, and other mishaps. The farm magazine *Breeder's Gazette* described 1904 drivers as "a reckless, blood thirsty, villainous lot of purse-proud crazy trespassers." Rural vigilantes struck back by strewing broken glass, dropping trees, and digging ditches across country roads; some even took potshots at drivers enjoying God's great open spaces.

John Burroughs, the early-twentieth-century naturalist, a good friend of both Ford and Edison, was wary of the car Ford sent him—for an entirely different reason—writing:

> I see what a fraud the car is—how much it has cheated me out of. On foot and lighthearted, you are right down amid things. How familiar and congenial the ground is, the trees, the weeds, the road, the cattle look! The car puts me in false relations to all these things. I am puffed up. I am a traveler. I am in sympathy with nothing but me; but on foot I am part of the country, and I get it into my blood. If it were not for Mrs. Burroughs I should hang up the car.

Eventually electric cars were outcompeted, farmers won over, and naturalists ignored, and the gasoline car triumphed. The last hurrah of the early electrics was a much publicized collaboration by Ford and Edison to develop a better, cheaper electric car, based on Edison's improved storage battery. (Edison had latently come around to see the potential advantages of a rechargeable battery.) They labored from 1912 through 1914, making high claims in the press for an electric vehicle that would retail for less than $1,000 and run one hundred miles on a charge; to prove their claim, they built two viable prototypes. Ultimately, however, the great geniuses of electricity and automotion could not pull the project off, even after Ford invested over $1.5 million of his own and ordered 100,000 of Edison's batteries.

It's not clear why the collaboration collapsed; some have claimed that oil interests pressured the pair or caused a mysterious fire at Edison's research facility in New

Jersey in June 1914. Others claim that Edison's batteries, like most modern examples, just could not compete with the energy density of oil. (Remember that there are almost six hours of microwaving energy in a pound of gas and therefore thirty-four hours in a gallon of gasoline; modern lead-acid batteries manage only four *minutes* of microwaving energy per pound.) In any case, world events soon took precedence. World War I broke out in October 1914. Ford was a pacifist and spoke out against the war, but when the United States declared war against Germany in 1917, Ford Motor Company dutifully provided engines and vehicles to carry men and equipment into the trenches of Flanders Fields. Edison and Ford remained good friends for the rest of their lives, but after 1914 they never worked on an electric car again.

Though electrics could still be seen on the road when the next world war broke out, Ford's better, cheaper, internal combustion automobile had won—and so had he; he became the wealthiest man in the world after Rockefeller passed on. In 1947, the year Henry Ford died, Americans traveled 371 billion vehicle miles.

Parallel Infrastructure

Ford's cars needed roads to run on—neither Burroughs's weedy tracks nor the steel railways of the locomotives would suffice. (And streetcars share space uneasily.) A road is by definition a reinforced surface designed for vehicles; it placates the topography, bridges waterways, pacifies the mud, seals in the soil, and enables faster, more comfortable trades of time for space. There were over 2.4 million miles of roads in early-twentieth-century America, but the vast majority—93 percent—were unpaved. Some were just a pair of parallel ruts through the countryside, which degenerated into quagmire as soon as it began to rain.

The Good Roads Movement was initiated by bicyclists, not motorists, in the 1880s. Bicycles were the craze at the time, especially in towns and cities. Mass excursions were organized for both men and women, who rode Albert Augustus Pope's Columbia safety bicycle, another transportation monopoly. The Columbia would look familiar to us today, with its system of gears, brakes, and equal-sized wheels; earlier bicycles, called ordinaries, had no gears, no brakes, and a larger front wheel than back. Bicycles were true auto-mobiles, requiring only a pair of legs and a desire to go; quiet, exhaustless, thrilling, they worked best where roads were smooth, clear, and dry.

Up until then, roads were mostly a local and private matter. Though Congress sponsored a toll turnpike in the 1830s from New Jersey to St. Louis, money ran out when the railroads began to expand. By 1904 America had only 204,000 miles of improved roads to match its 297,078 miles of railroad, and most road "improvements" were just grading, at best gravel or macadam (a kind of crushed gravel); only on rare occasions, mainly in cities, would someone lay some tar or stones to keep down the dust. Some of these improved roads were turnpikes, maintained by private companies for a toll, but what bicyclists and motorists really wanted were free roads funded by the government. After all, a gas-powered Model T was no good if the conditions forced you to proceed at a fraction of maximum potential speed; you might as well ride your horse. As if to make the point, the Good Roads Movement advocates traveled on a Good Roads Train.

For decades in the early twentieth century, pro-road and anti-tax lobbyists argued in state capitals and Washington about who should build roads and whether it was proper for the government to tax the people to pay for them. Road advocates proposed various routes with appealing names, and intrepid early drivers took off on transcontinental adventures to prove them workable: the Lincoln Highway, connecting New York and San Francisco; the Dixie Highway, connecting Chicago and Miami; the Lakes to Gulf Highway, from Minnesota to Texas; and so on. Congress passed a federal aid package for highways in 1916, offering to split the cost of road construction with the states, but the bill failed to require localities to coordinate road placement. Though there were more drivers than ever before—2.3 million registered vehicles in 1915, 17.5 million in 1925, 22.6 million in 1935—improved roads were scattered aimlessly in small strips across the map, reflecting the whims of local politics. The 1916 law did require states to develop professional highway departments, staffed by engineers, if they wanted to receive federal money. These engineers brought data, sharp pencils, and a heightened sense of the importance of roads to the task of preparing for a national system of highways to come, even if no one was bound by the plans they drew.

Federal money was needed to buy land for roads because there were no longer federal lands to give away, as there had been back in the railroad era. This time, the treasury would lead with tax incentives, taking from the people to give to the people. While the haggling continued over the proper role of government in transportation infrastructure, several states approved toll highways built to the standards engineers were devising for the new expressways. They also approved fuel taxes—a kind of excise tax—on gasoline and diesel fuel to help pay for roads, the first in Oregon in 1919. By the time the cheap oil window opened, all states had fuel taxes and then the Feds added one of their own, a penny per gallon, in 1932. In 2012, the federal excise tax was 18.4 cents per gallon on unleaded gas (with states adding another ten to thirty-three cents per gallon), collected at point of purchase.

Tax-resisters continued to fight road-spenders through the Roaring Twenties, the Great Depression, and World War II, as Americans continued to take to the car in larger and larger numbers. By 1945 there were 25.8 million registered motor vehicles in America driving on 1.7 million miles of improved roads. Bicycle lobbyists were long out of the picture, their role assumed by a powerful consortium of automotive, oil, rubber, and real estate interests, who carried the day by arguing that a federal roads

The Paving of America, 1900–2008

The United States entered the twentieth century with over two million miles of roads, mostly unimproved and largely impassable in poor weather. As the car took hold, we built additional roads, and paved many more. The Interstate Highway System, a generously funded mandate begun in 1956, set the pattern for late-twentieth-century road construction. The federal government provided 90 percent of the costs to the states, which eventually built over 47 thousand miles of limited access highways, creating the backbone to which hundreds of thousands of miles of smaller roads could connect.

Source: Compiled from US Federal Highway Administration (1967, 1987, 1997, 2011a) and US Public Roads Administration (1947).

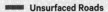

■■■■■ Unsurfaced Roads

■■■■■ Gravel Roads

■■■■■ Paved Roads

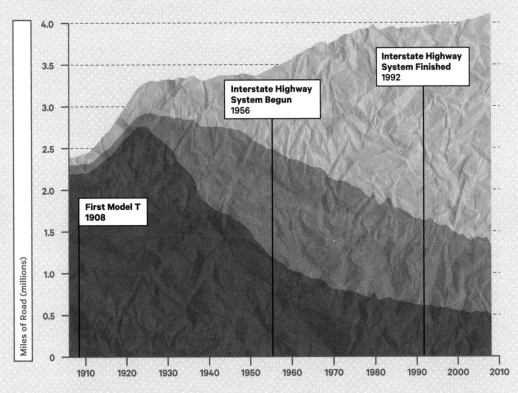

First Model T
1908

Interstate Highway System Begun 1956

Interstate Highway System Finished 1992

Miles of Road (millions)

program was necessary to provide jobs for veterans, familiar now with the jeep, truck, and jerry can, returning from Europe and the Pacific.

President Roosevelt signed a bill in 1944 authorizing a national interstate highway system, and President Eisenhower, after experiencing for himself America's terrible roads on a sixty-three-day transcountry odyssey in 1919 (which he later contrasted with the meticulous autobahns he found in fascist Germany in 1945), lobbied hard for the money to pay for it. In 1956 $25 billion was appropriated to build a true, coordinated, interregional highway system planned to cover forty-one thousand miles, the largest public works project ever attempted in the history of the world. We know it today as the Dwight D. Eisenhower National System of Interstate and Defense Highways, or the interstates.

The federal government agreed to pay 90 percent of the costs for right-of-way, labor, and materials through large block grants to states and local governments, which would provide the remaining 10 percent. States still determined the exact alignment (location) of the highways, but the Federal Bureau of Public Roads would make sure they lined up, and provided exacting design standards that had to be followed. Often the interstates ran near the old railroad lines, which were being abandoned or turned over for freight; in other cases, new interstate routes were carved out of city neighborhoods and farmland seized by eminent domain.

Federal investment in interstate roads was the death knell for intracity electric streetcar lines, which had been failing, despite increasing ridership through the Great Depression and during World War II; similarly, intercity passenger rail faded to just a shadow of its former self. Although the playing field was already tilted toward the motorcar by the cheap oil window and new government investments in roads, some in the auto industry thought it necessary to further increase the pressure on the struggling urban trolley lines. In 1947 General Motors, which had overtaken the Ford Motor Company as America's number one carmaker, was indicted as part of a consortium of oil and rubber companies involved in an elaborate scheme of buying out streetcar lines, replacing them with buses, and then pruning the bus lines back to nothing. Though not found guilty of trying to control transit lines, as is sometimes claimed, they were convicted of monopolistic practices and told to pay a nominal fine, but by that time, it didn't matter. The first era of the electric streetcar was finished by 1960. Few cities could afford both excellent public transportation and great roads, and in the cheap oil window, private transportation via the gas-powered automobile driving over free public roads won again and again.

In 1992, the Interstate Highway System was finally completed with the opening of the Glenwood Canyon stretch of I-70 in Colorado. The total cost of the project had risen, as it inevitably would—to over $114 billion—and extended to forty-sixty thousand miles of four-lane, limited-access, well-engineered expressway, the principal conduits

Vehicle Miles Traveled, 1900–2011

The history of twentieth-century America can be read in the small pauses and triumphal accelerations of the increase in vehicle miles traveled (VMT)—so it came as a great surprise when VMT seemed to plateau in 2007 at over three trillion miles, then decline slightly. From 1900 to 2010, Americans drove collectively 112 quadrillion miles, equivalent to 1.2 million astronomical units (the distance from the earth to the sun), or about 19 light-years.

Source: Compiled from US Federal Highway Administration (2010, 2011a).

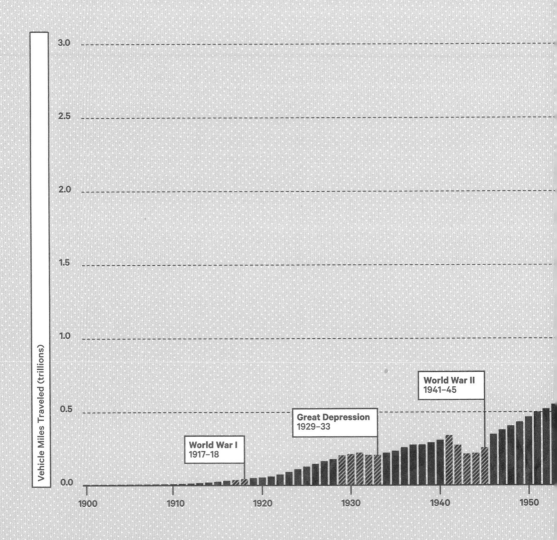

World War I
1917–18

Great Depression
1929–33

World War II
1941–45

Vehicle Miles Traveled (trillions)

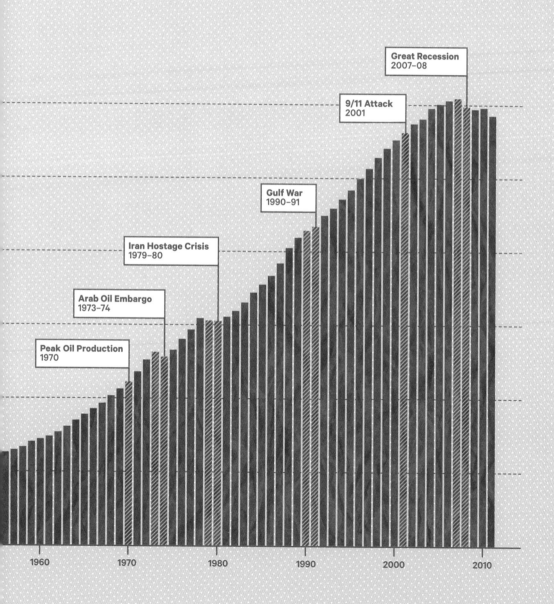

Great Recession
2007–08

9/11 Attack
2001

Gulf War
1990–91

Iran Hostage Crisis
1979–80

Arab Oil Embargo
1973–74

Peak Oil Production
1970

1960 1970 1980 1990 2000 2010

The Cost of Owning a Car, 2011

Motor vehicles are expensive to own and operate, costing the average citizen 29 percent more than food and 152 percent more than healthcare each year. Fixed ownership costs include vehicle purchase, interest, insurance, license, and registration, paid regardless of how much you drive; variable operating costs include fuel, maintenance, and tires, which depend on your mileage. Gasoline is only about one-fifth of motor vehicle costs (as of 2011), yet it looms large in our economic imaginations, because a car without gas is just an enormous, wasted investment.

Source: Compiled from American Automobile Association (2012 and previous annual editions) on owning and operating a car, and the US Bureau of Labor Statistics (2012b) on consumer expenditures by category.

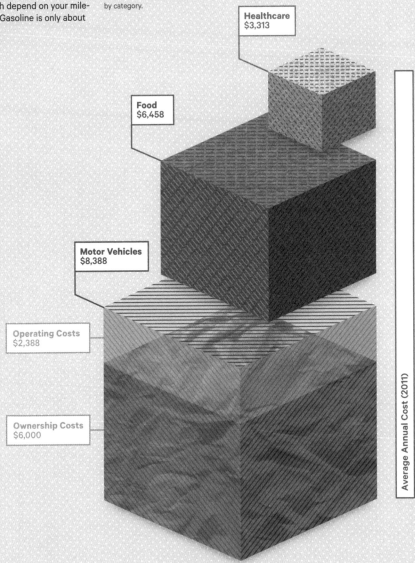

Healthcare
$3,313

Food
$6,458

Motor Vehicles
$8,388

Operating Costs
$2,388

Ownership Costs
$6,000

Average Annual Cost (2011)

of a network of main arteries and local streets that stretches into every neighborhood in America. Today US local, state, and federal governments manage over 8.5 million lane-miles on four million miles of paved roads. Another 380,000 miles of unpaved roads riddle our national forests and other public lands; the US Forest Service is the greatest road owner in the world. A century after the invention of the Model T, the farthest a person can get from a road in the conterminous United States—an area of over 3.1 million square miles—is just a bit over twenty miles.

Driving America

Given this history, it is tempting to think that cars, roads, and gasoline were inevitable, a manifestation of destiny, part of some kind of preordained plan, encoded in our cultural DNA, to ensure that everyone in America would drive. But that's the Siren song talking. Thinking that way requires neglecting all the players—the lobbies, planners, industrialists, inventors, and government bureaucrats—who were also part of the design to make us drive. It also means ignoring the generations of Americans who lived before the car was invented (presumably George Washington felt American even while riding his horse) and forgetting about the trains, bicycles, streetcars, and electric cars that once got us around. Car companies, rubber companies, oil companies, and their advertisers saturate the landscape with the useful myth: Americans wouldn't be American if we didn't drive.

Obviously America, and being American, amounts to more than a transportation mode, but I also hear the car's appeal to identity. Like personal flags, our automobiles convey more than our physical selves. They also declare in public our wealth (by what we drive), our personality (by how we drive), our sex appeal (by whom we drive with), and our beliefs (by what we affix on our bumpers). We drive automobiles to places that we have to go (work, shopping) and to places that we want to go (dates, vacation, second homes in the country). We spend on average 415 hours per year in the car, the equivalent of 2.5 weeks. Necessity is coupled with pleasure and isolation from the madding crowd.

We also bond with our cars simply because they are everywhere in America. During our long journey as a species on earth, human beings have developed a special and, arguably, unique ability to acculturate and identify with our surroundings. Boys in particular connect powerfully to moving things. I noticed it when my son was three; he latched on to identifying every kind of moving vehicle that crossed his path: Trains, planes, cars, and especially taxicabs were the greatest delight in his life, not the dogs, birds, frogs, and other living, moving things that my wife and I diligently tried to show him. It was the same for me. When I was a kid visiting family in Colorado in the 1970s, my grandfather would take my brother and me around the farm in his pickup. To this day, I associate the smell of dirt in the cab of his truck with being a man—not my grandfather's hat, his boots, his unnerving silences, his unending service to the family—but the musty, dusty smell of the cab of his Ford half-ton. Someday, I believed, when I'm a grown-up, I'll have a pickup truck just like Grandpa's.

Cars imbued with identity and associated with free movement are constant companions in a turbulent and fast-moving world; no wonder they mean so much. Even for environmentalists, cars are important. When I finally broke down and bought a car in college, a kind of rite of passage of its own, Brünnhilde became my own personal flag of identity. I loved that car: with its sticker of earth as a blue marble; its length, just right for me to unroll my sleeping bag in the back with room (potentially) for someone else; its faithful hum driving home late at night. On road trips friends and I would play a mix tape of road music on the cassette deck, leading off with Richard Wagner's leitmotif: Ride of the Valkyries. But the thing I loved most about my car was the feeling that I could go anywhere, anytime.

Not Driving

So we drive, conscious of the virtues and blurring the costs. Perhaps it is only later in life that one begins to realize that feeling free isn't the same as actual freedom. Freedom presupposes I'm making my own choices, but what if all the choices are determined by someone else? I might choose a color, model, stereo system, or bumper sticker for my car, but can I practically choose to do without a car at all? Not in today's America: We all have so far to go, we all have to drive. Our chariots of freedom chain us to the politics, economics, and perils of oil.

In fact there are only three groups of Americans today who get by without automobiles—the young, the dense urban, and the poor. The poor do not have the money to entertain this part of the American Dream because cars have become expensive. Long gone are Ford's penny-a-mile motorcars. The average American spends 21 percent of her disposable income on transportation, almost as much as food and healthcare combined. A study by the American Automobile Association (AAA) in 2011 found that Americans paid on average $8,388 per year per car, or $22.96 per day. Fixed costs were higher for large cars and SUVs ($21.75 per day), and lower for small cars ($11.76 per day). On top of those lump sums for insurance, license and registration, depreciation, and financing there are the variable costs for gas, maintenance, and tires, calculated at an average 19.64 cents per mile. So, if the average American drives an average 12,000 miles per year, then he or she is also spending another $2,388 per year in gas and maintenance, $6.54 per day, to go places; more when gas prices spike up, causing the prices of everything else to spike up, too.

If anything, the costs of driving are higher in cities because of less parking, pricier fuel, lower fuel efficiencies, higher insurance costs, and more repairs from stopping and starting on traffic-congested, potholed roads. Because driving in town implies so much hassle and expense, millions of urban Americans opt out of having a car altogether and find other ways to identify themselves—clothes, shoes, jewelry, tattoos, and witty conversation. Moreover, many city denizens enjoy the freedom of choosing between walking, bicycling, or taking the subway to get around town; they delight in the independence that public transportation provides with the same passion with which they disparage the system when it fails them. (Traveling underground, while made convenient by lack of traffic, is a less than ideal way to enjoy the urban tableau.) Density contributes to shorter trip distances, and shorter distances make walking or biking and public transit amenable, even preferable. Some recent data indicate that poor people gravitate to urban areas, despite high costs of living, because in town they can make trades of time for space more cheaply (anything less than $23 per day is a bargain, according to the American Automobile Association) and because cities provide the best chance to get ahead economically. However, even in urban America, the carless are the exception, not the rule. In New York City, with nation-leading density, a world-class public transportation system, and an expanding system of bike lanes, 44 percent of households (including mine) still own a car.

The future lies with the young. Many youngsters born in the twilight of the twentieth century are opting out of the car, at least until they settle down. First-time drivers are getting their licenses later than they used to, not only in America, but in Britain, Canada, France, and South Korea. Sixteen to thirty-four-year-olds in households with over $70,000 in income increased their public-transportation use 100 percent from 2001 to 2009, according to one study. Why? It may be that young people move around more and have families later than their folks did, or it may be, given the economic times, that they are less able to afford the car that for their parents was a rite of passage. Or perhaps it's just that being stuck in traffic isn't that cool anymore.

Nevertheless, cars have a lot of advantages when oil is cheap, roads are clear, and we have a long way to go. Most people like to get out of town to see the countryside. Others drive across states to visit family or take a new job. Ford's Model T and subsequent generations of automobiles, SUVs, and pickup trucks gave us the chance to trade time for space, to make choices, to enlarge the circle of our interactions. And where did we decide to go?

To the suburbs.

6

The Great American Expansion

Our property seems to me the most beautiful in the world. It is so close to Babylon that we enjoy all the advantages of the city, and yet when we come home we are away from all the noise and dust.

From an unknown correspondent, writing by cuneiform letter to the King of Persia, circa 539 BC (cited in *The Crabgrass Frontier: The Suburbanization of America* by Kenneth T. Jackson)

As I took up my new job at the Wildlife Conservation Society a decade and a half ago, my greatest fear wasn't traveling to far-off countries, encountering fearsome animals, catching tropical diseases, or grappling with hard scientific questions: It was finding a place to live. In the go-go market of the late 1990s, New York real estate was expensive and fiercely fought over; many apartments would be listed in the morning and gone by the afternoon. And I didn't want to settle for just anything.

Fortunately the Society came through and offered me an apartment on City Island, a small, mainly residential spot of former forest in Long Island Sound, connected to the mainland Bronx by a bridge. The Zoo was six and a half miles away, with free parking. To get to work, I could ride my bike in about thirty-five minutes; or take the bus, about forty-five minutes with one connection; or drive the Volvo, and it would take twenty minutes to an hour, depending on traffic. I liked City Island—though my apartment was small, I lived close to my colleagues, with whom I shared many interests; I had a small garden out back for flowers and my dog; the island offered views of sailboats at the end of every street; and the city provided a large park nearby with a beach and even a wildlife sanctuary. There was ample free parking. I had moved to the suburbs, New York City–style.

What People Want

I wasn't alone. One estimate had about 47 percent of Americans—somewhat over 132 million people—living in suburbs in 2009. There must be something there that people like. What might it be?

A study investigating American preferences in 2005 found out what every real estate agent knows: Home buyers want to be close to friends, family, and other folks similar to them; we like open views and proximity to natural areas and recreational opportunities; we value attractive and comfortable houses; we prefer safe neighborhoods (especially if sprinkled with good schools); and, given a choice, we want to live close to our workplace.

What goes unsaid in this kind of survey is that most of us also need a job and housing we can afford. Finding the solution to these competing demands—work, affordability, social comfort, openness, amenities, safety, and proximity—is not easy. And it is the main reason that we drive so much: Driving is the safety valve for the modern lifestyle. We will put up with a long commute if it makes the rest of our wish list come true.

For most of us, the landscapes in which we make our choices were not designed by us; they were shaped by nature first, and then remade by people who came before, who planted the land, laid the streets, and built the houses. Earlier generations also faced choices about housing, transportation, and energy, and our landscape expresses their solutions, just as our choices will set the context for the future. For most of American history, suburbs were rare, a province of the rich and powerful, available only to those who could afford to travel. The sprawling landscapes of today—the "crabgrass frontier"—are a relatively new invention, prescribed to cure a problem of the nineteenth century: overcrowding.

Density Dependence

The American nation swelled with people in the years after its founding, growing from less than four million souls when George Washington was president to over seventy-six million by the time Theodore Roosevelt was sworn into office in 1901. (The US population in 2012 was 314 million.) Many of the new Americans poured into the cities of the East Coast during the nineteenth century. These cities were renowned for their crowds, noise, and danger because of a phenomenon that biologists have long recognized called density dependence.

Population density is the number of people (or any other critter) living in a given area. It can be calculated for a neighborhood, city, county, state, or country—often with different answers depending on how the boundaries are drawn. For example, the 2010 census holds that the population density of City Island, where I live, is 6,917 people per square mile when calculated over the land area of the island; 1,204 when measured over the boundaries of its census tract, which includes the nearby park; and 1,263 when calculated over the area of its zip code. Population densities vary tremendously in today's America: Some counties house fewer than one person per square mile, especially out west, while New York County (Manhattan), the most densely populated place in America, is home to more than seventy thousand people per square mile.

Urban densities were once even higher. In 1910, 2.2 million people lived in Manhattan, an area of about eighteen square miles. That's a density pushing 122 thousand people per square mile overall, with some neighborhoods—like the Lower East Side—holding even more. As you might imagine, such enormous population pressures caused tremendous problems, and also enabled a few important pleasures. The pains and pleasures that change as density does are called density dependent.

One of the best attested facts of economic life is that vitality follows density. Having more people in a place creates a fortuitous aggregation of producers and consumers that reduces transportation and communication costs and creates profits. Innovation thrives in cities because there are more minds thinking in the same place, responding to each other, and responding to the diversity they see before them, whether by accident or design. That combination of diversity, serendipity, and fluidity of thought engenders creativity and generates economic activity. The case of the oil developer George Bissell, described before, is illustrative: He had his inspiration about drilling for oil not in rural Pennsylvania, but by chance in New York City, taking the shade under an awning. Same with other inventive minds: Franklin lived in Philadelphia;

Density Dependence

Wildlife populations vary over time, typically as a result of interactions that depend on density (the number of critters in given area). As a result, some populations bounce up and down; some roar and crash; others are quiet and steady; and then for one species, there seems to be only increase. Can it last? Only if the human population can manage the negative aspects of density dependence (e.g., war, disease, wastes, resource exhaustion) and amplify the positive ones (e.g., creativity, conviviality, culture, economic growth).

Source: Redrawn from MacLulich (1937), Turchin et al. (2000), and US Census Bureau (2012a).

Snowshoe Hares and Canadian Lynx, Hudson Bay Company

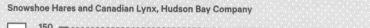

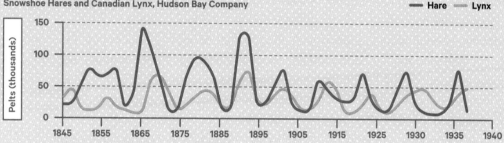

Lemmings in Kilpisjärvi, Finland

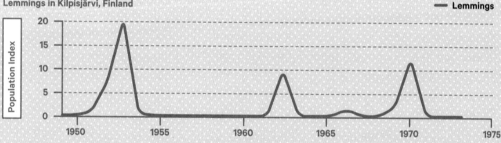

People in America

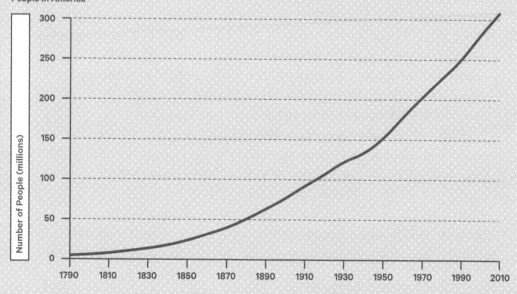

Joule in Manchester; Benz in Mannheim; Faraday in Oxford; Ford in Detroit; and Edison in Boston, New York, and Newark, before settling in the wealthy suburbs of Menlo Park, New Jersey.

Sadly, these positive aspects of density dependence can emerge only if the negative aspects can be controlled or mitigated in some way. Disease, for example, spreads more rapidly in dense populations, risking epidemics. Dense populations also produce concentrations of waste, which if not removed, can literally poison the nest or the city. Higher densities of people or other organisms can outpace a place's natural supplies of food and water. When this happens to animals, most move to greener pastures; human beings use trucks, trains, pipes, and aqueducts to bring additional goods into town.

In nature, population size is regulated by the positive and negative effects of density dependence and occasional shocks from the outside, like a flood or a fire. (Such calamities are called density-independent factors.) The population of some species oscillates up and down like a wave over time (for example, the population of the Canadian lynx chases the population of the snowshoe hare in the boreal forest); other populations ascend to dizzying heights, crash cataclysmically, and then recover (think lemming). Sometimes species populations seem to vary randomly, especially if density-independent factors dominate (think locust). And then there is the human population curve, which has been bending upward at ever-steeper rates since the city developed ten millennia ago. Ecologists grimly joke about whether the human population curve possesses a form unique in nature, or whether it's just a case of the lemming curve that hasn't run itself out yet. We shall see.

The negative aspects of density dependence, no more desirable for people than they are for bees and bunnies, were in full flower in late-nineteenth-century American cities. The photographer and journalist Jacob Riis gave us this startling description of life on the Lower East Side of Manhattan in 1890:

> In the tenements all the influences make for evil; because they are the hot-beds of the epidemics that carry death to rich and poor alike; the nurseries of pauperism and crime that fill our jails and police courts; that throw off a scum of forty thousand human wrecks to the island asylums and workhouses year by year; that turned out in the last eight years a round half million beggars to prey upon our charities; that maintain a standing army of ten thousand tramps with all that implies; because above all, they touch the family life with deadly moral contagion.

Crime surged alongside poverty in New York, Chicago, San Francisco, and other cities, fed by and feeding the corruption of city officials. Infant mortality rates rose by two-thirds between 1810 and 1870. Consumption, pneumonia, bronchitis, diarrhea, typhoid, and other diseases ravaged the families of tenement dwellers, creating terrible misery. These problems were compounded by the effects of having factories right next door. The factories required raw resources from the countryside to come in through the city, and they passed noise, heat, and light back out, also congesting the streets with people, horses, wagons, and railways, and filling them with the products they manufactured, as well as garbage, offal, manure, and other wastes. The clamor was unbearable. Soil and water reeked. Some of the dreck still festers underground today in brownfields, a pox on postindustrial cities ever since.

It was a terrible way to live, but most people had no choice. Cities were where the jobs were. The wealthy could and did buy their escape from the crowds; they lived in protected enclaves in town and frequently escaped to nearby country homes. The poor suffered in place. It took longer than anyone might wish, but in the closing years of the century reform finally took root, leading to less corruption, enforced construction standards, and investment in municipal sewage and garbage collection systems. Cities also started to plan their development through land-use zoning, bringing order to the chaos of urban spaces. These innovations all acted to control the negative aspects of density dependence.

Another way to deal with the woes of city living was to get out. Mr. Edison's electricity, Mr. Ford's automobile, and Mr. Rockefeller's oil meant that people of sufficient means—the emerging middle class—could vote with their feet and leave the city for more commodious areas nearby.

The First Suburbs

How exactly to define a suburb is a matter of debate among scholars, though most would recognize some combination of the following attributes: areas of low-to-moderate population density, single-family housing, more separation of commercial and residential land uses than in cities, and the now-ubiquitous commute, dependent on mechanized transportation. Today that mechanized transportation mode is the car, but the first suburbs were facilitated by the streetcars we met before. Powered by electricity, streetcars provided brief and affordable commutes to people living within a few miles of downtown. The first suburban communities were denser than our suburbs today, but less dense than the tenements in the central city.

Because these new streetcar suburbs on the edge of the city were built for people of varied means, they provided a variety of housing. Single-family homes were intermixed with apartments and mom-and-pop stores, all laid out within walking distance of the streetcar lines. Homeowners cultivated small yards with fruit trees and vegetable gardens, in contrast to the flowerbeds and alleys of trees marking off second homes for the wealthy, which were being rapidly surrounded. Many streetcar suburbs—in Dorchester, Massachusetts; Shaker Heights, Ohio; Highland Park, Pennsylvania;

Coral Gables, Florida; and West Hollywood, California, for example—are still intact. Walking their side streets away from the main driving thoroughfares remains a distinct pleasure.

Suburbs were a huge hit because they disconnected where people lived from where they worked, reducing the negative aspects of density dependence while maintaining the positive ones. Commuting into the city, workers (and employers) received the positive economic and cultural density benefits of the big city; commuting out in the evening, workers returned to quieter, safer, healthier, greener spaces. The streetcar, a technological marvel, was wildly popular.

As streetcars enabled better lives for workers, they also provided a new horizon for real estate development, one with ample scope for profit. Municipalities encouraged the growth of streetcars by granting long-term monopoly concessions that guaranteed a fare base. Fares were fixed by the government, for many years, at a nickel a ride. Owners capitalized on their monopolies by making land deals in outlying areas ahead of the track—purchasing "greenfields" (typically agricultural areas on the edge of town) and then subdividing and selling the land along the way. In the streetcar suburbs, most people bought lots first, then built or contracted to build their own new homes. Sears, Roebuck and Co. and Montgomery Ward made their names selling prefabricated houses to new homeowners by mail. As the streetcars reached further out, the amount of land available increased parabolically, according to Archimedes's old rule that the area of a circle grows by three times the square of the radius ($A = \pi r^2$).

Having purchased, subdivided, and sold farm fields as building lots, developers locked in a captive market for their transportation service. As the streetcar lines expanded, sometimes extending ten miles from downtown, nearby property rapidly rose in value, especially as others built farther out the line and envied those closer in. Land values had always appreciated close to town—people spend money to be near the action—but now this effect extended outward, generating new waves of value for developers, builders, and residents.

One potential sticking point of the Progressive Era building boom was that house building needed to be capitalized—in other words, someone had to pay up front for the wood, nails, and other materials, as well as the labor to build the new home. Most Americans didn't have that kind of money, and the culture of the time looked skeptically at borrowing, with its risk of long-term indebtedness. Banks viewed the American consumer with equal doubt, and so most turn-of-the-century mortgages were written with terms of only three to five years, allowing only interest to be paid during the term of the loan. Borrowers had to make large balloon payments at the end or renegotiate new deals every few years.

The citizens of streetcar suburbs solved the financial problem by providing the money themselves through newly invented neighborhood savings-and-loan

A Case Study of Manhattan and the Bronx, 1790–2010

Between 1890 and 1929, the population of Manhattan peaked at over 2.2 million people, then dropped by 20 percent, while the adjacent Bronx soared from 89,000 to 1.2 million souls during the buildout of the streetcar suburbs. After World War II, the population of both boroughs declined as cars enabled people to live farther afield, in the Great American Expansion into the automobile suburbs. Bereft of funds and population, New York City struggled through the 1960s and '70s with unrest and poverty. Since 1980 the population has started to grow and the city is thriving again.

Source: Compiled by Jackson (1995).

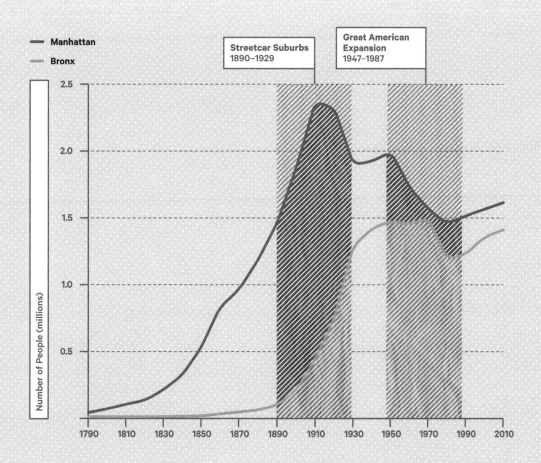

Why Banks Like Home Loans

Mortgage holders pay mostly interest at the beginning of the loan. A bank stands to make $123,610 on a loan of $150,000 to be paid back over 30-year term with a fixed, 4.5 percent interest rate (i.e., a conventional American mortgage). Relatively low, unwavering monthly payments and the mortgage interest tax deduction encourage Americans to gladly enter into three decades of indebtedness in the name of homeownership. The deeds, however, are held by the banks until the mortgage (derived from Old French, meaning literally the "dead pledge") is paid.

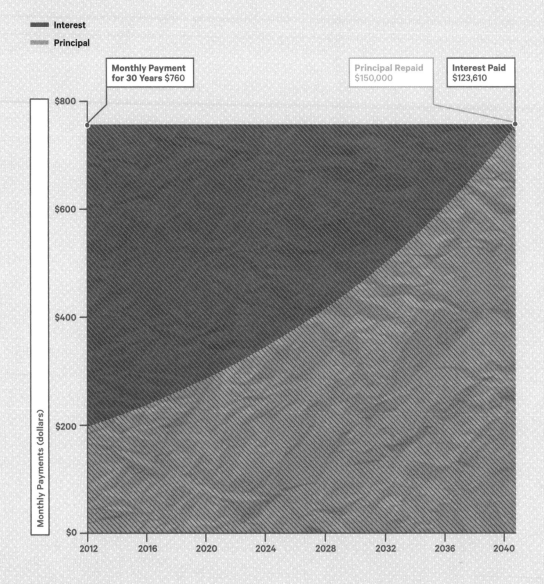

Interest

Principal

Monthly Payment for 30 Years $760

Principal Repaid $150,000

Interest Paid $123,610

Monthly Payments (dollars)

$800

$600

$400

$200

$0

2012 2016 2020 2024 2028 2032 2036 2040

associations. While these small, local institutions acquired a scandalous reputation in the 1980s, they played an essential catalytic role in developing American real estate in the 1910s and 1920s. Member-formed and -operated, S&Ls invented the self-amortizing mortgage, which allowed borrowers to pay off principal and interest simultaneously over a longer term, eventually set to thirty years. Since the loan came from the members of the association, who were themselves homeowners in the same area, neighbors helped neighbors own their homes, bolstering home values and a sense of community. One 1921 advertisement caught the spirit of the times: "The association has kept our boys' money safely invested, and they are $925 better off than two years ago. (These boys had formerly spent all their money for drink.)" Playing an officer in one S&L in Frank Capra's 1946 movie, Jimmy Stewart learned that it really is a wonderful life.

The streetcar suburbs set an important precedent by aligning the needs of government, business, and the public. They solved society's problem with urban overcrowding and generated a new way for business to make profits, while fulfilling the public's longing for space, greenery, and employment—all for the price of a streetcar token.

"Build and Build"

The streetcars created the pattern, but it was the automobile that would see the suburban model of real estate development come to dominate the American landscape. Through the long years of the Depression and World War II, a new economic model was incubating, one that would bring new wealth to millions of Americans. But one more critical adjustment was required: The streetcar tracks were too short, faster trades of time and space were needed, and in swept the car and government road.

The Levitt family of Brooklyn, New York, ushered in the new suburban age. Abraham Levitt, the father, and his two sons, Alfred and William, were real estate developers on Long Island through the long, slow years of the Great Depression and World War II, where they constructed houses for the wealthy. During the war, the military contracted Levitt and Sons to build 2,350 houses as temporary housing near Norfolk, Virginia. The scale of the project caused the firm to develop new techniques for mass production of buildings, modeling their efforts on the automotive production line. Then William was drafted into the Navy Seabees, where he oversaw barracks construction around Pearl Harbor, Hawaii. Seeing the end of the war on the horizon, William surmised that when veterans returned, they might want to buy their own private houses, homes he knew had not been built during the war or during the economic depression. Moreover, he knew that even before the war the Federal Housing Administration had been offering government loan guarantees for home construction, and the Federal National Mortgage Association (or Fannie Mae) had backed home mortgages, to jumpstart economic activity. And then he caught wind of the clincher: A new federal program was in the works to offer low-cost mortgages to returning soldiers. It would eventually be known as the Serviceman's Readjustment Act of 1944, or more colloquially, the G.I. Bill of Rights; it also provided support for college tuition and other benefits. William telegrammed his father back on Long Island: "Buy all the land you possibly can. . . . Beg, borrow or steal the money and then build and build."

In 1946 the Levitts found the bargain they were looking for on the old Hempstead Plains, about twenty-five miles east of Manhattan, in a Long Island community called Island Trees (named after a stand of white pines that from a distance looked like an island floating above the fields). The Hempstead Plains were once the largest grassland east of the Appalachians, a unique relict ecosystem of eastern tallgrass prairie with its own indigenous flora and fauna, "purchased" from the local Algonquin-speaking Indians by English colonists in 1643. Plowed and irrigated, those prairies made good potato lands, but three hundred years on, farming was failing in Island Trees because of an unintentionally introduced pest.

At first the astonished farmers were glad to sell to the Levitts for bargain rates of $300 per acre. (Others, catching on, would later hold out for ten times that amount.) Levitt and Sons eventually bought up 4,700 acres, razed the pines, and bulldozed the fields. They laid out neat and identical 60 x 100–foot lots on the model pioneered by the streetcar developers, but on a completely different scale: They built 17,400 mass-produced homes in two types ("Cape Cod" and "Ranch"), enough to house eighty-two thousand residents. Techniques of standardization inspired by William's military experience and the factory assembly line were used to streamline construction. Teams of carpenters, plumbers, and other craftsmen would go house to house, doing exactly the same thing over and over again, in a choreographed ballet of hard hats and hammers. They called their new development Levittown.

Levittown houses were nothing fancy, but they were cheap and abundant. Prices started at $7,500 (about $74,000 in 2011 dollars), which meant that returning GIs, who had left crowded cities and rural fields to fight in Europe or on Pacific Islands, could become first-time homeowners in the suburbs with nothing down and payments as low as $65 per month. Bendix washers, General Electric stoves, and Admiral televisions, many supplied by Levitt subsidiary companies, were included. Fathers would get in the car in the morning to go to work. Some would drive into the city; others parked and boarded the nearby Long Island Railroad. Midcentury mothers would stay home to cook dinner, organize scouting trips, and attend Tupperware parties. A new suburban culture was evolving. Billy Joel, the piano man, grew up in a Levittown home, as did Bill O'Reilly, the conservative talk show host.

It probably goes without saying that the Levitts did well, clearing an estimated $1,000 per home, making them the most successful home builders of their era and

The Number of American Houses, 1900–2011

From 1900 to 2010, Americans built housing at a furious rate, curing the ill of overcrowding, and then some. US Census Bureau data shows that 18.8 million housing units (out of a total of 113.5 million) were unoccupied for all or part of 2011. Some of those are second homes; others are foreclosed properties awaiting another willing buyer capable of taking a mortgage.

Source: Compiled from US Bureau of the Census (1975), US Census Bureau (2012a), and US Census Bureau and Social Science Research Council (1949). Prior to 1940, data were reported only for non-farm housing; after that for all housing units (HU). An HU consists of the separate living quarters of an individual or family. Apartment buildings have multiple HUs on the same property, and single family houses only one.

▬ **Occupied housing units**

▬ **Vacant housing units**

▬ **Data only for non-farm housing, 1900–38**

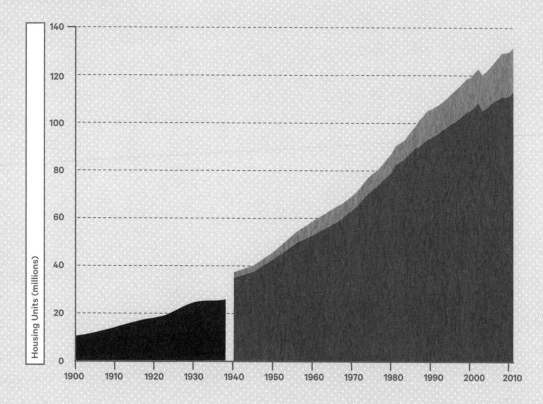

The Great American Expansion, 1940–2000

A nation sprawled. In 1900, 6 percent of the US population lived in suburbs. In 1990, 46 percent of the populace did, and the citizenry had grown by 172 million people. Between 1950 and 2000 we built over 62 million new homes, paved over 1.7 million miles of road, and swallowed over 598 thousand square miles of farm fields, forests, and meadows—an area larger than Texas, California, and Montana combined.

Source: Redrawn from Radeloff et al. (2010). The colors indicate areas with residential densities of 1,000 people per square mile or more after adjusting for household size; lower density developments are not shown. Cited statistics from US Bureau of the Census (1975), US Census Bureau (2012), US Federal Highway Administration (1967, 2011a), and Brown et al. (2005).

very wealthy. With so much money at stake, so much demand, and generous government assistance, it is not surprising that the Hempstead Plains were swept under a monoculture of single-family homes, along with a goodly portion of the rest of the region. Soon others would buy, divide, and plant new fields of houses near town, not just around New York, but from sea to shining sea. The Levitts would go on to build Levittowns in Pennsylvania, New Jersey, Puerto Rico, and on the outskirts of Paris, France. Bill Levitt, the "King of Suburbia," made the cover of *Time* magazine in 1950. The potatoes and other vegetables would just have to come from somewhere else.

The Great American Expansion

Because of idiosyncrasies in how the Census Bureau reports its population numbers, no one knows exactly how many middle-class folks emptied cities and rural areas to move into newly built suburbs after World War II. One study estimates that 46 percent of Americans lived in suburbs in 1990; in 1900 it had been approximately 6 percent. The US population grew by more than 200 million people during the twentieth century, and something like two-thirds of those people settled in the 'burbs.

After WWII houses grew like weeds in rings outside the cities, twenty, thirty, and eventually fifty miles from downtown. The greatest gains were ten to forty miles out, regions beyond the easy reach of streetcars but comfortably attained by automobiles running on new urban interstates and freeways. Construction peaked in the 1950s, but continued at phenomenal, if declining, rates right into the 1980s.

The great American expansion was underway. Between 1940 and 1990, the number of housing units in the United States grew from 37 million to 106 million, at a rate of 13.8 million per decade. In 1950 alone, foundations were laid for 1.4 million new houses. In 1973 annual housing completions peaked at 2.1 million units a year. As a result, the ratio of houses to people collapsed, from one housing unit for every 6.5 persons in 1900 to one for every 2.5 people in 1990, even as the population increased. Overcrowding was licked for good.

In midcentury America, the Sirens were singing in full chorus. The cheap oil window, based on production quotas in the United States and arrangements for oil in the Middle East, kept the price of gasoline continuously low. Motor vehicle registrations zoomed upward as cars were making faster trades of time for space on new roads; and those roads brought new land into development at three times the square of the radius. Develop those lands did, fueling the surging US economy with real estate deals, construction, and jobs. The American economy exploded. The gross domestic product rose 138 percent in 1946–70. The population was growing and people were spending more: Per capita inflation-adjusted personal expenditures rose 56 percent between the end of World War II and the peak of US oil in 1970. Inflation was generally low, and so was unemployment. It was, as the 1970s sitcom called it, Happy Days.

The Great New York Expansion, 1940–2000

Houses grew like weeds in twentieth-century America. This histogram shows the number of houses in rings, working out from the Empire State Building. Since 1940 the most houses have been built in the 10–20 and 20–30 mile rings, and the greatest increases were from 1950 to 1960 and from 1960 to 1970. Note how the largest gains in the number of housing units (measured by comparing the heights of adjacent bars) marches outward over time, ever farther from the city center, with a surge in construction from 1990 to 2000, bucking the historical trend.

Source: Analyzed from data in Radeloff et al. (2010).

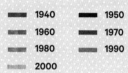

▬ 1940	▬ 1950
▬ 1960	▬ 1970
▬ 1980	▬ 1990
▬ 2000	

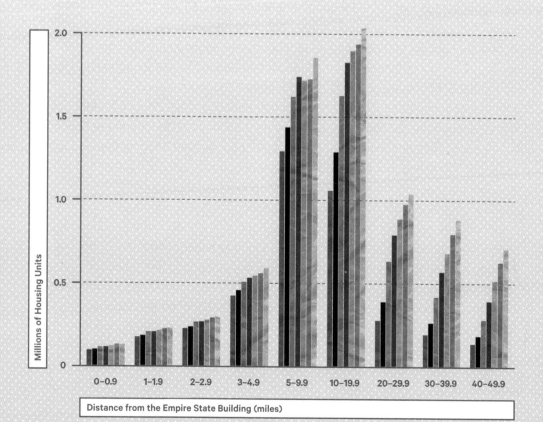

Millions of Housing Units

Distance from the Empire State Building (miles)

Figure No. 31
The American Dream

What drove the twentieth-century American dream? A simple and profitable formula: a farmer near town sells out to a developer, who lays the streets and builds the sewers, so that a builder can construct a house for a happy family to buy, with thirty years of debt.

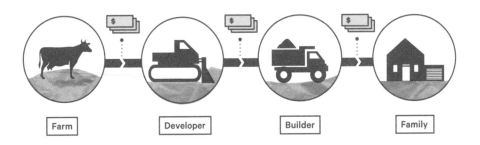

Farm — Developer — Builder — Family

The economy feasted on a delectable menu of oil and land. Each new house was a profit center. Developers snatched up as much land as they could and then subdivided and made money by selling to a builder, who then constructed a house and made a profit on selling it to the homeowner, who took a government-backed loan from a bank. The banks made a profit on fees and interest, which were further encouraged by mortgage-interest tax deductions to the taxpayer who signed up for thirty years of debt. The government also fueled the construction industry by bankrolling roads, sewers, lights, schools, police precincts, fire houses, and other infrastructure. The people moving into the new houses bought cars, furniture, new appliances, and other durable goods, which generated jobs in sales and manufacturing. The economy rode the back of the suburban expansion, which rode the back of cheap land and oil.

Despite the surging economy, joy did not reign everywhere. As the population was diluted across the landscape, the old cities began to rot, cut off from their source of power: people. The urban magic was bleeding away. The government launched some half-hearted efforts to staunch the wounds, but otherwise diddled, focused on the Cold War and the potential for nuclear annihilation (another good reason, some argued, not to congregate too closely in cities). Because people paid taxes mainly where they lived, not where they worked, the tax base of major cities declined at the same time that suburbanization was helping reorganize neighborhoods, in the city and without, by race and class, intensifying problems. People of color were actively discriminated against by a new set of red lines drawn on maps around "ethnic" neighborhoods, demarcating places where the banks refused to lend. Trapped, the poor inherited the inner cities, demanding that their elected officials address the reasons for their impoverishment even as municipal governments had less means to do so.

Where We Live Today

The Great American Expansion re-sorted the US population with respect to density (across the top) and the distance to work or school (going down). The number in each box is an estimate of the total number of Americans living at each combination of density and distance in 2009. (People living at zero distance work at home, or are unemployed or retired.)

Where Americans live is fundamental to the fate of the nation: It predicates economic growth, national security, and our health, welfare, and environment.

Source: Analyzed from 2009 data in US Federal Highway Administration (2011b).

Distance to work or school (mi)	Residential Density (people/sq mi)								Total
	Rural			Suburban			Urban		
	0 – 99	100 – 499	500 – 999	1,000 – 1,999	2,000 – 3,999	4,000 – 9,999	10,000 – 24,999	More than 25,000	
0	8,900,000	8,589,000	4,741,000	6,156,000	10,585,000	11,764,000	4,826,000	3,169,000	58,735,000
0 – 1	1,437,000	1,170,000	806,000	882,000	2,144,000	2,189,000	1,071,000	861,000	10,560,000
1 – 5	6,508,000	8,587,000	6,605,000	9,042,000	14,181,000	18,604,000	5,656,000	4,082,000	73,266,000
5 – 10	6,989,000	8,591,000	6,062,000	8,701,000	13,661,000	14,925,000	3,772,000	2,100,000	64,799,000
10 – 20	10,185,000	11,760,000	6,641,000	8,698,000	12,383,000	12,554,000	3,154,000	1,268,000	66,644,000
20 – 30	4,719,000	4,135,000	2,249,000	2,385,000	3,228,000	3,085,000	766,000	159,000	20,726,000
30 – 40	1,774,000	1,425,000	649,000	770,000	1,211,000	841,000	230,000	64,000	6,965,000
40 – 50	704,000	459,000	148,000	316,000	104,000	168,000	38,000	*	1,937,000
50 – 100	943,000	459,000	303,000	422,000	361,000	293,000	50,000	49,000	2,879,000
100 – 200	84,000	31,000	47,000	90,000	41,000	26,000	7,000	*	328,000
> 200	41,000	57,000	6,000	14,000	16,000	31,000	2,000	*	167,000
Total	42,287,000	45,264,000	28,258,000	37,477,000	57,916,000	64,481,000	19,571,000	11,753,000	307,007,000

* Not estimated.

The Housing Price Escalator, 1890–2012

For most of the twentieth century, housing prices were flat. Then after the close of the cheap oil window in 1971, nominal housing prices (what people pay; dark blue line) escalated dramatically, before flattening in the late 1980s, and then zooming again from 1998 to 2006. The inflation-adjusted line (light blue line) wavers more and shows a flatter trajectory, until 2000, when it too climbs and crashes. For homeowners, riding the housing price escalator allowed them to "adjust" for inflation since 1970, and in some markets, make significant gains, while renters and younger folk were left out in the cold.

Source: Drawn from indices compiled by Shiller (2006 and updates: www.econ.yale.edu/~shiller/data.htm).

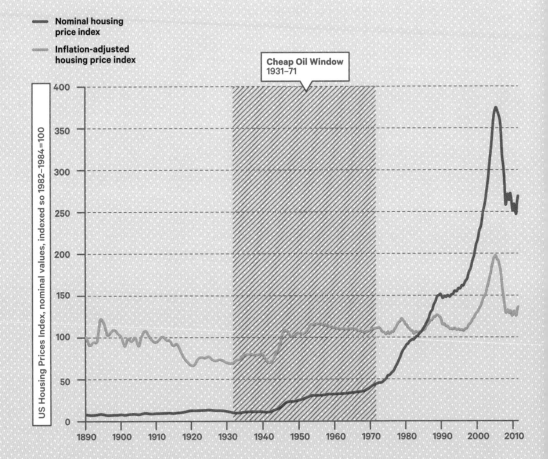

Legend:
— Nominal housing price index
— Inflation-adjusted housing price index

Cheap Oil Window
1931–71

US Housing Prices Index, nominal values, indexed so 1982–1984=100

Crime spiked and riots broke out in cities across America in the late 1960s and early 1970s, part of an era of tremendous social change; the same density that fueled innovation, it appeared, could also amplify and express anger, especially when ignited by violence. And urban decay, danger, and blight reinforced the desire of people with the means to leave to do so, deepening divisions.

Rural areas also transformed as American society reorganized itself around the suburb. Peri-urban rural areas were swallowed in suburban sprawl, as farmers took their last "harvest" by selling out to development, covering some of the most productive farmlands in the country with cement and shag carpet. The Santa Clara Valley in California, once known sweetly as the Valley of the Heart's Delight for its pleasant climate and miles of orchards and blooming farms, became Silicon Valley instead. A corresponding Silicon Forest soon took root in the rich farmlands and nurseries of Oregon's Willamette Valley and beside Washington's Puget Sound. Queens and Brooklyn had long been the breadbasket of New York City—the first and third most agriculturally productive counties in the United States in the 1870s, respectively—but now they fell under the relentless action of the asphalter; the last truck farm closed in Queens in 1938. Across the Hudson in New Jersey, the "Garden State" was soon better known for its refineries and shopping malls than its produce. In California the house where I would grow up was built in a torn-up walnut orchard.

At the same time, in areas more distant from the city new capitalization from the growing economy was destroying the small family farm. Capital sought new investments in agriculture with its now time-tested recipe of standardization, mechanization, and economies of scale. Large companies began to deploy huge, expensive machines to do the jobs that people and animals had handled before. Those machines needed vast swaths of land and fewer farmers to be profitable. The farmland had to be made as uniform as possible for monocultures of corn, soybean, and other crops, fed by new generations of chemical fertilizers and pesticides. The super-productive plants were transformed by super-sized factory machines into thousands of foodstuffs sold in supermarkets. The cost of food plummeted, and so did the farming way of life. The small, self-sufficient farm of diverse plants and animals—the American homestead at the heart of the Jeffersonian version of the American Dream—was virtually abandoned. Between 1890 and 1990, the percentage of Americans working to grow food and fiber dropped from 43 percent to 2.6 percent of the population, even as the population grew by 130 million.

Meanwhile, the suburbs were not turning out to be the paradisiacal gardens they were sold as. Cultural critics began to write about the monotony of living in communities that all looked the same, living with people who culturally were more or less the same, shopping in malls that all sold the same stuff. (For some folks, though, predictability and insulation were exactly what they were looking for in turbulent times.)

Despite homogenization of social groupings and increasing wealth, sociologists began to document dramatic declines in "social capital"—a measure of an individual's ability to rely on networks of others—as people moved to the suburbs, especially the television 'burbs of the late 1960s, like where I grew up. Americans largely stopped going to church, belonging to clubs, participating in service organizations. Many people seemed to fold into their homes, caring mostly about private concerns, and neglecting public ones. Backyards rather than front porches became the place where families gathered; citizens started getting gossip and news from the television, rather than at the diner or barber shop. Counterintuitively, for many, apathy and loss of meaning came to characterize what generations of Americans had been working and sacrificing to obtain. We were collectively richer and more powerful than ever before, and yet something vital was gone.

Lewis Mumford, the historian of the city, knew what was to blame, and he put the onus squarely on the automobile in words that not only make his point, but capture the spirit (and sexism) of 1968, when he wrote:

> As long as motorcars were few in number, he who had one was a king: he could go where he pleased and halt where he pleased; and this machine itself appeared as a compensatory device for enlarging an ego which had been shrunken by our very success in mechanisation. That sense of freedom and power remains a fact today in only low density areas, in the open country; the popularity of this method of escape has ruined the promise it once held forth. In using the car to flee from the metropolis the motorist finds that he has merely transferred congestion to the highway; and when he reaches his destination, in a distant suburb, he finds that the countryside he sought has disappeared: beyond him, thanks to the motorway, lies only another suburb, just as dull as his own. . . . In short, the American has sacrificed his life as a whole to the motorcar, like someone who, demented with passion, wrecks his home in order to lavish his income on a capricious mistress who promises delights he can only occasionally enjoy.

The Price Escalator

If everyday life promised less interest and relevance as America entered the 1970s, at least house prices were going up. In the years following 1970, coincidentally with the closing of the cheap oil window, the price of housing began to appreciate dramatically. This pattern is clearest in the time series assembled by economist Robert Schiller for his book, *Irrational Exuberance*, about speculative bubbles. For most of the twentieth century, nominal housing prices were relatively flat, with a small gain after World War II reflecting high demand and tight supply as veterans returned home (just as William Levitt had predicted). Then through the 1960s there was relatively little growth in prices, as the millions of new homes added to the landscape kept up with the demands of population growth and internal migration.

But in the 1970s, something extraordinary began to happen. Over the decade, nominal (non-inflated adjusted) housing prices surged, gaining 122 percent. Between 1980 and 1990, they rose another 73 percent. All told, over the two decades before 1990, the market value of household and nonprofit real estate in the United States rose by $6.3 trillion and the market value of *all* real estate increased by $12.1 trillion. (For comparison GDP of the entire economy in 1990 was $5.8 trillion.) It became an article of faith among politicians and economists that if oil shocks led us into recession, then housing would lead us out.

Americans had discovered a new attraction in the Siren song. A house was no longer just a place to live and raise a family: It had become an investment opportunity, simultaneously as buoyant as stocks and as reliable as bonds. And as prices increased so did the economy, the stock market, government budgets, and America's power in the world. Inflation was also on the rise, but for homeowners, gains in the price of property helped compensate, especially since housing comprised some 40 percent of the index used to measure inflation.

The price escalator worked well for my parents, who bought their suburban house in a former walnut orchard twenty-seven miles from downtown San Francisco for $29,000 in 1969. It appreciated in value every year my family lived there, and not just because Dad added a garage and an extra bedroom and Mom lavished care on the blooming roses out back and we were good people; all our neighbors' home values were escalating too. By the mid-1970s, some land-rich, cash-poor suburban Californians were reputedly being ruined by the tax bills on their real estate. Proposition 13, passed in 1977, attempted to rectify the problem by fixing the value of a house at its 1975 level for the purposes of taxation. Henceforth, new assessments would be limited and made only when a property changed hands or new construction occurred. That worked great for my folks, whose tax bill stayed the same even as the value of our house doubled, tripled, and quadrupled. On the downside Proposition 13 also meant I had no sixth-grade camp (because education in California was largely funded by property taxes and now there was no money for camp or other extracurricular activities; public education in the state declined in other ways, too), nor could I ever hope to buy a house in the town where I grew up: Walnut Creek has long since priced wildlife conservationists out of the market.

And then in the early 1990s, perpetual growth seemed to stall, causing concern inside and outside of government. Housing had proven a profitable economic model for everyone involved for fifty years: not only the farmer, the developer, and the builder, but since 1970, the home buyer who saw his and her personal wealth enlarged through the simple act of owning a home.

No one wanted it to end.

7
The Crescendo and the Crash

Change before you have to.

Jack Welch, former chairman of General Electric

Economies come from somewhere. To hear most commentators talk about it, you would think economic growth is something you just whip up by lowering taxes and exhorting people to work hard and spend more, but in fact economic systems, like ecological ones, depend not only on their internal arrangements, but also on what you put in and what you take out. No one expects a forest to grow without sunlight, water, and soil, and similarly economies don't grow without inputs of creativity, effort, and resources.

As we have seen, the American economic system was humming along until the 1970s, when oil shocks and inflation threatened to put it off the rails. Fortunately, just in time, housing prices started escalating too, which helped to compensate. Sure, more households had two breadwinners instead of one, more families were falling into poverty, and inequity was on the rise, but most Americans could still get to work and pay their bills. And then something important happened that is critical not only to its particular moment in American history—the late 1980s and early 1990s—but essential to understanding the economy and politics of the first decade of the twenty-first century as well: The housing market started to tank.

Clinton's Song

In the twilight of Reagan's America, in 1987, completion rates for new housing topped out at a seasonally adjusted rate of 1.8 million homes per year, then plummeted 40 percent over the next five years. In 1985 vacancy rates nationwide hit 11.5 percent, and then hovered at over 10 percent for ten years running, which meant that one in ten housing units in America stood shuttered. Over the previous forty years, the United States transitioned from a country with 46 million housing units for 151 million people in 1950 (a ratio of 1:3.3), to 69 million housing units for 203 million people in 1970 (1:2.9), to 102 million units for 245 million people by 1990 (1:2.5). Suddenly it seemed that everyone who could afford a house had one, and there were fewer takers for new construction. In 1990, as happens in markets when supply exceeds demand, prices began to level off, which, coupled with the brief oil shock caused by the Gulf War and the collapse of the savings and loan industry, set off the recession into which William Jefferson Clinton was elected president in 1992.

President Clinton and his Housing Secretary Henry Cisneros diagnosed the problem not as too many houses, but as too few homeowners. Clinton put it this way at a press conference in June 1995:

> [Back in 1975, in Fayetteville, Arkansas . . .] Hillary had to go away to some-
> where—I can't remember where she was going now, but anyway she was tak-
> ing a trip on an airplane, so I was driving her to the airport. And we drove by
> this wonderful old house. It was an old, old, very small house, and she said,

Housing Cycles, 1889–2011

Housing construction is a sensitive measure of the oil-cars-suburbs economy. When construction is up, the economy hums with growth, and in mid-twentieth century America, a humming economy meant prestige, power, and wealth. When construction declines, however, the economy wallows in recession, as happened repeatedly over the last forty years, each dip a little deeper than the one before. In the past, housing always led the US out of recession. Do we think the music can continue forever? Or do we need to learn to sing a new song?

Source: Housing start data on new residential construction compiled from Blank (1954) and US Census Bureau (2012a); recessionary periods as defined by the National Bureau of Economic Research (2010).

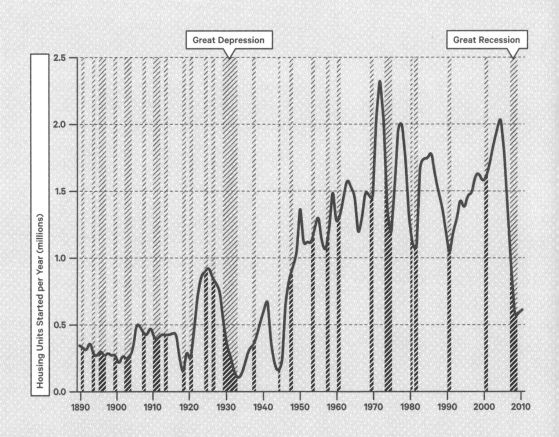

Homeownership, 1965–2012

President Clinton showed a graph similar to this one in 1995 when he announced his National Homeownership strategy, which included relaxing lending requirements by the government-backed mortgage-security corporations Fannie Mae and Freddie Mac. The strategy worked, in terms of fueling a boom in construction, and enhancing homeownership by a few percentage points, a policy that President G. W. Bush happily extended in 2000. Unfortunately it had more dire consequences in 2008–2009 when the global economy nearly collapsed with the bursting of the housing bubble.

Source: US Census Bureau (2012b). The 2012 figure is based on first two quarters only.

"Boy, that's a beautiful house." And I noticed that there was a little "For Sale" sign on it. So I took her to the airport, went back, and bought the house. And when she came home after the trip, I drove by the house. I said, "See that house you liked? I bought it while you were gone. Now you have to marry me." [Laughter] And it worked; twenty years ago this fall, it worked. Most people do it the other way around, but you know—[Laughter]

I still remember that home cost $20,500. It had about 1,100 square feet. And I had about a $17,500 mortgage on it, and my payments were about $176 a month, as I remember, something like that. And that was twenty years ago this fall that I signed that fortuitous contract.

Those prices aren't very much available anymore, but the objective for young people, with their futures before them and their dreams fresh in their minds, starting out their families, to be able to own their home and to start a family in that way, that's a worthy objective—just as worthy today and, I would argue to you, more important today than it was twenty years ago. . . .

[After showing a chart showing declining homeownership during the 1980s] . . . more and more American families [are] working harder for the same or lower wages every year, under new and difficult stresses. It seems to me that we have a serious, serious unmet obligation to try to reverse these trends. As Secretary Cisneros says, this drop in homeownership means 1.5 million families who would now be in their own homes if the forty-six years of homeownership expansion had not been reversed in the 1980s. . . .

Now we have begun to expand it again. Since 1993, nearly 2.8 million new households have joined the ranks of America's homeowners, nearly twice as many as in the previous two years. But we have to do a lot better. The goal of this strategy, to boost homeownership to 67.5 percent by the year 2000, would take us to an all-time high, helping as many as eight million American families across that threshold. . . .

And I can say without knowing that I'm overstating it, that if we succeed in doing this, if we succeed in making that number happen, it will be one of the most important things that this administration has ever done, and we're going to do it without spending more tax money.

So the Clinton Administration set about offering incentives to prospective homeowners, encouraging them to take on more debt by keeping interest rates low and requiring lenders to relax their standards for qualification, starting with government-owned housing banks Fannie Mae and Freddie Mac. It worked the Clinton miracle; more people began receiving loans and buying houses, and the prices of houses started rising again, which, as predicted, did wonders for the entire economy. Clinton left

Land Values, 1952–2007

The price of a house reflects not only the building and other improvements on the property, but also the value of land. Although the nominal value of residential and nonprofit property has increased hugely since the 1950s, the greatest gains have been in land, even after adjusting for inflation. Either the demand for land has increased, or buildable land is in shorter supply, or both.

Source: Redrawn from Davis (2009).

—— **Land as percentage of total real estate value**

▨ **Household structures**

▬ **Land**

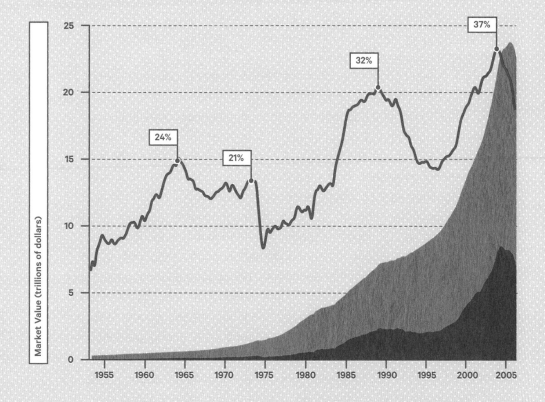

office with GDP topping $9.3 trillion in 1999 and a federal budget surplus. The next president, George Walker Bush, continued these policies into the new millennium, cut taxes further, and then worked with Federal Reserve Chairman Alan Greenspan to drop interest rates lower still. The housing market took off, then began to soar like Icarus toward the sun. Homeownership nationwide peaked in early 2005 at 69.2 percent. Long-time homeowners started to take out second mortgages for everyday expenses, with the expectation that rising real estate values would pay the bills. Banks offered a diverse menu of enticing loan offers, facilitated as a matter of government policy, and Wall Street came up with some novel innovations to repackage mortgages into derivative products that were sold into a world market eager to have a piece of ever-growing American prosperity.

The Sirens were pleased.

The Hidden Cost of Housing

Meanwhile something strange was happening with the economics of housing.

Property values have two principal components: the buildings or "improvements" on the land, and the land on which those "improvements" stand. In the 1950s the nominal (non-inflation-adjusted) price of land was about 10 percent of the overall market value of household real estate; the rest of the value came from the structures. Land contributed one part in ten of housing's value nationwide.

In the late 1950s and through the 1960s, land's share began to rise, reaching 20 percent of total real estate value by 1972. The recession of 1973–74 caused a short-term collapse in real estate, but housing came roaring back, leading the economy out of recession. Land valuation led the way.

Land values continued to increase relative to buildings through the double-dip recession of 1980 and 1981–82, and peaked in 1990 at over 30 percent of the value of a home purchase nationally. The recession of 1990–91 occasioned another dip, but not as deep as before, and soon land was gaining again over structures as the incentives for homeownership took hold and expanded the market. By 2005, 37 percent of the price of an average American house was for American land.

Housing prices were escalating to dizzying, even ridiculous, heights. *Chicago* magazine in its annual survey of housing prices in 2005 wondered out loud: "Can It Last?" That year nine Chicago neighborhoods or suburbs bragged of average property prices over one million dollars; in 2000 there were only two neighborhoods where that was true. Housing prices in Bakersfield, California, rose by 33 percent between the first quarter of 2004 and the first quarter of 2005; nationwide prices were up 12.5 percent in a single year, according to data from the feckless Office of Federal Housing Enterprise Oversight. Economists began speaking darkly of irrational exuberance and black swans, unpredictable events with huge implications.

Where Is the Buildable Land?

Land isn't all made the same as far as developers are concerned. "Buildable" land has to be flat, dry, and not set aside for other purposes. Ideally it will also be unbuilt. These are factors we can measure. In this analysis we show how much land was "buildable" and already "built" around New York City in 2006. Nowadays, to find appreciable amounts of buildable greenfields, one needs to look more than thirty, and in some directions even fifty, miles from midtown Manhattan—i.e., beyond the practical commuting horizon.

Source: Flat equals less than 15 percent slope from the National Elevation Dataset (Gesch et al., 2009); built and dry means "developed" and "not wetland, shore, beach, and open water," respectively, from the National Land Dataset (Fry et al. 2011); and set aside includes park and reserve lands, from the Conservation Biology Institute (2010).

██ Too Steep
(unbuildable)

▓▓ Protected
(unbuildable)

▒▒ Built

▒▒ Water + Marshlands
(unbuildable)

░░ Buildable

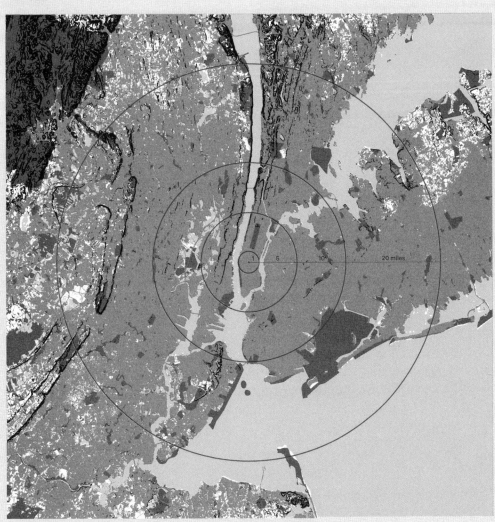

Here Is the Buildable Land (Not!)

For the fifteen most economically productive metropolitan areas in the United States, which together account for almost a half of the gross domestic product of the nation, we observed that most of the buildable land is already built, especially within twenty miles of downtown.

Source: Geographic analysis as described on p. 131; GDP data from US Bureau of Economic Analysis (2012a). Named cities suggested the locations of the relevant "downtown" centers.

Metropolitan Area	2009 GDP ($ millions)	Percent of US GDP	Percent of Buildable Land (miles from downtown)			
			1	5	10	20
New York City	$ 1,210,387	9.6%	100	99	99	95
Los Angeles	$ 730,941	5.8%	99	99	98	94
Chicago	$ 508,712	4.0%	96	80	71	57
Washington, DC	$ 407,463	3.2%	89	65	61	43
Houston	$ 363,201	2.9%	99	79	71	47
Dallas/Ft. Worth	$ 356,615	2.8%	100	89	74	41
San Francisco/Oakland	$ 335,563	2.7%	100	93	90	85
Philadelphia	$ 335,112	2.7%	99	91	73	49
Boston	$ 298,256	2.4%	100	91	80	61
Atlanta	$ 264,700	2.1%	98	78	72	51
Miami/Ft. Lauderdale	$ 252,647	2.0%	100	99	99	89
Seattle/Tacoma/Bellevue	$ 228,797	1.8%	97	89	75	45
Phoenix	$ 190,725	1.5%	100	89	77	45
Minneapolis/St. Paul	$ 189,801	1.5%	98	91	80	32
Detroit	$ 185,800	1.5%	100	99	94	66

Was it just exuberance? Was the housing bubble unpredictable? Inflation, at any rate, was reliably swelling: 1.6 percent in 2001, 3.4 percent in 2005, 4.1 percent in 2007. Inflation-corrected land prices, which had been flat until 1970, gained steadily over inflation from 1975 through 1990, sagged a bit in the early 1990s, and then increased 300 percent from 1995 through 2005. Pricier homes expressed more than the amount of money in circulation. Land itself was dear. Why was land becoming so expensive?

Land to Build On

Sixty years of increases in land values suggest that land was either becoming rarer, more desirable, or both. Land in America is a commodity like any other, subject to the laws of supply and demand. Maybe the problem wasn't actually a lack of home buyers per se—the Clintonian hypothesis—but rather a lack of desirable land on which to own a home.

Housing economists routinely dismiss land scarcity as having anything to do with real estate prices. After all, urban areas (including the suburbs), occupy only 2.6 percent of the country's land area. Though it is a well demonstrated fact that costs per square foot are highest in city centers (hence the skyscrapers), on a national scale there seems to be plenty of land to be had, as anyone can see flying across the country, peering out the window.

But land in general, anywhere, is not what is needed; rather, housing needs to be built where someone will want to buy. As we have seen, home buyers want affordability, social and environmental amenities, a job, and a reasonable commute. Developers need "buildable" land.

Topography, water, price, and regulation define what is buildable, which is to say that a potentially developable property has to be reasonably flat, not in a swamp or the tidal zone, on offer, and approved according to local codes. From the developers' perspective, it is also best if the property has not already been built on since greenfields, not brownfields, provide the greatest potential for profit and least risk for all involved. The question of land scarcity boils down to: How much buildable land has already been built?

That's not a standard economic question, but one that conservationists think about all the time (though typically in the inverse). So I enlisted help from my WCS colleague Kim Fisher to estimate the distribution of buildable land in the lower forty-eight United States. We used data derived from satellite imagery from the government, distributed over the Internet. These datasets are different from your ordinary maps; they represent the land as pixilated images, in which each pixel, approximately thirty yards on a side, represents a land-use class: developed land, agricultural fields, forests, wetlands, and the like. Comparative data nationwide were available for 1992, 2001, and 2006. Using other geographic information, we masked out three kinds of land that are unbuildable: places with slopes of more than 15 percent; parcels mapped as open water, beaches, or wetlands; and areas that are off limits for private development, like parks, nature reserves, and military bases. We then measured how much of the remaining buildable land had already been built on, at various distances from the centers of metropolitan statistical areas as defined by the Census Bureau.

Here is what we discovered: The top twenty most economically productive metropolitan areas in the country—which provide 52 percent of the nation's economic output—are finished within five miles of downtown, that is to say, on average 92 percent of the developable land was already developed back in 2006. Five to ten miles out, 81 percent developed. Ten to twenty miles out, 57 percent developed. These numbers do not account consistently for city parks and gardens, which sometimes show up as potentially developable. They also don't include what the satellite can't see: zoning regulations that require minimum-sized lots, owners unwilling to part with their land, and customers priced out of the market.

Thus in 2006, near the height of the most valuable real estate market in American history, developers would have to look fifteen to thirty miles from downtown—in some cases, fifty miles—before finding appreciable quantities of greenfields to turn into new homes around America's most productive cities. Beneath the speculative bubble, buildable land had become scarce, and thus expensive. As with easy oil, there were no more easy houses to be had.

The Practical Commuting Horizon

Fifteen to thirty miles doesn't sound like so far if you have a car that can exceed sixty miles per hour, but commuting, as most of us know all too well, is not only a matter of distance, but also a function of how many other people are between you and your destination. Vehicle occupancy in personal motor vehicles averaged 1.5 persons per trip in 2009. The average American household consisted of 2.5 people who owned 1.9 vehicles, with 20 percent of households owning three or more cars or light trucks. Each person travels on average thirty-three miles per day. The average population density in the suburbs varies between one thousand and four thousand people per square mile. The inevitable conclusion: Being farther out means not only more miles to drive, but more traffic to wade through.

Traffic congestion is one of the most hated aspects of modern American life, and for good reason. The average commuter sat in traffic for fourteen hours per year in 1982; twenty-five years later, she had the pleasure of spending thirty-six hours per year idling in a smoking steel box among her fellow citizens. The average American commute exceeded twenty-five minutes one way in 2009, and extreme commutes of an hour or more were no longer unusual, especially if there was an incident on the roads. To add insult to injury, gas has become more expensive, on an inflation-adjusted basis rising from $1.35 per gallon in 1998 to $3.46 per gallon in 2012.

America has built out beyond its practical commuting horizon, a distance that varies from place to place, and from person to person, but as a rule of thumb can be estimated by how far you can drive in thirty minutes at rush hour. (In the Bronx, that's about six miles.) And who bought those houses on the periphery? Primarily new homeowners, people who couldn't qualify before, young families, people of color, and other lower-income folks that Clinton and Bush hoped would carry the old American Dream forward another generation. Why wouldn't they? It worked for their parents and grandparents and neighbors; maybe it would work for them, too.

The Crash

Or it might not. Conventional explanations of the financial collapse of 2008 focus on speculation in the markets, particularly the sky-high price of housing in years after 2001. In the fall of 2008, $12.4 trillion in American net worth evaporated in a few weeks' time.

Market speculation was no doubt aided and abetted by the greed of Wall Street and the lax oversight of government minders, which allowed the "securitization" of fundamentally insecure mortgages, that is to say, mortgages that people couldn't actually pay back. Those securities were sold unscrupulously or, at least ignorantly, by firms like Goldman Sachs and Deutsche Bank into a world market, prettied up with ratings provided by professional evaluators, like Standard & Poors and Moody's, whose computer models turned out to be flawed, but whose formal seals of approval made a lot of money for all concerned. As a result the bad loans spread far and wide, contagiously infecting a much larger pool of good mortgages. Few on Wall Street, Main Street, or anywhere else stopped to ask where all the money was coming from as long as they knew some of it was going into their pockets.

Thus it came as quite a surprise to all of us when the emperor had no clothes. This story is told in replete detail elsewhere, so I won't repeat it, but in brief: The world economy nearly collapsed. It turned out that lending standards—where borrowers demonstrate a reasonable probability of being able to repay loans offered on reasonable terms—are useful. It also became obvious that banks and other investment houses should have transparent ways of keeping their books and avoid selling high-risk investment products with "safe" AAA ratings that they don't understand. Many people, especially young people, learned that trusting the system, as currently configured, might not be in their best interests.

A recession begun by the summer of '08 oil shock turned out to be a whole lot worse when the banking, automobile, and housing industries all nosedived in succession. The first two received government bailouts; the third was too big to save.

Economic Vital Signs

Dow Jones Industrial Average

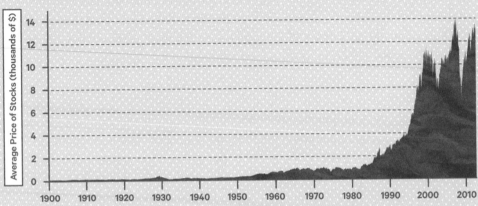

Employment

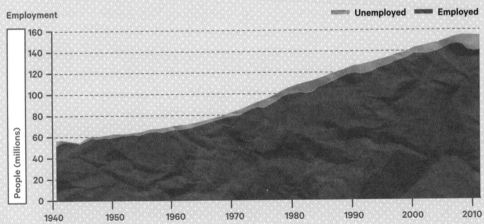

Legend: Unemployed, Employed

Federal Government Net Income

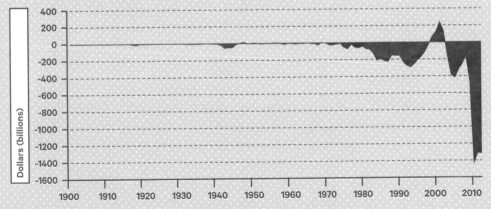

Here is what we are told to pay attention to in the economy (left): the stock market, the employment numbers, and the federal debt. Here is what we should also be paying attention to (below): the money supply, the distribution of income, and the amount of carbon dioxide in the atmosphere.

Source: See Notes.

Monetary Base

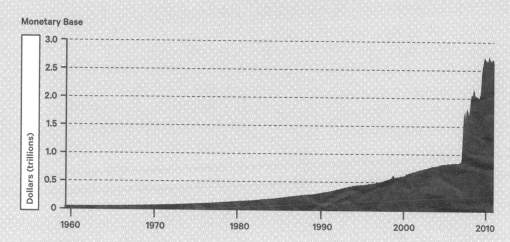

Income Inequality

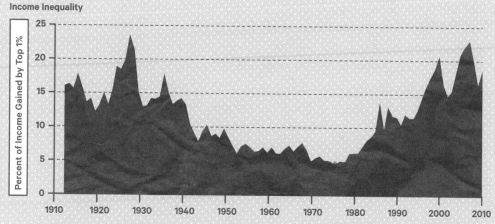

Carbon Dioxide

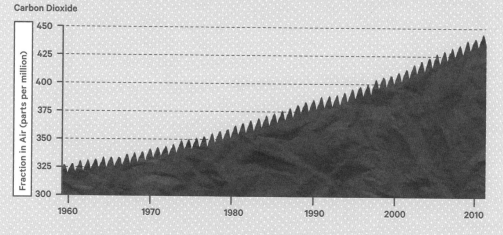

8.7 million Americans lost employment and 9.9 million American homes lost owners via foreclosure. Three million families fell into poverty between 2000 and 2011. Vacancy rates rocketed above 14 percent. New home construction practically ceased. Homeownership returned to levels last seen in 1997, the same percentage as in 1980, when Jimmy Carter was president. As of mid-2012, the real estate market remained mired in its most depressed state in seventy years, though as the book was going to press the housing market was starting to perk up again.

Meanwhile the federal government and the Federal Reserve Bank have taken extraordinary measures to stimulate the economy. President Barack Obama entered office with a budget deficit of $1.5 trillion, two foreign wars, and a 3.7 percent drop in total GDP from fall 2008 to spring 2009. He pushed an $819 billion stimulus package through Congress, while the Fed, under Chairman Benjamin Bernanke, dropped short-term interests to essentially zero (0.15 percent in April 2009).

Since that wasn't enough, through a monetary maneuver known as quantitative easing, the Fed bought financial assets from banks on the open market, paid for with money created de novo, increasing the total amount of dollars in the world by 86 percent between September 2008 and January 2009, and then again by 36 percent between November 2010 and July 2011. The plot of the US monetary base looks like a broken leg, pointed the wrong direction; when I first saw it, I literally felt sick to my stomach. Nothing in ecology or economics is supposed to look like that. Milton Friedman's words kept echoing through my mind: "Inflation is always and everywhere a monetary phenomenon in the sense that it is and can be produced only by a more rapid increase in the quantity of money than in output."

Dear reader, we have a problem that goes far beyond lending standards and banking regulation, beyond stimulus and monetary policy, beyond inflation and unemployment, even beyond oil, cars, and suburbs, as such. At the heart of the Siren song lies a false presumption made in ignorance a long time ago about the land where we live.

The American Presumption

The presumption is that there is always more to be had. More oil, more land, more air, more water, more soil, more nature, so that it doesn't matter how much you consume and appropriate, there is always more to be had over the next hill, in the next valley, on the frontier—a frontier that no longer exists.

The American presumption is older than the nation itself, formed in arrogance and ignorance during the time of European discovery and settlement, and then codified into our system of laws and economic practices. When the country was first explored, sciences like geology and ecology were still hundreds of years away and few understood how land was created or how the qualities of land varied and were sustained. To foreigners testing the shore, the people who could best speak on behalf of the land were the people already living here, the "Indians" or Native Americans, who them-selves had arrived some twelve to forty thousand years before. When they learned of European conceptions of land and right, they were shocked and confused; it hardly made sense to conceive of land as property given to only one person for his or her exclusive dominion in perpetuity, yet here were these strange people, speaking in tongues, who said it was so.

Equally bewildered, Europeans of the time did not recognize how nature sustained clean water, fertile soils, an agreeable climate, and other benefits, despite the clear and ample demonstrations before them; nor did European kings and assemblies properly acknowledge the rights of indigenous peoples, though some token payments were made. The conquerors of North America marched in and claimed the land, with force as necessary, and then as time passed their governments sold or gave it away. William Penn received most of what is today Pennsylvania and Delaware in 1681 to pay off a debt King Charles II owed to Penn's father; nothing was said of the thousands of people and the billions of creatures who inhabited the land already. New York State (or New Netherland as it was first known) was given as a monopoly to the West India Company by the independent staatholders of the Dutch Republic in Amsterdam who knew nothing of the nature of the Hudson River Valley, except it provided beaver felt for hats. The Pilgrims arrived in Cape Cod Bay, and not having a legitimate royal charter, wrote their own compact to plant a colony and permanently preserve it. Florida and California were settled after land grants from the Spanish crown; Texas was wrested from the Mexicans, who themselves derived claims from the Spaniards, who took what they wanted from the Aztec, Navajo, Comanche, and other nations.

Eventually those European landlords sold their lands to other people who sold the land to other people who sold the land to . . . you and me. All property claims, including mine and yours (only excepting some very few held by Native Americans on their ancestral homelands) extend back to these original appropriations. Each taking repre-sents an assumption that there will be more for someone else, for the future, for nature. John Locke, the English philosopher, said as much writing the essays that laid the modern justification for private property in 1690:

> Nor was this appropriation of any parcel of land, by improving it, any prejudice to any other man, since there was still enough and as good left, and more than the yet unprovided could use. So that, in effect, there was never the less left for others because of his enclosure for himself.

Locke, as we can surely see three hundred years later, was wrong. Enclosing the land means someone else can't have it; improving land, whether by extracting oil or building a house, alters its natural productivity. Appropriations and improvements prejudice everyone who comes afterward, and the larger community of nature that gave the land life and form in the first place. As Locke says:

> Whatsoever he removes out of the state that Nature hath provided and left it in, he hath mixed his labor with it, and joined to it something that is his own, and thereby makes it his property.

Taking from the state that Nature hath provided is the basis of private property, land for sale, and oil to burn. And what if Nature hasn't provided enough?

Like the sea breaking against a lee shore, the presumption of unlimited resources swells the Siren song from Rockefeller straight through to 9/11 and the Great Recession. Always more oil is why we valued our domestic petroleum so little when other countries found theirs so dear, even after the cheap oil window closed, and why we now fight oil shocks and oil wars. Always more distance is the false promise of the automobile to carry us beyond the practical commuting horizon and to build ever more highways, bigger houses, and sprawling suburbs. Always more land is the patter of the real estate agent offering a better deal one exit farther along the interstate and the banker offering the next great mortgage security. Always more to be taken without recognition or responsibility is the nightmare that is swallowing the American Dream whole.

It is the deadly end of the Siren song.

> And Odysseus said:
> "Friends, up to this point,
> we've not been strangers to misfortunes.
> Surely the bad things now are nothing worse
> than when the Cyclops with his savage force
> kept us his prisoners in his hollow cave . . . .
> I think someday we'll be remembering
> these dangers, too. But come now, all of us
> should follow what I say. Stay by your oars,
> and keep striking them against the surging sea.
> Zeus may somehow let us escape from here
> and thus avoid destruction. You, helmsman,
> I'm talking, above all, to you, so hold
> this in your heart—you control the steering
> on this hollow ship. Keep us on a course
> some distance from the smoke and breaking waves."

The Crescendo

Putting the prices of oil and housing on the same graph with inflation and the money supply for the last forty years highlights the perils of our time. During the cheap oil window, oil was inexpensive, housing abundant, and inflation practically nonexistent. After the cheap oil window, the economic system has had multiple heart attacks, each worse than the one before. And the cure, in terms of loosening the money supply and relaxing financial regulation, may turn out to be worse than facing the root of the problem—which is the American fantasy that oil, cars, and suburbs can last forever.

Source: Oil prices from Federal Reserve of St. Louis (2012a), housing prices from Shiller (2006 and updates), consumer price index from the US Bureau of Labor Statistics (2012a), and monetary base from Federal Reserve Bank of St. Louis (2012b). All are indexed to 100 as of January 2000 for purposes of comparison.

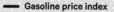

- **Gasoline price index**
- **Housing price index**
- **Consumer price index**
- **Monetary supply index**

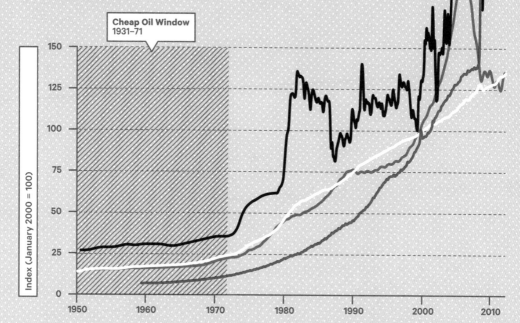

Part II
Terra Nova

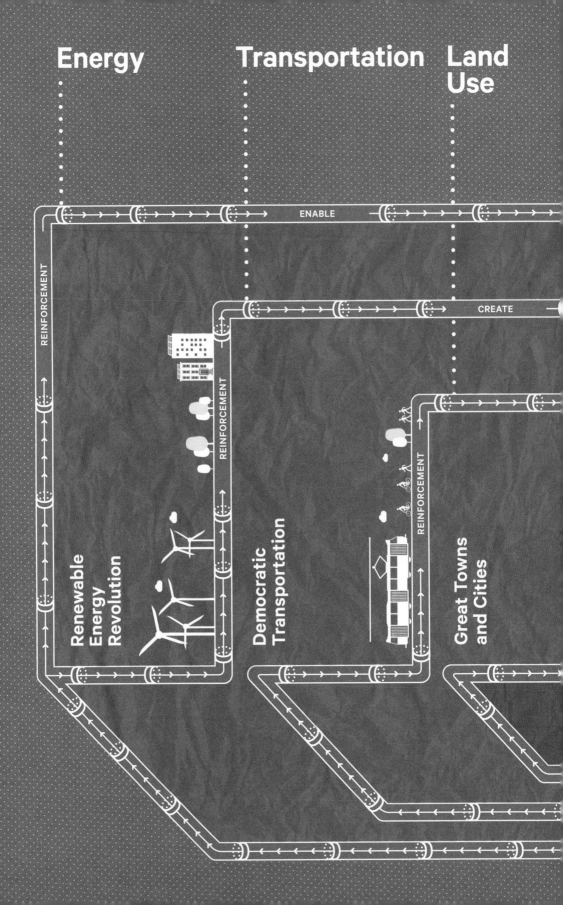

Nature

Figure No. 41

What America needs now is a virtuous positive feedback loop. Valuing the land will help encourage great towns and cities, which in turn will enable non-fossil fuel dependent transportation, which then will ignite a renewable energy revolution. And as we come to depend on America's advantages in wind, sun, heat, and ingenuity, they will reinforce electrified transport, better communities, and the natural capital of the nation.

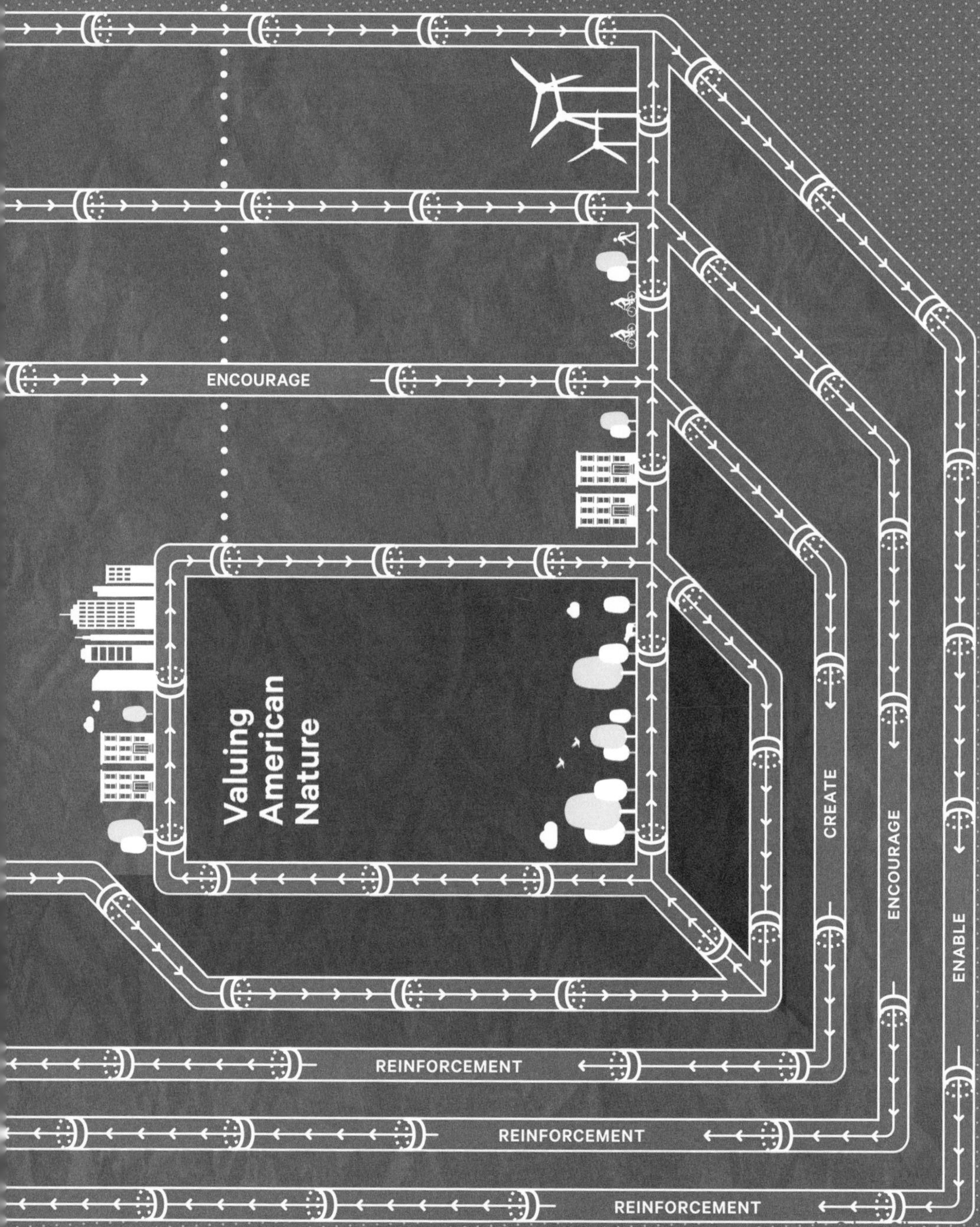

ENCOURAGE

Valuing American Nature

CREATE

ENCOURAGE

ENABLE

REINFORCEMENT

REINFORCEMENT

REINFORCEMENT

8
Holding Council

To help the task of fulfilling
and creating the vision:
the Grandfather Spirits of the
Rock, Fire, Wind, and Water.

From the Lenape creation myth of Turtle Island

The animals ran and the birds flew. The rain fell for days and nights without stop, until at last only a small patch of ground was left showing on top of the highest mountain where a small cedar tree grew. It was now apparent that all of the seven island mountains of the world would drown.

Seeing this Nanapush, the Strong Pure One, the Grandfather of Beings and Men, picked up the animals and birds and tucked them into his shirt. He approached the cedar tree and spoke to it before starting to climb. Even as he reached the top of the tree, rising waters were already lapping at his feet. Then great Nanapush began to sing and beat on his bow-string, which served as a drum. As he sang, the sacred tree began to grow, and it kept on growing as the waters continued to rise.

After a long time like this, Nanapush grew tired of singing his peace song to the unrelenting storm, so he threw down the branches he had plucked off the tree as he climbed. On this raft he carefully placed all the creatures he had saved, then he climbed aboard himself. The tree and the mountain peaks drowned; only Nanapush and our animal brothers remained alive, floating on a vast sea.

Nanapush decided a new earth should be made, a task he could well perform through the powers granted to him by the Creator. So he held a Council with our brothers, and they agreed to help him. Their duty would be to get some soil from the submerged old earth.

Mitewile'un, the Loon, offered to go. He dived and stayed down a very long time. When he came floating back up to the surface, he was dead. So the great Nanapush breathed upon the unfortunate Loon, and his life was restored. Kuna'moxk, the Otter, dived down, but he also drowned and was revived. Then Tamakwa, the Beaver, tried, failed, and was brought back.

At last Nanapush turned to Tamask'was, the Muskrat, and told him that he must try very hard to reach the old Earth. The little Muskrat stayed down twice as long as any of the rest, and he came to the surface exhausted, but alive! And in his mouth he carried some of the precious mud from the lost world.

Great Nanapush was pleased, and so he blessed the Muskrat, promising him that his kind would never die out.

So it begins, with these words from the Lenape creation myth, told long ago in the land that would someday be called New Jersey. I like Nanapush. He had an optimistic turn of mind. When the world was drowning, when the mountains went under, when all the small creatures were crowded with him on a raft made of sticks, he didn't despair; instead, he called a council, considered the facts, made a plan, and asked everyone to contribute. It was hard and there were failures—as the loon, the otter, and the beaver discovered—but perseverance carried the day. I don't know that I would have predicted the muskrat as the deliverer, but there it is: Sometimes solutions come from unexpected places.

Now we need to hold council together, too, standing as we do over the body of the failed twentieth-century economy. The economy based on oil, cars, and suburbs is done. If not dead, it stumbles like a zombie, propelled on by ancient urgings but no longer capable of growth or life. We have more houses than we can use and more infrastructure than we can take care of. VMT peaked in 2005, and the best the carmakers can do now is replace the auto fleet we already have since everybody who wants a car (and can afford it) has one already. And oil, as we all know, is expensive, if not continuously so; each oil shock provides another cycle of recession, downsizing, and foreclosure.

And it is not only our economy that suffers from the Siren song; our national security is also endangered. Dependence on foreign oil antagonizes people who don't like our soldiers on their lands; it props up dictators and compels us to support them against our better interests; and it imposes a terrible tax on the nation, paid in the lives of men and women lost and ruined; machinery, weapons, and fuel expended; and prestige squandered on the global stage. We fight wars we don't have to and leave the debt to future generations. People die, and schools and bridges go unbuilt so that we can commute. There can be no real national security as long as we cling to suburbs that require cars that require oil.

It is time to imagine a new way of living in the new world. I don't have all the answers, nor can I, nor any one of us, get there alone. But I can tell you this: We can get there together. The American experiment is not dead. Now is our time to demonstrate that a democracy can not only defend itself against external perils, but can look long and deep into its soul and muster the courage to change from within. Finding the wisdom to apply the knowledge we have is the challenge of our generation.

Here is what we know.

A Self-Reinforcing Cycle. Going forward, we can no longer think of oil-cars-suburbs (or more generally, energy-transportation-land use) as separate entities; rather, it is clear they form a system of interacting parts, a self-reinforcing cycle. The energy in oil enabled rapid, individualized transport, and the speed of the automobile and cheap land made possible sprawling development. Because energy, transportation, and land use are mutually reinforcing, for as long as our land use is extended and diffuse, we are chained to the explosion car and its fossil fumes and company masters, and the death, war, and terrorism that this dependency entails. No solution that addresses only one sector will work—but a solution that addresses all of them can't help but succeed.

Happenstance and Flexibility. No one planned the Siren song; it unfolded through a combination of happenstance, invention, crisis, and adaptation. Imagine that evening when Mr. Rockefeller first saw the bright light of Mr. Edison's electrical contraption and realized his market in kerosene illumination would eventually be snuffed out. Within a matter of years, the automobile came along as the next great absorber of oil's magic. Imagine Bill Levitt standing with the other Seabees in the 1940s, watching Quonset huts going up in neat rows and thinking of Long Island potato fields. He glimpsed new opportunities and wealth to be grasped, especially when backed by government programs of a grateful nation. Similarly, as we build a new model for American life, we need to look for lucky breaks and unexpected connections, be the inventors and producers of our future, and adjust to new conditions even as we create the conditions necessary for adjustment. These are traditional American strengths— creativity, earnestness, and optimism—born out of our experience as pioneers and not forgotten yet on the sidewalks of our cities and towns.

Physics Matters. No matter how ingenious we may be, however, there are certain laws we must respect, like the laws of physics. You can sing, you can pray, you can cajole, you can legislate to your heart's content, but gasoline is still going to have more or less thirty-four hours of microwaving energy per gallon, and few other materials have that punch. Since the amount of affordable oil is diminishing and we won't be making more any time soon, we should save its energy pop for things that really require it— like airplanes—and focus on lower-energy solutions for getting groceries and going to work. Physics offers some suggestions, as well as drawing some bright lines around what is possible and what is not. We should take heed.

Incent and Deliver. Physical laws can't be broken, but every other law is created by human beings and therefore can be re-created by human beings. People can change, and do. One of the most striking lessons of the Sirens' work together is how oil, cars, and suburbs appealed with the right notes at the right times to orchestrate the American body politic over a century and a half. From the grand diversity of our throbbing needs and desires, a remarkable unison developed around the Siren song, and thus we layered a continent with asphalt and linoleum. Going forward, we will also need incentives—and by that, I mean profits as well as patriotism and good sense. We need a new solution to the old American Dream of a better life for each generation, and that solution needs to be coordinated by signals from the people to the people.

Economies of Scale. If physics and society can be aligned and activated according to their own laws and internal motivations, then we might look to achieve economies of scale. Economies of scale are cost-savings that result from making more of something, usually through greater efficiency. Such savings occurred several times during the Siren song: in the expansion of the railroads, the vertical integration of the oil companies, the build-out of the streetcar lines, the development of the suburbs, even— bizarrely—in the mass securitization of insecure mortgages. Economies of scale make significant changes possible, and we'll look for them again and again.

Extreme measures. Finally, we should take note of the extreme measures necessary to sing the Siren song: Armed force, oil diplomacy, price manipulations, secret deals, government loans, tax breaks, inflationary monetary policy, saturation advertising, interstate highways, eminent domain, and land giveaways were all used to lure, hustle, and corral Americans into the current economic model. More, we have been extreme in what we gave up to chase the Sirens: Not only did we lose lives, homes, health, community, and hope, we've endangered species, paved farmlands, altered weather, fought wars, and diminished American power and prestige. Fortunately such extremes will not be necessary in the forthcoming suggestions. The present isn't always a good guide to the future, especially when building a new world. Ask the muskrats.

All is not lost yet, so take heart: We have many, many advantages still. We are a large nation with a diverse and abundant, if sorely used, land base; and the land community is resilient given time and attention. Under our land, we still have some oil and natural gas (a riff on the Siren song, already being sung), and we also have vast windy plains and coastal waters, sunny deserts, warm rocks, and tall wet mountains. We have a lot of roads, cars, and other mechanical things; factories and houses too, and many terrific towns and cities. We lack neither technology nor innovation nor the desire to succeed. We have a large, diverse, friendly, if politically fractious, population: over 314 million souls, drawn literally from every nation on earth, united in a shared belief in democracy, tolerance, and freedom. Many of us are educated and able, and though our population is aging, we possess a tradition of hard work and the closest thing to a meritocracy the world has ever seen, despite conspicuous inequities. Even after the financial crash, our capital markets are the envy of the world. There is a lot of money in the hands of a few who might be convinced to help. And now we have experience and knowledge to temper our wild ways.

Indeed the foreclosure crisis, the Great Recession, the slow-motion recovery, the divisive political rhetoric, the frustrations of the Tea Party and the Occupy movement, and the high cost of gas might be blessings in disguise. They are signals that we need to develop another way to live, a new basis for the economy, an evolved form of the American ecosystem, a better way to live in the old New World.

A Muir Web

This complex network shows what nature is made of: lots of connections tying together many parts. Each sharp point on this diagram represents a different species (a mammal, a bird, a plant, etc.) or a natural resource (the water, air, sun, etc.) or an ecosystem (a forest, meadow, wetland, etc.). One of the nodes represents human beings. And inside the human node? Another complex network, formed of our economic, social, cultural, and intimate relationships. Everything that it means to be human is embraced by what it means to be part of nature.

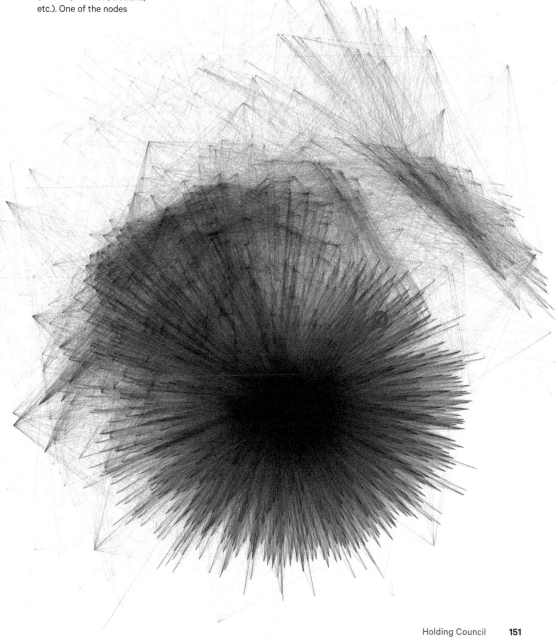

9
Gate Duties

We abuse land because we regard it as a commodity belonging to us. When we see land as a community to which we belong, we may begin to use it with love and respect.

Aldo Leopold, *A Sand County Almanac* (1949)

A few years ago I was immersed in a study of the ecology of Manhattan before it was a city, at the moment of its European "discovery" by Henry Hudson in September of 1609. At a certain point in that work, it became helpful to document all the habitat connections for all the species that once lived on Mannahatta, as the island was then known. The result was a diagram, rendered by a graduate student on a computer, composed of thousands of thin gray lines that formed a cloud of interconnections among the species and elements of the ecosystem. We called that cloud a Muir web, after the famous naturalist John Muir. Each gray line connected two nodes, which stood for plants and animals, kinds of rock and types of soil, forests and meadows and the like. One of the thousands of nodes represented Nanapush's Lenape (the Real People, as they knew themselves), who hunted and fished in the same precincts where skyscrapers would one day stand. I pinned a photocopy of the diagram on my office wall and looked at it from time to time. We are one of many, whispered the Muir web.

Sitting at my desk one day, I was staring at the human node in the ecological web, thinking how small it was, how contingent it was on all the rest, and then my eye happened to alight on the human footprint map—the garish chart of global human influence we'd made years before—dangling on the wall nearby. The contrast startled me. Humanity is also large and powerful. We have marked the world with such ferocity that most of the globe is inflamed by our presence. No other species has so changed our planet. We are one unlike any other, trumpeted the map of the human footprint.

I used to think that the human footprint is a direct consequence of human arrogance, and perhaps it is, but I've become increasingly convinced that the human footprint is more about ignorance than anything else. We have so insulated ourselves within our node that we have forgotten how we belong, in Aldo Leopold's memorable phrase, to land as a community.

Perhaps our forgetfulness is a result of preoccupation. After all, the Muir web's human node isn't just a dot in the network of life; it is a container of networks. Inside it lie our connections to family and friends and work; history, art, and culture; economic transactions; social systems; our understanding of the world. Our ignorance and probably our arrogance are manifestations of a kind of self-absorption, a mistaking of the part for the whole, a set of narcissistic blinders that keeps us, and our American society at large, from really seeing how the world works.

Oil-cars-suburbs demonstrates the absorption well enough. Yes, we tell ourselves climbing into the car in the morning, we are polluting, causing traffic congestion, chancing accidents, requiring roads, fomenting wars, and the rest of it—but don't you understand? I really have to get to work! The businesspeople chase the bitch-goddess success; the politicians fight over who is responsible for the price of gas; the oil masters rub their hands with glee; and those who can't speak—the community of nature and the future—drown in the indifference of the suburban street.

We need to see the community to which we belong. It needs to be made real in our terms, in the human language of dollars and cents on which decisions are made. Then other forms of relationship will ramify across the land. New dreams will be dreamt. But before we dream, dollars and cents means we need to visit with Professor Adam Smith.

Adam Smith's Pins

An eccentric, comical archetype of the absent-minded professor, Adam Smith once fell into a tanning pit because he was so distracted in conversation. In 1776 he published a book whose influence on American society is arguably as great as the Declaration of Independence. In *An Inquiry into the Nature and Causes of the Wealth of Nations* (usually shortened to *The Wealth of Nations)*, Smith observed that wealth comes from the combination of three "factors of production": land, labor, and capital. By land, he meant the soil, but today we would define it as all the natural resources that come from nature into the economy, including oil, natural gas, and other fossil fuels, minerals, metals, water, wood, wildlife, even sunshine and an amenable climate: the goods and services of nature. Labor is the time and effort of human workers, some of whom are skilled at particular tasks—such as manufacturing, computer mapping, book writing, oil drilling, and so forth. Capital is money that can be used to buy the raw materials, hire employees, and obtain the tools and machines that enhance the labor of human beings.

In Smith's day, land there was, green, wet, and rolling around his native Glasgow. Labor there was, too, though mostly unskilled. But he lacked capital—capital was the limiting factor. Smith observed that fortuitous combinations of land, labor, and capital (what he called an "enterprise," or what we might call a "business model") can lead to new products generated at lower costs than they were before. Sold into the competitive market, those products generate greater profits—profit being the difference between the costs of creation and the price paid by a willing buyer—in the form of new capital, which can be invested in other enterprises, thus inducing greater demand, more supply, lower costs, and additions to the general welfare. Smith's system of self-perpetuating relationships is now called capitalism.

Smith's genius was to realize that taken across a large number of buyers and sellers, these transactions could arrange a system for the production of the goods that society wants in the quantities that society wants, subject only to the constraints of supply and demand and the human capacity for innovation—without anyone ever issuing an order of any kind. It is really quite remarkable when you think about it. It is hard to imagine anyone effectively planning for the trillions of transactions that occur in the US economy each year, let alone devising a way to effectively track them. Moreover, Smith surmised that his system of prices would create incentives for initiative and reward innovation, sending signals for the system to change as circumstances changed. Nature works that way too. No one tells the forest how many trees to grow or the bees how many flowers to pollinate, yet an ecosystem coheres and evolves. The economic system also adapts, meeting need with substance and flexibility, as long as all the costs and all the benefits enter fairly onto the market floor.

Smith illustrated his idea by conceiving of a pin factory down the road from the university. Without capital, each worker could make just a few pins per day. With capital, the factory owner could buy tools and machines to streamline the efforts of the workers, making each laborer more productive as he or she specialized in one step of the pin-making process. These efficiencies, enabled by a loan from an investor, would allow the factory to make many more pins per day, and over time, enable the owner to pay back the loan with interest.

What is left out of the parable of the pin factory? Although Professor Smith neglects to notice it, we will: In the process of making more pins, the pin factory would also use more steel. And that steel has to come from somewhere, such as another enterprise, one that mines iron from the ground and makes wires for making pins. And where does the mine get more iron from? Smith, the economist, has no answer.

Here we need a scholar of a different kind. An ecologist would note that there are also fortuitous combinations that create things anew in the community of nature. Like business enterprises, ecosystems are diverse and take many forms; they even have a conserved currency we have met before, called energy. Living things and non-living things are connected in ways not completely unlike a factory (greener perhaps, with less concrete, more soil). Given energy, space, and time, natural ecosystems produce sources of wealth in the form of land. In fact, land—the community and its products, nature's resources and its services—is one of the essential factors of production. Smith's pin factory can have workers and capital, but without iron, no pins.

It continues to astonish me, and no doubt it would Professor Smith, that the mainstream of American economic thinking has so completely forgotten about the iron to make pins, the land to make wealth, even the plankton to make oil. Robert Solow, a remarkable MIT economist, won the Nobel Prize in 1987 for determining that seven-eighths of US economic productivity gains between 1909 and 1949 came from technological progress, as opposed to increased inputs of labor or capital. Land, natural resources, and the cheap oil window never entered his model, yet without them there would have been no economic phenomena for his study: He had blinders on. His colleagues fill American economic journals with similarly clever but parochial findings. In short, much prizewinning, hugely influential twentieth-century economic dogma is based on less ecological understanding than a third grader takes for granted.

Banished to the wilderness, a small group of economists (the "ecological" economists and their cousins in the agricultural college, the "natural resource" economists) have been howling into the wind about how to fix this fundamental flaw in the way the economy is conceived. Their answer: Put the economic household back into the ecological house.

The Household and the House

Economics, from the Greek *oikonomia*, is the study of the household; ecology, derived from the Greek *oikos,* is about the house itself. The classical social view of the economy is of a black box containing two groups of people. You can label them as Producer and Consumer, or Employee and Employer, or Investor and Business, or Government and Citizens. A pair of slightly curving arrows connects the two groups, closing a circular cycle. The arrows reflect exchanges from donor to acceptor. The consumer gives money to the producer in trade for a product (e.g., dinner at a restaurant or a new electronic gadget). Employers give money to workers in exchange for their labor. Investors provide capital in return for dividends and interest returned from business later on.

Inside the economic household's black box, ideally every time an exchange takes place, profit is generated. A successful firm, like an oil company, has a product that a consumer wants to buy, like gasoline. The price of gasoline includes the costs of its production and a profit margin for the producer. The profit margin is important because it is the main motivation for the producer to go to all the trouble to haul a barrel of oil from the other side of the earth, refine it into gasoline, and sell it at the corner gas station—no mean feat—but of course, a priori, no one knows what the profit margin should be. The sellers, such as ExxonMobil, Chevron, and BP, want to make as much money as possible, and buyers want the lowest prices and the greatest convenience. Thus in a competitive market with many sellers and buyers expressing their preferences through what they will and won't buy, the price—and the profit—is determined by where supply meets demand.

Economic growth occurs when the cycle in the black box spins faster than it did before, with more exchanges, larger exchanges, or both, as measured in money. The gross domestic product of a nation is the sum total of these market exchanges. Economic growth is considered desirable because each successful exchange generates profit for the seller who finds a willing buyer. More profits mean more wealth, and in Smith's terms, more opportunities for innovation and improvement as money is reinvested back into the cycle. Economic growth also implies more consumption, which is generally thought to be in the consumer's best interest (after all, he agreed to the exchange) and for the worker who made the product or provided the service (because now she has more to do, which translates into steady employment and higher wages).

Economic growth is also pleasing for the government, which gains financially from each exchange through taxation, either based on the value of the transaction (as in a sales tax) or through the profits generated (as in the income tax). Taxes act as a kind

Figure No. 43
The Economic
Household

An illustration like this one appears at the beginning of most elementary economics textbooks: the cycle of exchange between consumers and producers. Consumers pay money for products, from which the producer ideally realizes a profit and the government, sales-tax revenue. This black box has analogous forms for investors and producers, employers and employees, and government and citizens. Economic growth occurs when the cycle spins faster than it did before.

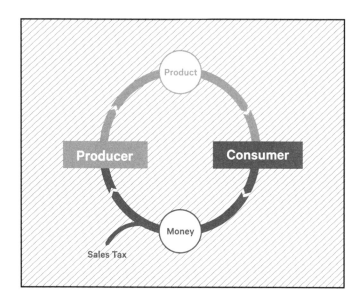

of friction on the economic wheel, slowing it and throwing off sparks of protest as a result, but they also generate the revenues that enable government to fulfill its responsibilities with greater speed and verve. Thus for its own sake, and for the sake of the people it serves, a democratically elected government typically works to make the economy grow as fast as possible. In fact everyone wants the circle to spin faster and faster: business leaders, politicians, poverty advocates, conservation organization fundraisers, and broad swaths of the general public.

Outside the household, in the house where nature dwells, the exhortations for ceaseless economic growth appear a bit less benign. You will not be surprised to learn that for each turn of the economic wheel, natural resources are consumed. Since those natural resources might have been deployed in a cycle of nature other than the economic one, their loss is an opportunity cost charged to the natural world. Some commodities depend directly on those natural resources, like gasoline from crude oil; others less so, like librarian services. As the aggregate economy spins faster and faster, more inputs are necessarily fed into the economic maw—and more wastes come out the other side. Burning gasoline creates air pollution. Librarians toss out printouts and break pencil leads. These wastes we leave for nature to assimilate with nary a thanks.

Because nature bears these costs, the market does not see them, and therefore can't hope to act on them. As we have seen, the oil company does not pay the cost for using oil faster than nature can generate it, but only for the "toil and trouble" (in Smith's memorable phrase) of getting the oil out of the ground as fast as possible. Nor does the librarian pay for the air pollution emitted by his car on the way to work. In Adam Smith's economy, if the seller doesn't pay all the costs of production, then the product will be supplied at an artificially low price and/or generate artificially high profits. If the buyer doesn't pay all the costs of consumption, then demand will be unintentionally— inefficiently—high, and thus the buyer will use too much, possibly to her individual detriment and surely to the detriment of nature and some future consumer who might also have wanted to consume. Free becomes free to waste.

When supply and demand are out of balance, the usual spurs for innovation to find better, more efficient enterprises—the kind that made the professor so justly proud— go missing. America adopted cars whole hog in the 1920s, not because they were powerful and sexy and it was our cultural destiny to do so, but because the energy that animated them was so outrageously cheap. Smith's invisible hand inevitably directs the economic blind man to the "free" candy on the table.

Gatekeeping

Fortunately our friends in the wilderness have a better idea. Imagine the economy as a city and around the city is all of nature. Inside the city it's much the same as the economists' black box: The buyers and sellers are trading in the market, workers are rushing to work, investors are evaluating enterprises, and the government is attending to its duties on behalf of the public. Now let's imagine there are two gates to the city: the front gate, where all the energy and materials from the land community enter, and the rear gate, where the wastes of consumption return to nature. At each gate there is a gatekeeper monitoring what goes in and what goes out.

In 2010 the front gate of the US economy received 1.1 billion tons of crude oil. The rear gatekeeper reported 2.4 billion tons of carbon dioxide pouring out, as well as 1.4 billion tons of waste from extracting and refining American oil; 30 million tons of plastics discarded into American landfills; 567,000 tons of herbicides and pesticides applied to American crops, lawns, and gardens; and 125,000 tons of pharmaceuticals flushed out with the sewage from American hospitals. And that is just from using oil.

As Robert Solow said, seven-eighths of economic growth is a matter of technological progress.

A Better Way to Tax

I give Solow a hard time, but it's not his fault; Harvard and Columbia never trained him to see beyond the social ties that bind human interactions in our little node on

In Nature's House

The textbook representation of the economy neglects to note that the economic household is part of nature's house. Nature provides goods and services on which the economy depends and receives the wastes the economy no longer wants.

Because nature cannot negotiate in terms the economy understands (that is, with prices), these inputs and outputs are undervalued, causing waste, inefficiency, and over-indulgence, the costs of which are borne by the poor, the environment, and the future.

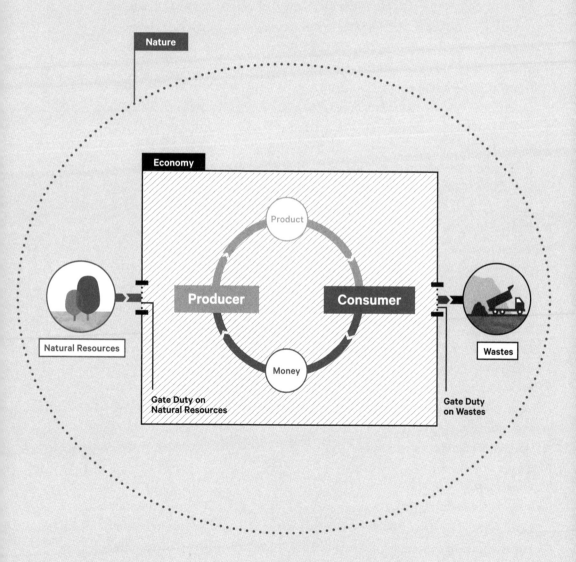

the edge of the world's vast ecological network. But some economists, to their credit, have seen further: Arthur Cecil Pigou was one. An English economist born into Queen Victoria's empire, Pigou originally went to Cambridge to study history, then shifted into moral philosophy, like Adam Smith a century before. From theories of ethics he turned to investigations of economics, eventually becoming a fellow of King's College, where he supported another brilliant upcoming economist, John Maynard Keynes, in the early decades of the twentieth century. Professor Pigou loved the mountains, where he liked to climb to take the view. It is unclear whether history, morality, or the out-of-doors eventually opened his eyes, but in his academic studies he became concerned with economic costs the economy didn't see. He called them externalities.

Externalities are costs or benefits of an economic interaction that are not transmitted through money, so they are *external*, and thus invisible, to the normal economic relationships. Pollution is a kind of externality on people and ecosystems because people get sick and ecosystems decline but at no cost to the polluter (except what regulations impose). Too much money in circulation is an externality paid in future inflation. Government support for education is considered a positive externality since a society that invests in its children receives skilled workers and engaged citizens later on, but the generation making the investment rarely sees the direct benefit of what it supported. Technological innovation, art, and science are also positive gifts to the future. Externalities are typically "paid" or "received" by nature, the poor, and posterity, and more often than not in human history, the payments have exceeded the receipts.

Pigou's recipe for dealing with the cost of externalities was to charge for them in the here and now, as a way of limiting their negative qualities and amplifying their positive effects; later economists called his notion Pigouvian taxes. Many modern economists believe this sort of taxation, rather than regulation or outright proscription, is the right way to tame the less desirable features of the capitalist economy. Gregory Mankiw, noted Harvard economics professor and advisor to presidents, keeps a list of members in what he calls the Pigou Club: economists, politicians, speechwriters, and journalists who sign up simply by endorsing Pigouvian taxes publically. These taxes, which I will frame as gate duties, are the foundation on which a new American economy can be built, especially when put together with other A-list ideas from Professors Leopold and Smith.

Simply stated, we need to signal to the economy that it is part of, not separate from, the land community. Here is how.

Gate Duties

First, we reform the tax system to value natural resources taken from nature and place costs on wastes released back to nature; it's a system of duties payable at the front and rear gates of an economy situated within its environment. Gate duties will make it more expensive for natural resources to enter the economy, but once those resources are in the scope of our economic interactions, gate duties will encourage producers and consumers to keep them circulating for as long as possible, by discouraging waste and encouraging recycling and reuse. Gate duties express the simple fact that natural resources are not free; they have costs to nature and to the future and to us today, particularly when we use them inefficiently.

Second, we release or remove taxes that act as friction on the economy's internal arrangements. The goal is to place taxation on what we don't want (i.e., wasteful consumption), and take away taxation from what we desire (i.e., innovation and productivity). By moving taxes from the heart of the economy to its peripheries, where labor and capital meet the land, we improve the landscape of opportunity and benefit, without changing the absolute amount of taxation. In fact, over time taxes will decrease as a direct result of choices people make.

Third, we make sure the new policies and incentives apply to everyone, including nonprofits and the government. The government in its various forms—local, state, and federal—is such a significant portion of the economy, it cannot be exempted if we are to make real change. Besides, it wouldn't be fair to excuse the public concern from the burdens of the private citizen. The government can tax itself as easily as it taxes us.

Finally, we need to show how these changes make other adjustments possible in our land use, transportation, and eventually, our energy systems. Describing how these new systems of relationships will unfold is the topic of the next three chapters.

Taxes and the Law

Before detailing how a system of gate duties might work, let me say that no one likes taxes or discussing tax policy, me included. (My parents are certified public accountants, and I can tell you that dinnertime conversation was a significant factor in my becoming an ecologist.) But taxes are important. Government levies are critical because they enable a democratic people to act as a group to "provide for the common defence, promote the general Welfare, and secure the Blessings of Liberty," and so forth. Taxes are also an essential way for the group to signal the economic system, especially a market economy, what its goals might be. As we change the tax environment for Smith's pin factory, and all other factories, to do their work, they adapt without anyone actually telling them how to change. Nudged, incented, compelled perhaps, at least they are free to choose how they will adapt to the new world.

The only alternative to taxes—the usual one in our litigious society—is to appeal to the coercive power of the law to enforce the rules, which means that some activities are completely okay, some are acceptable in certain circumstances, and others are completely illegal. We can all agree that obviously immoral acts, like murder and rape, should be outlawed. In other cases, where the morality is less clear, regulations are more hazardous. For example, how much pollution can be acceptably dumped in a river? Some would say none, not ever; others would allow some dumping, subject to the welfare of frogs and fishes and the amount of stink; yet others might argue that being a sewer for pollution is what rivers are for, so dump away.

Gate duties address pollution and other uses and abuses of natural resources by requiring people to pay for what they use and abuse, rather than proscribing an absolute level of behavior. If you are going to pollute the air and the water, then you should pay for it. If you are going to extract natural resources out of the ecological commons, you should pay for that too. When Colonel Drake drilled America's first oil well, he paid for a lease, but he never paid for the oil he found, even though that oil took a hundred million years to form. Neither Bissell nor Rockefeller had anything to do with the long sequence of geological events that led to the black gold lying beneath Oil Creek, yet they made fortunes. And the nameless landowner who leased the spring on his property only had rights to that spring through a chain of appropriations, all perfectly legal and fundamentally ignorant about how the world really works.

Who should speak for nature and the future? They can't speak for themselves, so either we continue to ignore them as our predecessors have done (with consequences we now well know), or we speak for them ourselves, through the government. The government represents not just those of us alive today, but the continuity of our national union over time. And the government has a mechanism that the economy, which distributes the resources and flushes the wastes, monitors intently: taxation. We should tax not simply to raise funds for people today, but to protect the dreams of the future.

Entry, Exit, and Use

Gate duties come in three forms: entrance duties, on extraction of natural resources; exit duties, on releases of wastes back into nature; and ecological use fees, assessed on development of the land. We should set these duties to reflect the relative costs of natural resource use, of which there are at least three:

First, the cost of renewal: Natural resources exploited more quickly than the rate of their natural formation (such as oil) and wastes produced faster than nature can absorb and reprocess them (carbon dioxide) should carry higher duties than resources (wind energy) and wastes (water vapor) used within the capacity of natural systems to create and renew them.

Second, the cost of scarcity: Naturally scarce resources (oil) entering the economy should carry higher entrance duties than more abundant ones (timber).

Third, the cost of connection: Natural resources with more and stronger connections to the rest of nature (water) should have higher duties than natural resources with fewer and weaker connections (oil). Analogously, wastes that are more poisonous or more difficult for nature to assimilate (spent nuclear fuel rods) should be taxed at a higher rate than wastes that can be accommodated more easily (banana peels).

Beyond that, setting the amounts of gate duties need not be complicated. We establish them as we do other fees and taxes in the economy, through a sense of fairness, practical effectiveness, and shared goals. The secret CPAs whisper at night is: There is no scientific method to set taxes any more than there is a right answer to the amount of pollution to dump in the river. Taxation always represents a value judgment, which is why tax policy is so often in dispute. Value judgments are what democracies are for: We all get a say, and after we have spoken, we all agree to play by the rules. And over time we adjust the rules as circumstances warrant.

Laying out a full course of action on gate duties will require a book three times as thick as this one and a tenth as interesting, but let us imagine as an illustrative example levying a federal entrance gate duty of $1 on each gallon of crude oil extracted from the ground. This duty would be paid by the oil company in Alaska or the Gulf of Mexico or wherever else the oil is taken domestically (on private land or public), or for foreign oil, at its point of entry into the United States.

What would the company do with this cost? It would, of course, pass it on to the refiners, who would pass it on to the gas station owners, who would pass it on to you and me. Gas prices, we can be assured, would rise by at least forty-eight cents (the proportion of a gallon of crude that is refined into gasoline), if not a little more.

Gas stations would collect another fee from consumers in the form of exit duties on the combustion of gasoline. Let's imagine exit duties set at three pennies per pound of carbon dioxide returned to the air via automobile tailpipes. Combustion of a gallon of gas produces about twenty pounds of carbon dioxide (recall that combustion combines hydrocarbons with oxygen from the air, adding to the mass). So the exit gate duty for gasoline would add another sixty cents to the price at the pump. Gas that is at $4 per gallon today would cost a bit over $5 a gallon in the future.

What would be the effect of this added expense? We know that when prices go up, consumption goes down. For example in the summer of 2008, when spiking oil prices drove the price of gasoline over $4.10 a gallon, July sales dropped *four million gallons per day* compared to the same month the year before. You and I would no doubt grumble to find that gas prices go up, but we would probably also use less of it. We would fill up less often, sending a signal of reduced demand to the oil companies, and we might decide differently about driving to the corner market for milk; walking and biking would seem like better options. Hybrid cars, electric cars, even streetcars would all become more desirable. Over time we would grow ever more motivated to seek long-term solutions to reducing our dependency on gasoline, which would enhance opportunities for the nation as a whole to be creative and free, rather than grappling with the tiresome political, economic, and geological happenstances of the distribution of oil.

What Nature Provides

Nature is large and generous, but as houseguests, we have overstayed our welcome. Witness the enormous wealth flowing from nature into the US economy each year. Gate duties are a way to give something back while also activating a more sustainable and growth-oriented form of the economy. For example, a $1 tax on each gallon of crude oil and three cents on each pound of carbon dioxide emitted would produce enough revenue to replace all sales taxes (see text). Duties on other natural resource flows could allow us to lift corporate income taxes forever. Gate duties encourage us to be creative rather than greedy with what nature provides.

Source: US Energy Information Administration (2012a), Kelly and Matos (2010), Kenny et al. (2009), and Smith et al. (2009). All data for 2009 except timber (2007) and water (2005). Note the different units of measurement and processes of replenishment.

Natural Resource (units)	How Resource Forms in Nature	Annual Rate of Natural Formation	Rate of Extraction from the US	Rate of Consumption in the US
Crude Oil (barrels/ year)	Geological transformation of biomass (mainly plankton)	Negligible	1,956,765,000	5,206,450,000
Natural Gas (thousand cubic feet/year)	Geological transformation of biomass (mainly plankton)	Negligible	26,057,000,000	22,910,000,000
Coal (short tons/ year)	Geological transformation of biomass (mainly plants)	Negligible	1,074,900,000	997,500,000
Gold (metric tons/ year)	Stellar nucleosynthesis	0	223	254
Copper (metric tons/ year)	Stellar nucleosynthesis	0	1,110,000	1,600,000
Aluminum (metric tons/ year)	Stellar nucleosynthesis	0	1,727,000	3,250,000
Iron Ore (metric tons/ year)	Stellar nucleosynthesis	0	26,696,000	25,700,000
Uranium (pounds/ year)	Stellar nucleosynthesis	0	3,708,358	49,400,000
Water (million gallons/ year)	Evaporation followed by precipitation over land	1,724,000,000	149,891,657	149,891,657
Fish/Shellfish (pounds/ year)	Aquatic food webs	*	7,867,000,000	12,997,000,000
Timber (thousand cubic feet/year)	Photosynthesis	26,744,366	21,719,000	25,072,000

* In 2009, 43 of 176 US fisheries were considered overfished.

If that sounds good, just wait. The entrance duty on oil extraction applies not only to gasoline. Because the duty is assessed when oil comes out of the ground, the cost is transmitted to everything made from that oil—plastics, chemicals, drugs, etc., would all become more expensive in proportion to how much they depend on oil (by a few pennies per unit in most cases). Additional costs would also come from exit gate duties laid on the wastes generated by plastic makers, chemical manufacturers, pharmacologists, and the consumers of their products. Cheap plastic bags and forks will not be so cheap; bringing a canvas bag to the store or keeping cutlery at the office for lunch will make more sense. Your apples may have lower pesticide residues and your water fewer endocrine disruptors.

Beyond petroleum, entrance gate duties should also be assessed on other natural resources—coal, natural gas, gold, silver, water, soil, timber, grass, ducks, wild turkeys, and so on. And every household and office and factory should pay exit gate duties on garbage and sewage and emissions, a function of the amount of waste produced. Recycling and reuse will suddenly be profitable like they never have been before.

As we lay these costs on the market, no need to race; let's move slowly and steadily, because the last thing we want to do is induce an oil shock or some other kind of shortage by not having given everyone sufficient time to prepare. The schedule and the rules need to be clearly defined, well explained, and announced (perhaps by Hollywood stars) with ample notice. And gate duties need to apply to everyone. No one, not the government, the nonprofit, the corporate titan, the clever lawyer, the lobbying industrialist, or the humble citizen can be exempt, just as no kind of natural resource can be omitted, including the most important one, the land itself.

Ecological Use Fees

Land serves not only as a place to drill for oil and fill with garbage but also as the place where we live and work. As we have seen, the massive expansion of the suburbs arose in part because the economy valued strip malls over farmland and suburbs over forests. Greenfields, until recently, haven't been valuable at all, except as a means to the end of development.

Development occurs because it creates capital, which feeds the economic cycle, according to Professor Smith's formula. But development, when it means building roads, factories, farms, and suburbs, decreases the ecological capacity of the land. Extracting natural resources is an obvious kind of ecological reduction. Trees removed for tract houses are no longer available to birds.

Less appreciated is the ecological opportunity cost incurred by development. Opportunity costs are what you pay for not doing something. Going to the baseball game on Saturday means I can't go to the movies at the same time, an opportunity cost paid in pleasure forgone; investing my savings in a new sports car means I can't invest that same money in my retirement account, an opportunity cost paid in future income.

Ecological opportunity costs are what nature loses when the land is changed from a more ecologically productive form to a less ecologically productive one. Development almost always entails improvements, typically through land conversion, changing an ecosystem of some kind to another that's considered more desirable: pasture, row crops, ballfields, roads, buildings, parking lots, and the like. Those kinds of improvements for economic purposes are opportunity costs charged to nature. Small costs, infrequently applied, are no big deal, because nature is large and generous, but when development mounts up over large areas, over regions and continents, then we need to take notice. Depleting one of Smith's three economic fundamentals means depleting the economy. A healthy economy requires a healthy balance between capital, labor, and land.

In the process of development, nature, the future, and the public lose clean air, drinking water, fertile soil, wildlife tourism, and so forth; these natural freebies are called ecological services. One estimate places the value of global ecological services, even in our nature-poor world, at over $33 trillion per year, which is almost twice as large as the total monetary economy of all nations combined. But these services rarely enter our economic calculus.

We can recoup these losses through use fees, however. Think about these new fees as a replacement for the property tax. What is the property tax, anyway? It is an assessment of the economic value of a piece of real estate (the land and its improvements), which swings up and down as the market does. Normally assessments are made when a property is sold, or as if a property were to be sold. Ecological use fees are similar to the property tax, but instead of taxing increases in economic values, we tax decreases in ecological service. Ecological use fees mark the "rent" owed to nature.

How might we do that? Although it may seem unimaginable, I can promise you that at some time in the past, the land where your home sits today was once a viable natural ecosystem of some sort, perhaps a combination of ecosystems, determined through interactions between the climate, the soils, species, and other participants in nature's cycles. It might have been part of a forest, a meadow, a wetland, or a desert; it might have spanned a streambed (which is why your basement floods when it rains), or rested deep in a piney wood. The land under my place on City Island was probably once an oak and chestnut forest cresting a small hill not far from a rocky beach. Those former ecosystems represent a baseline of ecological performance for your land. You can think of them as the best nature can do without our interference.

To learn about these ecosystems, we refer to maps of potential vegetation and ecological region type, which are available nationwide, created by tireless ecologists working on your behalf. Some municipalities have even more detailed maps, based on studies in a subdiscipline called historical ecology. Even in downtown Manhattan, we know where and what ecosystems once lay under Madison Square Garden and One World Trade Center.

Now imagine the map of the developments on your property—the footprint of the building, the extent of the driveway, the layout of the gardens, as if seen from above. (It's easy enough to see them using a mapping service like Google Maps.) My current abode covers about 1,000 square feet, and the rest of the lot, another 1,500 square feet, is covered by a deck, flower gardens, and a chicken coop.

Under this form of gate duty, your use fee will be based on the ecological deductions made by the buildings and other improvements on your land, literally subtracting the area of development from the area of ecosystems that once occupied the same space. The geographic assessment is not hard with modern mapping software, satellite imagery, and aerial photography. You could do it with a little time and an Internet connection.

Pricing out these ecological changes, as with the entrance and exit gate duties before, is a topic for further investigation, but the guidelines outlined above can apply here as well, in an analogous form:

First, the cost of renewal: More intensive kinds of development, which limit nature more, should be assessed at a greater rate than less intensive kinds of development with less impact (for example: Parking lots ≈ buildings > agriculture > pasture > woodlot).

Second, the cost of scarcity: Development of less common ecosystems should be assessed at a higher rate than more common ecosystems, which is to say, rates should vary with the proportion of an ecosystem remaining within a jurisdiction, increasing as development proceeds.

Third, the cost of connection: Development of richer and higher-functioning ecosystems (in terms of species and ecosystem services) should be assessed at a higher rate than less species-rich or lower performing ecosystems (deserts < grasslands < forests < wetlands).

Ecological use fees can be combined with economic assessments of the traditional economic-taxing kind or stand on their own. They can be assessed by government jurisdictions according to different systems of ecological potential; a local government might take a more fine-grained view of the nature within its purview, whereas state or federal governments might use regional or even continental assessment methods.

Paying for ecological values rather than only economic ones will encourage denser development. From an ecological perspective, a parking lot is not so different from a skyscraper; they are both, more or less, a lost cause. For the developer, however, they are quite different. A skyscraper is worth a lot more than a parking lot, with more rentable space. At the same time, the new residents of the apartment tower will each have lower taxes, because they are sharing the burden of a single fee with more people. These effects on the real estate market will set up important social and economic consequences explored in the next chapter.

Ecological Use Fees

Imagine (**a**) a small, narrow lot covered by an oak forest and a stream, and then imagine (**b**) a house constructed on that same lot. (**c**) A simple geographic overlay allows us to calculate the area of forest and stream replaced by structures, pavement, and garden. (**d**) The ecological use fee for supplanted forest and stream depends on the proportion of each type already developed within the city boundaries. (**e**) For example, since 70 percent of the streams have been destroyed, the use fee rate is $5.50 per square foot of stream. (**f**) Applying those use fees and a rate adjustment depending on the kind of development (structures > pavement >> garden), enables an estimate of the total ecological use fee assessment to be made for the property (**g**). Municipalities can adjust the fee schedules to balance revenue and conservation goals.

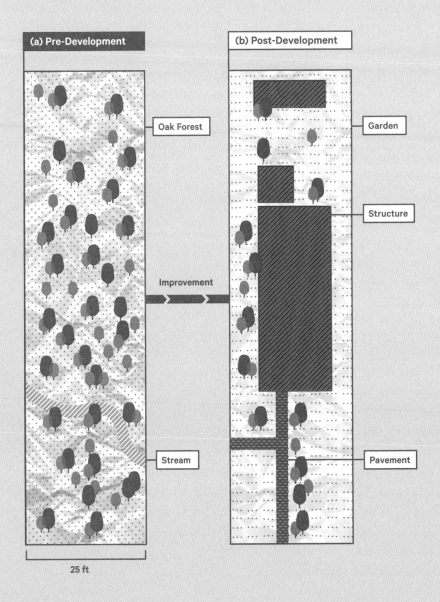

(a) Pre-Development

(b) Post-Development

Oak Forest

Garden

Improvement

Structure

Stream

Pavement

25 ft

(d) Ecological Use Fees Depend on Level of Development in Jurisdiction

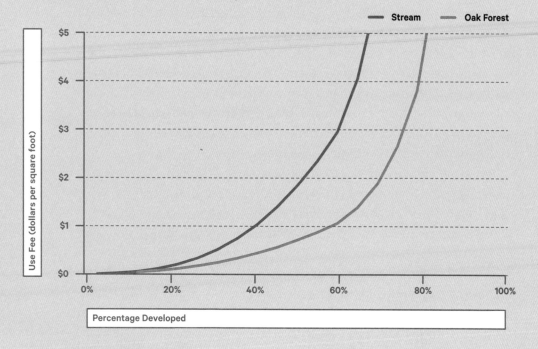

Pre-Development	Post-Development			Total
	Structures	Pavement	Garden	
(c) Analysis of ecological uses by property area (square feet)				
Oak-Hickory Forest	1,375	45	2,000	3,420
Stream	–	5	75	80
Total	1,375	50	2,075	3,500
(e) Ecological use fee ($ per square foot)				
Oak-Hickory Forest	1.75	1.58	0.58	N/A
Stream	5.50	4.95	1.82	N/A
(f) Ecological use fee annual assessment ($)				
Oak-Hickory Forest	2,406	71	1,155	3,632
Stream	–	25	136	161
Total	2,406	96	1,291	(g) 3,793

Meanwhile, on the failing edge of development far from downtown, ecological property taxes would encourage owners of poorly performing economic developments to restore part of their land. It may be less expensive to undevelop than to sustain lousy development over the long run. To lower the tax burden, a property owner can return land to higher performing states of nature and pay lower (or no) fees as a result. Where human infrastructure cannot succeed economically—whether it is a failed strip mall or an abandoned manufacturing plant—the land should be returned to a successful ecological state or the owner continues to pay the higher tax rate. As my son would say, "It's your mess—clean it up."

Boundary incentives will further increase the impact. A boundary incentive is a small reduction in the property tax when two similar land use classes meet along a shared edge. If your neighbors have built their house on the side of their property, then you might as well do the same, clustering development; similarly, if you leave forest on the back edge of your property, your neighbor should be encouraged to do so as well. Birds and bees and ecological processes are actually not persnickety about who owns what property, and larger patches of nature, all other things being equal, perform better than patches of nature scattered across the landscape. Similarly, concentrations of development, because of the positive effects of density on human interactions, also have benefits to the community in terms of economic productivity, community, and transportation.

Ecological use fees, like entrance and exit gate duties, will spark innovation across the American landscape as the economy inevitably adapts. To the extent that any piece of built infrastructure can be shown to perform as well as natural ecosystems—in terms of the costs of renewal, scarcity, and connection—it can and should be assessed at a lower rate. Architects, developers, and construction companies will be encouraged, as a matter of economic gain, to discover new ways for us all to live more gently and productively on the land.

We also now have a mechanism to harvest some of the economic benefits of the ecological services nature provides. Rather than regulating land use in a nearby watershed, a city might now pay upstream landowners a small premium to retain forests on their land, and thus protect the water supply without expensive filtration. A factory might pay for forest restoration to capture carbon that would otherwise be charged to its account, and thus drop its taxes. A community may decide to set aside some area of an endangered ecosystem because it will keep everyone's taxes down.

As with the other forms of gate duties, the government cannot be exempt from these ecological use fees. The government is a significant developer of the land, as a paver of roads, layer of dams, and constructor-in-chief for the infrastructure considered necessary to a functioning economy. Each of these projects takes away from the ecological commons while creating more obligations to maintain over the long run. Even the White House, a lovely home office, detracts from the hardwood swamps that formerly graced the Potomac River shores and slowed the floods. A 2009 report puts the cost of US public infrastructure requiring repair at over $2.2 trillion, while costing taxpayers $129 billion per year in broken cars, accidents, and lost time. So the government will pay, as we all will pay, so that public uses of public land are made more efficient.

Tax Shifting

Gate duties are a mechanism to reform the relationship of the American people to American nature and thus create the conditions for a new kind of American life and a new form of economy, less dependent on natural resources, more dependent on human capital and creativity, and better able to sustain long-term and growing prosperity. (The moral, aesthetic, health, safety, and security benefits of taking care of the land are important but secondary benefits.) And because new additional funds will flow into the government's coffers for its needs, such as they are, we can happily abandon some other kinds of taxation collected today.

That is, we should think of gate duties as tax shifting, rather than tax raising. Like most people, I would prefer to pay less taxes than I do today. But I also recognize that some level of taxation is necessary to pay for the military, police, fire protection, water works, and national parks that I need. So if I'm going to pay, which I must, then I want to pay taxes that work for me and my posterity, not intrusive, inefficient ones.

Sales taxes are high on my list of inefficient taxes to jettison as quickly as possible. Levied by local and sometimes state governments pursuant to state law, they add 6, 8, or 10 percent to prices of taxable commodities, while giving nothing productive back to the transaction itself except the indirect benefits of the government services. Sales taxes are intrusive because they make every transaction in a vast and complicated economy potentially accountable to the government, and within that mountain of paperwork there is ample room for mistakes and mischief. Sales taxes are a burden for all involved.

Moreover, sales taxes cascade, piling onto goods repeatedly as they move up the chain of production. A refiner may pay a sales tax when it buys oil from a wildcatter; the independent gas station owner may pay a sales tax when he buys gasoline from the truck; and I will pay a sales tax when I buy the gasoline at the pump. (Thus the sales tax also incentivizes vertical integration, where internal transfers within a business go untaxed, benefiting larger companies over smaller ones.)

Imagine how it would look if taxation occurs only at the beginning and end of the economic chain, where nature and the economy meet, not at various points along it. You'd go into a shop and the price of the item you wanted to buy would be its entire price, with no additional tax. If you operated the shop, you would not have to report your receipts to the government each month. The costs of making the things the shop sells would be built into the prices, with a small increment added due to its ecological

The Improvement of America

On a national scale, we can estimate the amount of development that has already occurred by over-laying the human footprint (as shown here) with maps of the different biomes or "ecological regions" of the country, like forests, grass-lands, shrub lands, deserts, and wetlands. This kind of overlay could drive a reformed and radically simpler property tax system based on a reckoning of the ecological values of our lands and waters, within an economic framework.

Source: Redrawn from data in Sanderson et al. (2002) and Olson et al. (2001).

A Better Way to Tax

Gate duties on land–called ecological use fees–rapidly add up. The contiguous United States covers approximately 2.9 million square miles, of which about 5 percent has been heavily influenced by human activ- ity, 50 percent moderately influenced (including most agricultural and mining lands), and 41 percent lightly touched. Application of ecological use fees mea- sured in pennies per square foot over these areas, calibrated to reflect differ- ences in ecological value and development, could yield potentially billions of dollars of revenue for the government, allowing us to cut taxes hampering other parts of the economy.

Ecoregions	Level of Development (based on human footprint)				Totals
	No or Minimal Development	Light Development	Moderate Development	Heavy Development	
Land Area (square miles)					
Desert	5,786	124,770	41,592	2,000	174,149
Forest	16,724	365,350	1,044,663	107,803	1,534,539
Grassland	3,824	395,871	369,043	21,916	790,654
Shrubland	14,289	322,438	107,831	9,713	454,270
Wetland	210	3,147	3,114	696	7,167
Total	40,833	1,211,576	1,566,242	142,128	2,960,779
Hypothetical Ecological Use Fee Schedule (pennies/square foot)					
Desert	0.0	0.1	0.4	1.0	NA
Forest	0.0	0.2	0.8	2.0	NA
Grassland	0.0	0.2	0.8	2.0	NA
Shrubland	0.0	0.1	0.4	1.0	NA
Wetland	0.0	0.3	1.2	3.0	NA
Hypothetical Ecological Use Fee Proceeds ($ millions)					
Desert	$ 0	$ 3,478	$ 4,638	$ 558	$ 8,674
Forest	$ 0	$ 20,371	$ 232,988	$ 60,107	$ 313,466
Grassland	$ 0	$ 22,072	$ 82,307	$ 12,220	$ 116,59
Shrubland	$ 0	$ 8,989	$ 12,025	$ 2,708	$ 23,721
Wetland	$ 0	$ 263	$ 1,042	$ 582	$ 1,887
Total	$ 0	$ 55,174	$ 332,999	$ 76,175	$ 464,348*

*For comparison, property tax revenues collected by local and state governments in 2009 were $472,834 million.

value. The shop manager and government agent would have a lot less paperwork. There could be fewer auditors because they would have less to do.

Another tax I would enjoy axing is the corporate income tax, collected from US businesses. Although it seems like a good idea to soak the oil companies when they have yet another banner year, taxing them doesn't do the job; they just pay the bill and pass on their expenses to you. Better for everyone to clean out the subsidies in the tax code supporting oil companies and other natural-resource extractors, and require them instead to pay the real costs of the natural resource they are extracting. What profits they take can be taxed when distributed to individuals through dividends, salaries, and bonuses. Progressively lowering and eventually eliminating business taxes, for small firms and large ones, would give all businesses a boost at the same time that expenses for precious or polluting raw materials are increasing on the margin.

Picking up our example from before, the imaginary $1 per gallon oil duty and the three cent per pound duty on carbon dioxide could produce a combined $532 billion in revenue, based on 2010 consumption. In comparison, the sales taxes collected nationwide in 2010 by local and state government, and from excise taxes on gasoline collected by the federal government, amounted to $511.2 billion, $20 billion less. Gate duties on other natural resources and other forms of waste could compensate for letting go of the corporate income tax system, which in 2010, collected $387.5 billion, across all government levels.

Property taxes in 2010 amounted to $430.6 billion nationwide, mostly flowing to municipalities and in some cases state governments. An ecological use fee schedule pegged at five cents per square foot for stream, three cents per square foot for wetland, and one or two cents per square foot for forest, meadow, or desert lands could raise more than the current property tax regime does.

If the problem with the current economy is that gas and land are too expensive, it may seem counterintuitive that the solution is to make these resources more expensive still, while other items become less costly. The reason to adopt gate duties is that they force a decisive break with the old way of doing business. Much as gas ignites in an internal-combustion cylinder only after a spark, so society will change only if we catalyze it to change; witness the extreme measures of the past. Gate duties, and the rearrangements described in the chapters to come, are the spark. In the resulting explosion, there will be dislocations and costs, even some sacrifices, but there will also be change. Through effort and perseverance, we will obtain a revitalized economy that builds, rather than destroys, the future.

Aldo Leopold said it, and no doubt Adam Smith and Albert Pigou would agree, when we see land as a community to which we belong, we begin to use the land with love and respect.

10
Moving to Town

Forget the damned motor car and build the cities for lovers and friends.

Lewis Mumford, *My Works and Days* (1979)

Many years ago, before I moved to the city, I had a job in the wilderness. I took a summer position working for the botanist at Sequoia National Park in the Sierra Mountains of California. We intended to catalog a sample of the plants that lived there, which entailed casting random darts at the park map (using the computer), then hiking into the backcountry to that point and identifying every tree, shrub, grass, and herb we found within a radius of 17.8 meters (equivalent to an area of one-tenth of one hectare, or about a quarter acre). If our circle landed in a conifer forest, we might get thirty or forty species, but if our circle happened to catch the edge of a stream, we would get three or four times that many. Water was the key to productivity and diversity.

In town between these scientific backpacking expeditions, I shared a house with an economics major (also a summer tree counter) and one night we fell to talking streams, which in the manner of collegiate conversations, somehow morphed into a discussion about economic fundamentals. My housemate declared there are three sure ways to make money in a competitive economy: (1) Make something first; (2) Make something better; and/or (3) Make something cheaper than everyone else. He said the long-term problem with the economy is that most people are not innovative enough to come up with something totally new or skilled enough to be the best at whatever they're doing, so many folks fall back on the strategy available to all of us, that is, Strategy #3: Make something more cheaply, by cutting costs for labor, using inexpensive ingredients, hiring machines to do the work, avoiding taxes, etc. My friend said it doesn't matter if you sell a few things for a lot of money, or sell a lot of things for a pittance. Either way, you make money. The modern Chinese economy, Walmart, and the plastics industry have all been built on Strategy #3.

I asked how that relates to a stream. And he said something that has been lodged in my memory ever since: "Cities are our streams."

Cities Are Our Streams

I chewed on that idea for a long time. It didn't make sense when I heard it because like most kids who grew up in the suburbs in the 1970s, I thought of cities as places where poor people lived. Downtown Oakland didn't seem like a place with a lot of money. Blighted city centers contrasted with the blooming productivity of the countryside in California, rapidly being bulldozed for cul-de-sacs and new houses. I knew exactly what cities did to streams: They obliterated them.

I began reading Lewis Mumford, Jane Jacobs, Max Weber, and other philosopher-historians of the city after I moved to the city. I learned that urban areas had a surprising track record of promoting innovation and productivity, back to the first walled towns of Mesopotamia, Egypt, and China. And why? Everyone seemed to have their favorite answer. Weber and contemporary economists like Edward Glaeser and

Richard Florida write about lower transaction costs due to proximity and sociality. They describe urban "agglomeration economies" enabled by increased specialization, greater division of labor, and economies of scale.

Jane Jacobs and other urban planners saw cities as places with tightly intermixed private and public spaces that promoted a uniquely dynamic way of life. The apartment overlooking the street allowed everyone in the neighborhood to keep an eye on the kids playing outside; the lively street in turn created a shared, vital public domain, which partly mitigated the apartment's tight quarters, encouraging people to turn outward, toward the city and the community, rather than inward, as suburban families do, toward the backyard. Cities also tend toward diversity—of culture, of income, of ideas—which in the right conditions can promote humility and tolerance, while generating the disconcerting, useful contrasts that often underlie new thoughts. New notions emerge when ecologists share houses with economists. It matters who you might bump into in the street and that enough people are out to have someone to bump into. Intensity, diversity, and density make cities engines of progress, economic success, and serendipity.

The results are remarkable. Consider New York City. The economy of the New York metropolitan area in 2010 was estimated to produce $1.28 trillion in value. That's more economic productivity than any one of these countries: Mexico, South Korea, Netherlands, Turkey, or Saudi Arabia. It's more than the economies of Malaysia, Portugal, Singapore, Hong Kong, and Egypt combined. And it's not just the biggest cities that produce. The metropolitan area of Austin, Texas, generated $86 billion itself in 2010, more than sixty national economies, including oil-blessed Oman, Azerbaijan, Gabon, and Bahrain.

To the arguments of Weber and Jacobs, we can add another beneficial quality of urban life described in David Owen's 2009 book, *Green Metropolis*, and a few others: Cities are more environmentally efficient. Shocking, I know: How can living in a leafy suburb possibly be harder on the environment than the burdens a city imposes on its land, air, and water? But the statistics do not lie. The average New Yorker uses about two-thirds of the amount of electrical power as the average American, and produces a third as much carbon dioxide. Because cities are dense, buildings tend to share walls, which share heat, lowering energy bills. Public transportation is more practical, and walking more likely, because distances are shorter. New Yorkers also use less water per person (by 74 percent) and generate less garbage (by 45 percent) than the average American and yet seem to enjoy, despite the traffic, noise, and attitude, a reasonably high standard of life.

Cities with enough density can promote creativity and resource efficiency at the same time: a win-win solution to both economy and ecology, obtained, counterintuitively, in town.

Density 5K

I write *town* because the benefits of urban life apply not only to the largest cities in the land, but to middle-sized cities and small towns too, even some suburbs, if they can build up enough density and a population of sufficient size to unleash the positive aspects of density dependency, while keeping the negative aspects in check.

Better, denser communities mean that the vast majority of Americans have a lot to gain. If we divide the current US population into sixths, then according to the government about two-sixths (one-third) of us live in localities with less than one thousand people per square mile; about three-sixths of us (one-half) live in towns and cities with one thousand to four thousand people per square mile (that's the suburbs), and the remaining one-sixth inhabit higher-density urban places.

The island where I live in the Bronx has about 4,250 people per square mile on a land area of about 0.4 square mile—it's dense, but small, probably too small to be economically viable on its own at current density, which is why most people commute off the island for work. By contrast, Manhattan Island, just down the road, has ten times the density and plentiful jobs as a result, but it also has rents that make most folks cringe. Though Manhattan's economy and culture are impressive, not everyone can live there.

How much density is enough to make a place great? The answer is not obvious (and depends, of course, on the secondary question: Dense enough for what?), but I suspect that many of the economically productive and environmentally beneficial aspects of density begin to kick in at around five thousand people per square mile. Depending on how you measure it (do you include parks—I do—or don't you?), density 5K is where the tradeoffs between amenities and congestion begin to balance; it's also about the density where public transportation starts to pay for itself. A threshold of five thousand people per square mile is less than the current density of Manhattan, and also less dense than Keizer, Oregon (5,145); Park Ridge, Illinois (5,261); Coral Springs, Florida (5,387); Newport Beach, California (5,402); and Woonsocket, Rhode Island (5,749). Density 5K is denser than most suburbs, but not unimaginably so. It's a density we can manage.

Minimum Population, 10K

Density alone is not enough; communities also need to be of a viable size to thrive socially and economically. Hunter-gatherers once lived in settlements of extraordinary concentration, sometimes at a rate pushing 100,000 people per square mile, even though there might be only fifty people living together amid a large wilderness; they chose to place their wigwams side-by-side and leave the surrounding countryside un-populated (by people). But fifty people are not going to be enough to support grocery stores or fire departments, the kinds of services that are essential to modern life.

What Density Looks Like

It's difficult for most people to visualize what different population densities look like. These studies from suburban Staten Island, NY, illustrate the building footprints in census tracts of different densities. Note that as density increases, open areas also may, as buildings become larger and taller. At the bottom are three examples of different arrangements of 5,000 people per square mile.

Source: City of New York Department of City Planning (2011).

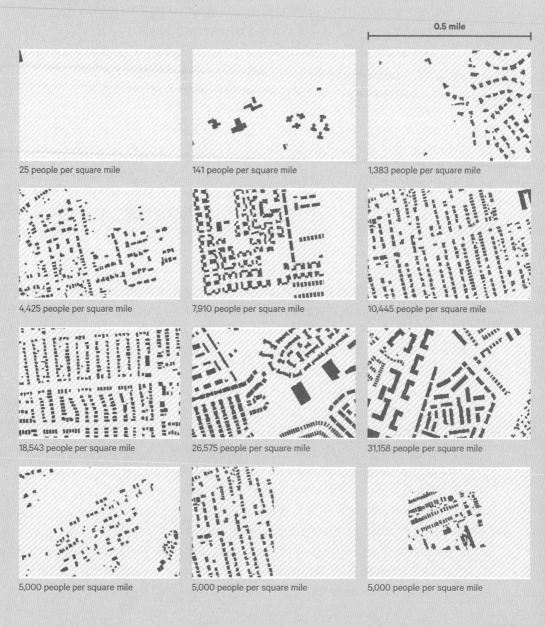

0.5 mile

25 people per square mile

141 people per square mile

1,383 people per square mile

4,425 people per square mile

7,910 people per square mile

10,445 people per square mile

18,543 people per square mile

26,575 people per square mile

31,158 people per square mile

5,000 people per square mile

5,000 people per square mile

5,000 people per square mile

To address the question of size, we need to invoke a power law. Power laws are very common in nature, describing the statistical distribution of earthquakes, the craters on the moon, and the foraging patterns of sharks, among other wonders. Cities follow a special form of the power law named after the American scholar George Kingsley Zipf. Zipf's Law says that for cities of a given size (x), the frequency of cities of that size will be approximately the reciprocal of the size (1/x). That is to say, relatively speaking, a city of a million people tends to occur approximately once in a million times and a city of ten thousand people will occur once in ten thousand times. Small cities are much more frequent than large ones. Zipf's Law has been demonstrated in a wide variety of countries and across many different historical circumstances; though explanations of the power law vary, the law itself is not in doubt.

Zipf's Law means that not all cities can be New York and Chicago; there will always be more small places than large ones. But it also means we can't have only suburban sprawl; we need cities too. Lots of chances, represented by the diversity of places to live, means there are more opportunities to experiment and specialize in the search for what makes a community not only viable, but great.

Specialization is important. The best cities figure out what they are good at and then attract specialists to them. They promote skill and ingenuity over cheapness, and over time, may discover new capabilities. Think of college towns that focus on education and then spin off new businesses; or technology hubs, like Silicon Valley, that reinvent the computer chip, then the telephone, then the social network. Natural factors—like the proximity of mountains or natural resources—also can promote specialization and repeated success: like Aspen, a ski resort that now sponsors festivals of ideas, or Miami Beach, a sandy strip that now specializes in parties.

The flip side of specialization is diversity—large cities excel at diversification, but all towns, as they concentrate, generate a necessary variety of occupations to fulfill the basic requirements of the people who live there. The trick for US towns and cities is to find an area of focus and build on that expertise, so that the community draws in more people to provide the food, healthcare, education, and other services that make a viable economic whole.

In endangered species conservation we often seek to restore (at least) the "minimum viable population" of a species, that is, the smallest population of a kind of plant or animal that can sustain itself indefinitely. We need the analogous concept for towns and cities. One 2012 study examined communities in Indiana, varying in size from a population of one to 252,000 (Indianapolis was excluded) and found that the threshold size at which essential establishments and services reliably developed was approximately eleven thousand people. Other studies have attempted to define efficient communities and optimum community sizes, but become entangled in the density-dependent effects. Because it's not clear that Indiana is representative of the country as a whole, and because it's easier to remember, let's be conservative and aim to restore economic communities of at least ten thousand (10K) American people.

Moving to Town

Powerful incentives led to the de-densification of America. Equally powerful ones will be required to reverse the trend and build the minimum density and community sizes to create a new revitalized economy. I have five ideas in mind about how the new American urban economy might come about.

The first idea comes from the last chapter. Gate duties are the foundation stone on which the new economy is built. Increased costs for gasoline and other fuels will literally change the cost of distance, and that will change where people decide to live. In the future, gate duties will provide an incentive to live close to work and rewards for living in more efficient ways. Ecological use fees will enhance those benefits by reducing taxes for those who choose shared living situations (e.g., apartments and condominiums) over sprawl.

Second, we must play favorites and support communities trying to grow into economic viability. I propose we ask existing municipalities to designate "New Town" districts that meet certain economic productivity criteria: (1) densities over five thousand people per square mile, (2) populations of at least ten thousand, and (3) revised zoning regulations that encourage neighborhoods to mix work, residence, and shopping, discourage and eventually ban automobile use, and promote walking, biking, and public transit. (No more free parking! Mandatory sidewalks!) New Town districts will automatically be eligible for a defined and substantive portion (50 percent?) of all federal and state support for transportation, education, infrastructure, and small business loans. Government will be required to establish offices and bureaus in New Town districts wherever possible. The creation of the districts will be completely voluntary—no community will be required to establish one—but communities that do will be immediately rewarded for their efforts. And no new money need be involved; we'd just be spending existing funds preferentially into neighborhoods that commit to helping the country into a twenty-first-century growth-oriented economic model based on urban vitality.

Third, employers need to be engaged in helping their employees find quality housing near the office. A simple strategy is the "location-efficient" mortgage, which modifies lending standards to favor home mortgages close to the applicant's workplace. Loan standards for commercial loans could include a stipulation requiring a certain proportion of workers to live near the business. If it's easier to get a mortgage by moving closer to the office or by moving the office closer to home, then colocation of home and work will spill over into other decisions.

Fourth, a more aggressive suggestion is to use what Google knows. Every year, each

Five Ways to Increase Density

Density drives economic growth, lowers environmental impact, and fosters diversity, life, and meaning. Here are five ideas to encourage density: Gate duties encourage density by lowering costs for con-sumers and residents willing to live closer together. New Town districts reward neighborhoods willing to commit to productive density and population targets. Location-efficient mortgages encourage home-owners to buy near work, and the home-to-work payroll adjustment nudges employers to care where their employees live and employees to care about living close to employers. Finally, the Superfund for Real Estate provides a way out for people living at eco-nomically inefficient densities, while returning some modicum of capital back to the land community.

Gate Duties

● Less expensive mortgage

More expensive mortgage ●

New Town District

● Jobs! Loans! Investment!

Oil shocks. Recessions. Foreclosures ●

Location-Efficient Mortgages

● Less expensive mortgage

More expensive mortgage ●

Home-to-Work Payroll Adjustment

● Lower payroll taxes

● Work

● Higher payroll taxes

Superfund for Real Estate

● The good life

● The good life

Neighborhoods of Different Size

Density also adds up.
This study in geometry and
multiplication shows for
neighborhoods of three
different sizes—with radii
representing distances associated with walking (0.5
mile), bicycling (2.5 miles),
and streetcar use (5.0
miles)—how many people
could live together at
different densities. Actual
neighborhoods will vary.

Streetcar Neighborhood

Biking Neighborhood

Walking Neighborhood

Radius (area in square miles)	0.5 miles 0.785 square mile	2.5 miles 19.635 square mile	5.0 miles 78.540 square mile
Density (people per square mile)	Population	Population	Population
1,000	785	19,635	78,540
5,000	3,927	98,175	392,699
10,000	7,954	196,350	785,398
25,000	19,635	490,874	1,963,495

of us reports to the Internal Revenue Service both our home and workplace addresses. Google (or other mapping websites) can easily calculate the distance in between. We should give employees a rebate on their payroll taxes when that distance is small. What's more, the employer should receive a rebate, too. For example, if a company has a worker who lives within five miles, then both employer and employee receive an automatic rebate of, say, 10 percent of their payroll taxes for the year. And where would the money come from? From new fees levied on those who live more than ten miles from work *and* on the people who hire them; the scale slides upward the greater the distance. With a "home-to-work" payroll adjustment in place, employers will aim to hire locally, and workers will seek employment closer to home. Second homes should be included in the calculation as well, with a separate assessment made for each one, according to the proportion of time spent there. (This might make things tricky for those with homes in Malibu, Jackson Hole, and Manhattan, but then, they can probably afford it.) The unemployed and the retired are exempted.

I recognize that these policies mean that many of us will have to move. But moving house is nothing new for Americans; we are a remarkably mobile people. Census data show that between 2009 and 2010, 12.5 percent of the American population relocated—that's more than 37.5 million people. The average American moves twelve times over the course of his or her life. Most of these moves are short—usually within the same county—and each one provides an opportunity to reshape the distance between home and work. Businesses move, too: A report on large corporations found that about five percent relocate their headquarters each year. Thus, with a little encouragement, a dance will begin in which employers and employees alike look to minimize the distance and maximize the tax savings for each other through the logical expedient of moving closer together. Being together will then generate other positive dynamics.

I will give you a specific example. The Bronx Zoo, where I work, is located in a central city neighborhood it shares with the New York Botanical Garden, Fordham University, and an old Italian enclave of terrific restaurants and delicatessens centered on Arthur Avenue. Yet few of the zoo's fulltime staff live nearby; the local schools have problems, and safety is an issue late at night. Instead, most of us drive to work from safe neighborhoods with better schools in New Jersey, Manhattan, Long Island, and the northern suburbs. What the Bronx needs—what America needs—are incentives for people to live near where they work. My guess is that if the staff were given incentives to live within walking distance, the schools would be fixed, safety would be restored, we would all get to work on time, and the Bronx would become the new Brooklyn.

Not everyone will win in the sorting that is about to take place. The combination of gate duties, New Town districts, and home-to-work adjustments may also trap some people in lands and buildings they no longer need or want, like those who bought

McMansions on acre lots a hundred miles from where they work. Those people are going to need help from society at large.

They need the fifth idea: a debt forgiveness plan, or Superfund for Real Estate. Property owners will apply to have the outstanding interest on their loans forgiven (a cost charged to the banks generously bailed out in the recent collapse) and a one-time cash payment from the government. The cash will come from a small surcharge on capital gains (say 2 percent, which would yield, based on US capital gains in 2010, $216 billion per year), thus returning a proportion of capital back to the land. The landowner will receive a payment calculated according to an average ecological use assessment estimated over the time the owner held the property, multiplied by the length of tenure (which means higher payments for people who have lived on the land the longest). Owners will agree to restore the land to its highest potential natural state, repay the principal on any outstanding loans, and retain title, with the option to "improve" again after an ecological recovery period of thirty years, the same term as for a home mortgage.

These policy measures (gate duties, New Town districts, the location-efficient mortgage, the home-to-work payroll adjustment, and the Superfund for Real Estate) combine carrots and sticks to re-create the American landscape. These incentives don't need to be permanent features of American life, but they will need to apply for a generation or perhaps two. After discussion, design, and legislation, these measures should be implemented incrementally but steadily—so that people may anticipate and respond in a timely fashion. It took a century to create the landscape we have today and even longer to become the people we are. Evolution takes time.

Where We Come From

Other suggestions for restoring great American communities arise from remembering where we come from, and I don't mean your hometown or even the country of your immigrant ancestors. The cities that will work best for people will be the ones designed as habitat for a very special species, people.

Once upon a time, our ancestors evolved from ape-like primates that lived for millions of years in the open savannas and woodlands of Africa. For a long time, we lived there and nowhere else; then seventy thousand years ago early human beings began migrating out of Africa to other parts of the world, eventually supplanting or absorbing two other species that had preceded ours in colonizing the world, *Homo erectus* and *H. neanderthalensis*. Our archaic predecessors were distinguished from other apes by shortened upper bodies perched on long legs, suitable for a new kind of bipedal locomotion called walking. We evolved opposable thumbs and big, color-loving eyes oriented forward out of our heads. As our brains grew bigger relative to our body size, our bodies became weaker. A lump of bones and tissues—the larynx—slipped down our throat;

we started talking. That's what made the difference. Our species, *Homo sapiens,* which literally means "wise human," succeeds because we can communicate with each other.

For the great majority of human history, people physically like you and me lived in small bands of friends and family, hunted animals, and gathered plants from the fecund lands and waters in our immediate environment. We cooked our food over fires set for the purpose, a technical innovation developed to mitigate small teeth and sensitive stomachs (or perhaps enabling the same). We were good at hiking and running when the occasion required it, even over great distances, and we learned to pause with our backs to cover, a predilection we retain today. We needed to survey the land to find resources, but we also made sure to have a place to hide in case predators or other humans came near. It may be why golf course developments with expansive views over the links appeal to so many, even though only a fraction of residents actually play the game. We like to see, and we like to hide.

By fifty thousand years ago, human beings had developed all the signs of recognizable humanity, what anthropologists call behavioral modernity: language, fine tools, burying the dead, figurative art, game playing, musical instruments, and self-ornamentation. These early speaking, skilled, mournful, artistic, playful, singing, decorated people did not live in cities, but they did live in groups. Like most primates, human beings are intensely social, at least in comparison to most other animals, many of whom live solitary lives. In the 1990s Robin Dunbar, a primatologist at University College London, observed a correlation between primate group size and the ratio of the volume of the neocortex (the thinking part) to the rest of the brain; on the basis of this correlation he hypothesized that an optimal group size for human beings was about 150 people. He then scoured the literature for the sizes of human groups and showed that across twenty-one different hunter-gathering cultures, the average group size was 148.4 at the band or village level—a remarkable concordance.

Dunbar suggests that 150 is about the largest number of people we can get our heads around at a time; as groups get larger, social cohesion and the usual bonds of kinship break down—we can't keep track of the web of relationships that bind the group together. Instead, formal relationships and shared allegiances, rather than talk, become required to maintain the group. In Paleolithic times, groups that became too big would divide to find their own way in the wide world. (Modern people, as a direct consequence of our numbers, have had to devise different solutions: democracy, the law, Facebook, and cities.)

Sociality, density, and communicative powers enabled humanity to flourish and spread over six continents. We told stories and shared visions of the future. Working together we domesticated plants and animals to our purposes, and those domesticates eventually led to food surpluses that created the conditions for the development of larger settlements, towns, and then cities. And as we have seen, when cities developed, we embarked on the achievements of civilization: Religion, the military, money, and the market economy soon followed.

Cities for People

Other inventions are now needed. Fortunately after ten thousand years of living in cities, we finally have a sense of what makes human communities work well—and there's a large and growing movement afoot to do better. A new generation of architects, planners, real estate developers, sustainability mavens, and urban farmers are leading the way. Some look to nature for models, curious about how light, air, water, soil, plants, and even bees and chickens can live in the same spaces with us, or above us, on green roofs. Others look to older, pleasant cities, like the streetcar suburbs we met before. These small cities lived in the middle ground of density and reaped the benefits. Some writers even peer overseas, to livable towns in Europe, where people seem to have realized that there is only so much land and space, so perhaps they should make the best use of what they have.

A classic in this vein is Jan Gehl's short, lively book *Cities for People*. Gehl, a Danish architect and urban designer, writes that the highest priority for retrofitting cities should be the life of the place. That is, we should decide what we want our city to do, and then we should create spaces to make it happen. Copenhageners like to ride their bikes, and so their city facilitates their delight with extensive cycle paths and bike share programs. (City bikes for borrow have large yellow smiley faces on the wheels.) Architecture traditionally emphasizes the buildings, which are a means to an end, but which ought not be an end unto themselves: Life is the end. Quality of life is what we should be designing cities for.

People are not giants. Most of the time, we experience the city from ground level, with heads five or six feet above the pavement, so our settlements should be attractive from that perspective, not King Kong's. The best urban landscapes cater to the perceptions of pedestrians, with long, unobstructed views, good lighting focused down (away from the stars), and multitudinous wonders, including other people, at eye level. Good cities instill the desire to stroll. For reasons that may have to do with our evolution, walking and thinking seem to be deeply connected. St. Augustine reputedly advised his students stuck on a problem to walk: "*Solvitur ambulando*," he intoned, "It is solved by walking." Charles Dickens, 1,500 years later, wrote: "If I could not walk far and fast, I think I should just explode and perish." (Fortunately for us, he found room and time to get out.) Abraham Lincoln was a walker; so was Mahatma Gandhi. No less an authority than Mad-Eye Moody, of J. K. Rowling's Harry Potter books, advises the precocious wizard: "Nothing like a nighttime stroll to give you ideas."

While out, people need places to interact with others, often along edges. Social ecotones, or "soft edges," are private places facing public ones, like a stoop of a brownstone,

Cities for People

Great cities cater to how people see and use the environment. They provide opportunities for interaction that foster safety, conviviality, and serendipity, and they provide what we need, whether transportation, shopping, employment, or recreation, within easy walking distance.

Source: Inspired by figures in Hester (2010) and Gehl (2010).

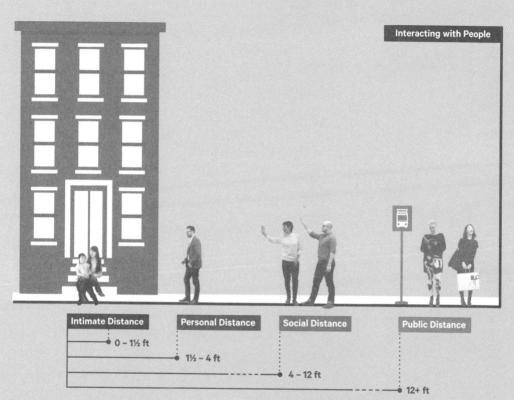

Interacting with People

Intimate Distance	Personal Distance	Social Distance	Public Distance
0 – 1½ ft	1½ – 4 ft	4 – 12 ft	12+ ft

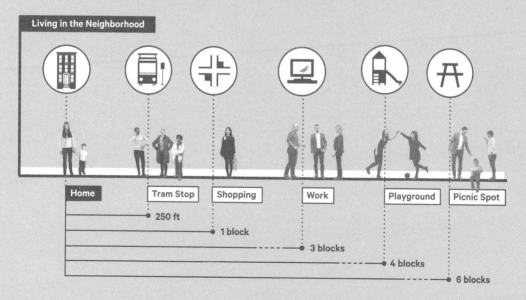

Living in the Neighborhood

Home	Tram Stop	Shopping	Work	Playground	Picnic Spot
	250 ft				
	1 block				
		3 blocks			
			4 blocks		
				6 blocks	

How Far We Travel

Americans are prodigious travelers. In 2009 the average American traveled 36 miles per day and 13,148 miles per year, but more interesting is how travel habits vary with density and distance to work or school. Travel distances drop off with densities above 10,000 people per square mile and when people live within five miles of work. Imagine the travel savings that might be achieved from some fraction of folks moving back to town.

Source: Santos et al. (2011); US Federal Highway Administration (2011b).

Distance to work or school (mi)	Residential Density (people/sq mi)							
	Rural			Suburban			Urban	
	0 – 99	100 – 499	500 – 999	1,000 – 1,999	2,000 – 3,999	4,000 – 9,999	10,000 – 24,999	More than 25,000
0	18,066	14,980	11,066	11,622	10,918	10,001	6,449	5,195
0 – 1	16,298	11,835	14,781	10,550	8,783	8,782	11,454	5,388
1 – 5	15,461	15,902	13,173	14,056	12,996	10,159	6,977	5,482
5 – 10	19,207	16,882	15,275	14,068	14,827	13,123	13,782	10,660
10 – 20	19,798	20,168	18,747	26,465	20,687	16,791	13,826	11,610
20 – 30	21,852	28,589	21,083	22,807	23,312	26,687	21,157	50,281
30 – 40	26,233	28,976	28,564	27,039	32,953	27,967	31,814	21,539
40 – 50	27,084	29,051	31,090	28,230	33,394	26,432	16,876	*
50 – 100	32,338	39,602	27,170	34,052	29,384	24,310	19,966	27,796
100 – 200	41,719	34,692	32,831	60,567	27,852	24,030	23,675	*
> 200	53,810	38,249	*	22,171	36,435	28,647	*	*

* No estimate

or a small garden in front of an apartment, or a street-level café. These private-public places invite people walking by to stop and talk, and give other people the opportunity to tend their own space while keeping an eye on the world at large. Doorways are another example. Portals to life on the street, they are also places to loiter: discrete, safe, intriguing niches. Shop windows and flowers, porticos and arcades, plazas and fountains all generate interest and life, while creating soft boundaries.

Around town, we need to provide opportunities to rest. Sitting actually brings vitality to the city, since the presence of other people is reassuring; it tells us a place is safe and comfortable enough that we might want to stay, too. The best sitting/leaning locations cover your back, provide a view, and have a favorable microclimate, so architects and planners must pay attention to sun and wind. Buildings cast dark, cold shadows, and bend the breeze; if a place is too cold or too hot, it won't be used. Moveable, foldable chairs, like the ones in Bryant Park in midtown Manhattan, are a nice solution to changing weather, allowing individuals to choose where they want to sit and with whom.

On a larger scale, interactions in the city are not just among its residents, but also with the idea of the city itself. The best cities are made of more than buildings; they are made of people and ideas. Randy Hester, a professor of landscape architecture at the University of California, Berkeley, says all the best places to live have a unique sense of themselves, a natural boundary, and a center. Think of your favorite town or city. What makes it distinct? How do you know you've left? Where is its heart? If you have ready answers, then you know what makes that place great.

And so, the best idea for restoring American urban communities is to build ones where people will want to live.

"Forget the Damned Motorcar"

One word you will not have read in this brief account of quality cities is *car*. Humanity went carless for the better part of four hundred thousand years and did just fine. Cleopatra, Michelangelo, William Shakespeare, Adam Smith, and Abraham Lincoln lived in cities and never drove an automobile. They didn't need one, and neither would you if we could arrange our lives the way our ancestors did.

Mumford had it right. The problem with cars is not only that they enable sprawl, but that even in otherwise dense, interesting cities, they take up so much damned space. Standing on a sidewalk, a person occupies about four square feet of land; most cars take up eighty square feet, twenty times more. Suburban zoning regulations commonly require three parking spaces for every thousand square feet of office space. Because a standard parking space measures 330 square feet, regulations require a one-story building to have as much parking lot as floor space; a three-story building requires asphalt at a rate of three times the footprint of the building itself. In most American cities, 30 percent or more of the lucrative land area of our American cities is dedicated to automobiles and their needs, counted in millions of square feet of streets, garages, and parking lots. People with cars need to park. People and cities free of cars will also be free of lots of wasted space. They will have parks instead.

Lost development opportunities are only part of the automobile burden. Cars also limit our interactions with each other. Our senses evolved for maximal effectiveness while walking at about two or three miles per hour, which is in sharp contrast to the complicated behavior called highway driving. Tailgating, road rage, and other poor behaviors of the American road are at some level an indication that the evolutionary process hasn't caught up to our high-speed lifestyle. It's impossible to see facial expressions or body language at seventy miles per hour; without these clues, we are poor judges of intent. Because we can't perceive intent, we often think the worst and occasionally act out, setting up cycles of bad behavior and poor decisions that litter the road with tragedy, car components, and body parts.

In town, cars seem to magically fill whatever space is provided. They clog the avenues and side streets, packing the intersections and bringing everything to a grinding halt. And what for? It turns out that an enormous amount of in-city traffic—some studies have counted up to 70 percent—is composed of people circling to find parking places. It is a kind of madness.

We all know the consequences: hold up, stand still, gridlock, traffic jam, congestion. Then when the flood of steel boxes moves, the clamor is amazing: horns, squealing brakes, growling engines of explosion cars; the rising roar of motorcycles; the sirens of ambulance and fire engine and police cruisers. The noise spills out over the sidewalks and cafés and into doors and gardens, stifling conversation and ruining a thought. Pollution collects along the roadways, too; even with modern controls, the products of combustion have to go somewhere, typically into trees and the lungs of passersby.

And for pedestrians, cars are agents of oppression. The streets—up to a third of the city's area—are forbidden to the walker and only uncomfortably share space with bicycles and buses. Crosswalks force the walking public to the ends of the blocks, where we stand and wait for an authoritarian system of automated lights, or risk tickets, life, and limb by jaywalking.

If cars on a lonely country road are a manifestation of American freedom, in American cities they are shackles, rendering us *urbanis in vincoli,* cities in chains. We've got to do something about that.

Journey to Work

The journey to work—the commute—is key to travel behavior. For most people, the distance from home to work determines their choice of vehicle. Once people commit to cars, every other transportation decision is made on top of that large investment. Living a little farther out adds a lot to the total distance traveled, suggesting that modest gains in proximity of home and employment could have large benefits in terms of distance, mode choice, and therefore, oil dependence.

Source: Analyzed from data from 2009 in US Federal Highway Administration (2011b).

Total distance (million miles)

Work trips (millions)

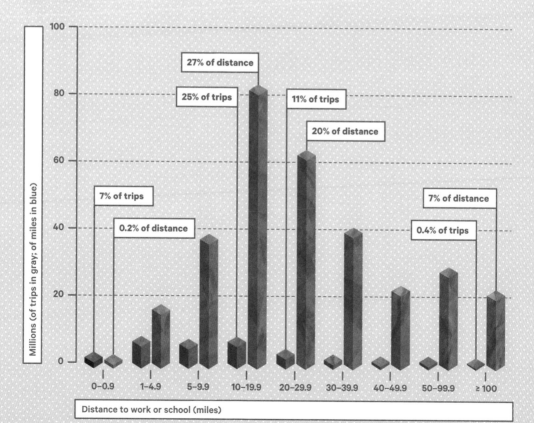

7% of trips

0.2% of distance

27% of distance

25% of trips

11% of trips

20% of distance

7% of distance

0.4% of trips

Millions (of trips in gray; of miles in blue)

Distance to work or school (miles)

0–0.9 1–4.9 5–9.9 10–19.9 20–29.9 30–39.9 40–49.9 50–99.9 ≥ 100

11
Roads to Rails

STELLA: He smashed all the lightbulbs with the heel of my slipper.

BLANCHE DUBOIS: And you let him? Didn't run, didn't scream?

STELLA: Actually, I was sorta thrilled by it.

Tennessee Williams, *A Streetcar Named Desire* (1951)

When we begin to value the land for what it is and build cities worth living in, density develops, and density makes things happen. Some of those happenings are economic, in the sense of improved productivity; others are environmental, in terms of fewer resources consumed. Density also has a lot to offer in terms of our trades of time for space.

You will remember that past transportation revolutions have been rooted in land. The railroad companies were encouraged to expand west by massive giveaways of public land; the streetcar operators were given monopolies to encourage their development; and the automobile industry received the greatest gift of all—roads—carved out of the public domain, bought or appropriated from private citizens. Many people and innumerable beasts were hurt in the process, so that other folks could be whisked on their way. Such radical efforts were necessary to make twentieth-century transportation feasible, affordable, and widespread in America.

A similarly radical approach is required today, but without all the giving and the taking. It's simple. We just need to decide to make better use of the land we all already own together: the public roads. Our roads today suffer from an identity crisis. We want them to provide thoroughfares for private cars, routes for public transit, spaces for parking, lanes for bicycles, sidewalks for pedestrians, access for people with disabilities, space and light for buildings, drainage for storm water, and even room for trees and flowers! Take a look out your window—the streets are contested territory, trying to be all things for all people.

The suburbs at least did this part right: They were decisive. Streets were for cars, not for bikes or pedestrians or anything else. Sidewalks were to be narrow, ornamental, or nonexistent, since it was assumed people would be driving. Public transportation was not a priority, because everyone has a car or two or three. As suburbs expanded, zoning codes mandated off-street parking for houses, offices, and mini- and jumbo-malls, which like medieval castles surrounded by moats of asphalt, are best approached on a trusty steed: the motorcar.

Though decisive, these choices were all decisively wrong from the perspective of energy efficiency, national security, and long-term economic productivity. Let's see what we can do to make them right again.

A Brief Physics Lesson

In choosing how to use our precious street space, we need to begin with the laws of physics, rules of the universe that explain how and why different kinds of transportation use different amounts of energy. Better streets will move more people and use less energy. Lower-energy forms of transportation will be easier to supply with fuels other than oil; denser cities will require more efficient ways of moving. How much energy and how many people is a matter, at least initially, of physics.

Speed vs. Energy

Speed takes energy: to get going, overcome friction from air and ground, and compensate for inefficiencies. A car traveling 20 mph uses about the same amount of energy as a train traveling 75 mph, on a per-person basis. Notice how the x- and y-axes vary when comparing modes; the bottom plot compares total energy use per person for cars, bikes, and trains.

Source: Redrawn from MacKay (2009).

 Energy to overcome rolling

Energy to get up to speed

Energy to overcome air resistance

Energy lost through inefficiency

Total energy consumption

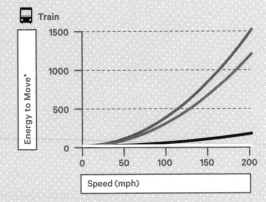

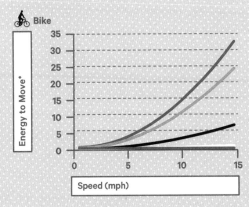

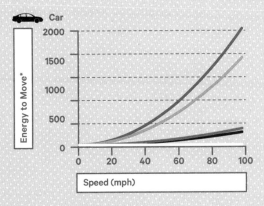

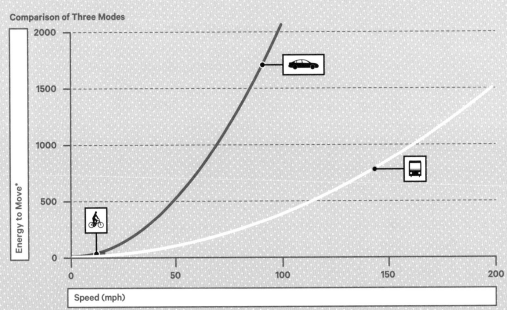

* Minutes microwaving per passenger

Recall that energy is "that which changes the physical state of a system"; physical state includes your geographic location. In a frictionless vacuum, the energy applied to accelerate an object would be all that is ever needed; once in motion an object would never stop. Sir Isaac Newton showed three and a half centuries ago that the energy of motion—the kinetic energy of an object—is one-half its mass times its velocity squared ($\frac{1}{2} \times m \times v^2$). This means heavier objects require more energy in proportion to their weight; faster objects require four times as much energy to double their speed. Thereby Newton gave us the first two rules to increase transportation energy efficiency:

Rule 1: Be lighter.

Rule 2: Go more slowly.

Note: Rule 2 matters four times as much as Rule 1.

The energy to put a vehicle in motion is lost when we stop at a red light or to let a pedestrian cross. It hasn't disappeared in a universal sense because energy is always conserved, but for our immediate purposes, it is gone, turned into manifestly less useful heat, vibrations, and brake squeal. The amount of energy required to get back up to speed is the same as what was lost, which suggests for efficiency:

Rule 3: Minimize starts and stops.

Note: Rule 3 explains why most cars make better mileage on the highway than in town.

Since we don't live in a vacuum, moving requires additional energy to overcome friction. Friction for most vehicles comes from two sources. One is rolling resistance from tires scraping along the ground. It is a function of gravity, the vehicle's mass, tire design, and the road surface. Different materials scrape differently: An inflated tire rolls with 6–7 percent less friction than a poorly inflated one, enough to affect your gas mileage; steel wheels running along steel rails, in contrast, roll along with 400 percent less friction than an inflated tire. Since less friction means less wasted energy, we have:

Rule 4: Slide, don't scrape.

Note: Rule 4 explains why trams are so successful at moving heavy loads.

The other source of friction is air. Air resistance describes how much air gets pushed around as a vehicle moves through it. It is a function of the vehicle's cross-sectional area, drag coefficient (which measures its aerodynamics), and speed. Think Camaro vs. Lincoln Navigator: The Camaro tries to slip through the air, while the Navigator just busts through. In either case, the air resistance increases with the velocity cubed ($\frac{1}{2} \times \rho \times d_c \times A \times v^3$, where ρ is the density of the air, d_c is the drag coefficient, A is the cross-sectional area of the car, and v is velocity or speed), which means that doubling your speed requires eight times more energy, assuming no wind.

Rule 5: Be sleek.

Note: Rule 5 is why racecars and jets are streamlined.

Transportation Energy

We have choices about how we move, and those choices dictate how much energy, and in what form, is required. Compare energy consumption per vehicle and per passenger, at usual and maximum occupancy, for walking, bicycling, skating, various automobiles, and shared transportation modes. Bicycling is the most efficient way to go, requiring just 2.4 minutes microwaving per mile, at typical speed. To do the same, a Cadillac would require 121 minutes microwaving and a Prius, 44 minutes (but of course, cars move faster). Electric cars (like the Nissan Leaf) are the most efficient per vehicle-mile. Streetcars and the New York subway win on a passenger basis at usual occupancy. At maximum occupancy electric rail systems rival bicycles, while traveling four times faster and transporting hundreds more people.

Source: See Notes.

Mode	Fuel	Representative Vehicle	Curb weight (lbs)	Occupancy (ppl per vehicle)		Energy Consumption (minutes microwaving per mile)		
				Usual	Max	Per Vehicle	Per Person (usual)	Per Person (max)
Personal Travel Modes								
Walk	Food	Eric	not given	1.0	1	6.6	6.6	6.6
Bicycle	Food	700c Men's Schwinn Varsity Road Bike	34	1.0	1	2.4	2.4	2.4
Roller skates	Food	Chicago Bullet Speed Roller Skates	6	1.0	1	3.7	3.7	3.7
Automobiles								
Neighborhood Electric Vehicle (NEV)	Electricity	GEM e4	1,290	1.7	4	22.2	12.8	5.6
Subcompact Car	Gas	Mini Cooper Convertible	2,535	1.7	2	68.0	39.1	34.0
	Electricity	Smart Fortwo Electric Cabriolet	2,150	1.7	2	23.4	13.4	11.7
Small Passenger Car	Gas	Ford Focus	2,907	1.8	4	75.0	41.0	18.8
	Gas (hybrid)	Toyota Prius	3,042	1.8	4	43.5	23.8	10.9
	*	Chevrolet Volt	3,781	1.8	4	40.2	22.0	10.1
	Electricity	Nissan Leaf	3,354	1.8	4	20.4	11.1	5.1
Large Passenger Car	Gas	Cadillac DTS	4,009	1.8	5	120.9	66.1	24.2
	Gas (hybrid)	Lincoln MKZ Hybrid	3,752	1.8	5	55.8	30.5	11.2
	Diesel	BMW 335d	3,362	1.8	5	91.9	50.2	18.4

* Gas / electricity (plug-in hybrid)

Mode	Fuel	Representative Vehicle	Curb weight (lbs)	Occupancy (ppl per vehicle)		Energy Consumption (minutes microwaving per mile)		
				Usual	Max	Per Vehicle	Per Person (usual)	Per Person (max)
Automobiles								
Muscle Car	Gas	Chevrolet Camaro	3,780	1.7	4	114.5	65.8	28.6
	Gas	Porsche 911 Turbo S Coupe	3,200	1.7	2	114.5	65.8	57.3
Sport Utility Vehicle	Gas	Subaru Outback Wagon AWD	3,386	2.1	5	90.6	42.3	18.1
	Gas (hybrid)	GMC Yukon 1500 Hybrid 2WD	5,636	2.1	8	103.6	48.4	13.0
Pickup Truck	Gas	Ford F-150 Pickup	4,580	1.7	3	121.2	69.7	40.4
	Gas (hybrid)	Chevrolet Silverado 15 Hybrid	6,057	1.7	5	103.6	59.6	20.7
Shared Transportation								
Motorbus	Diesel	Denver RTD bus	24,000	8.3	77	544.0	65.9	7.1
Trolleybus	Electric	San Francisco Muni	31,500	15.6	83	301.8	19.4	3.6
Streetcar	Electric	Seattle Streetcar	66,200	25.7	140	478.8	18.7	3.4
Light Rail	Electric	San Diego Trolley	77,200	67.6	200	733.8	10.9	3.7
Heavy Rail	Electric	New York City Subway	72,000	247.7	1,000	2,646.4	10.7	2.6
	Electric	Bay Area Rapid Transit (BART)	56,000	147.6	800	1,860.0	12.6	2.3
Commuter Rail	Electric/ Diesel	NJ Transit Commuter Rail	100,000	244.8	1,000	6,274.3	25.6	6.3
Jet Planes	Jet Fuel	Boeing 737-700	114,000	122.0	149	4,189.4	34.3	28.1
	Jet Fuel	Boeing 747-700	525,000	429.2	524	15,933.2	37.1	30.4
Ferry	Diesel	King County Water Taxi	102,000	34.8	150	6,514.5	187.3	43.4
	Diesel	Staten Island Ferry	5,588,000	522.1	4,427	55,247.2	105.8	12.5

Putting these five rules of physics together, as David MacKay does in his book on sustainable energy, means that the break-even point between rolling resistance and air resistance for heavy, rubber-wheeled vehicles like cars is about 15 miles per hour. Below 15 miles per hour your car's weight and speed matter most in how much energy it expends. Above 15 miles per hour, shape and, especially, speed matter most. For an average car, energy consumption bends upward more stiffly as speed increases, which is why back in the 1970s, the Nixon administration introduced national speed limits of 55 miles per hour or less. These tradeoffs also present a design problem for automakers: How do you make a car efficient both in town and on the open highway? The answer is, you can't really. But you can make different choices about how you travel.

In town, where motion is dominated by low speeds and frequent stops, you can save energy by choosing a mode of transportation that is lighter (Rule #1), rolls with less resistance (Rule #4), and moves less rapidly (Rule #2). Walking, bicycling, and in-line skating all suggest themselves, rather than automobiles. Personal modes move a minimum of mass (our bodies plus the bike or skates) at low speeds, with little rolling resistance and smaller cross-sections. Though some of the energy is wasted in the inefficiency of our legs and backs, we don't mind: We call it exercise. Biking beats out walking for efficiency because the small gain in vehicle mass is more than compensated for by the increased efficiency of the bicycle's gears and pedals, making biking fast and fun, especially on paths uncluttered by pedestrians or motorcars.

Out of town, where higher speeds are required and stops are less frequent, vehicles make more sense. For fast-moving objects, like cars, energy loss is dominated by drag from pushing the air around. Under these conditions, your vehicle's weight matters less than its shape, so you can save energy by making your mode more streamlined (Rule #5) and unhelpfully—by moving less rapidly (Rule #2). Since making better trades of time for space is the point, especially over longer distances, the least you can do is split the energy use. More heads per cross-sectional area, like on a train, dramatically lowers the per-capita energy expenditure. The very best way to improve the fuel efficiency of your car is also the easiest way: Share with someone else.

Car pools are the only practical way to make up for the notorious inefficiency of internal combustion engines. Although it's been over 120 years since Benz sold his first motorwagen, automobile energy efficiencies remain stuck in the 18–25 percent range, not so different from you riding your bike. (Both you and your V6 are turning carbon-based chemical energy into motion.) Cars weigh more than people, so on a per-passenger basis, their energy efficiency drops even more. Consider that if you weigh 200 pounds and drive a run-of-the-mill 3,000-pound car, then your weight is just 6.25 percent of the total mass moved. If the energy to move you is consumed at 20 percent efficiency, then only 1.25 percent of all of the energy in all of the gasoline in your car is used to move you down the road. Energy loss accelerates as you do.

Electric motors for electric vehicles do a better job. Electrical engines typically obtain 80–95 percent efficiencies, because they are lighter and because electromagnetism skips the explosions and attendant hot gases, noise, and vibrations of combustion. But there's a catch. Electric motors need a constant supply of electrons to turn the wheel. Those electrons come from either a power cord connected to a power source, which is sending them in real time, as in streetcars, or they supply them on-board using a rechargeable battery. As Edison and Planté discovered in the nineteenth century, batteries are heavy because of the metals (like lead) required to hold the charge. Conventional lead-acid batteries add to the weight of the vehicle, which requires more energy to move because it's heavier, which requires a larger battery, which adds to the weight, etc. This ugly feedback loop leads to rapidly diminishing returns, and explains why, a century after Edison and Ford gave it a go, we are still struggling to make a speedy, long-distance, affordable electric car (though we will consider a few modern takes on the Electrobat below). The physical truth is a pound of gasoline holds 350 times more energy than a pound of lead soaked in sulfuric acid. (Lithium-ion batteries, the ones in your laptop, do better—gasoline:lithium-ion, 118:1—but are more expensive.)

SUVs zooming down the expressway at 70 miles per hour break every rule of energy efficiency, but manage to do what they do by relying on the remarkable energy density of their fuel. Aircraft, heavier and airborne, are even more dependent. Thus, if we value the ability to fly across the country, or to another continent, we might want to save our energy-rich oil for air travel. Back on the ground, we need to find a better way to trade time for space.

A Better Car

A curious fact about cars is that most of them are designed to carry more than one person. At maximum occupancy (four to eight people per vehicle), modern cars are actually reasonable in terms of their energy expenditure: They use only 300–500 percent as much energy per person per mile as someone walking or bicycling, but go on average a lot faster. As we all know from counting heads during the morning commute, most trips in personal motor vehicles are taken by lonesome drivers. Add some carpooling trips and family errands, and the overall average vehicle occupancy for personal automobiles in America works out to 1.59 passengers per trip (in 2009).

At this kind of occupancy, a car's energy efficiency, never great, collapses: A solo driver in a Ford Focus uses 600 percent more energy per person per mile than a pedestrian; a Camaro spends 1,000 percent as much. Thus, if you are going to drive, please share.

Hybrid cars are more energy efficient by making the best of a bad situation: They have two power trains, one electric and one internal combustion. They use a battery to start the car and run at low speeds; at higher speeds where more energy is required, or when the battery is drained, the gasoline engine takes over. Most hybrids also have regenerative braking that recaptures about 20 percent of the energy of slowing and stopping and shunts it back to the battery. (Gas cars can't have this feature because brakes can't regenerate gasoline, just electricity.) Despite the extra pounds required by the extra machinery and battery, hybrid cars are typically twice as energy efficient as internal-combustion-only automobiles of the same model. The problem with hybrids, beyond their purchase price, is that they still require gas as their sole energy source. Though more efficient, they are just a lighter version of oil's chains.

Better automotive energy efficiency can be obtained from a plug-in hybrid. At the time of this writing (summer 2012), there was only one such vehicle for sale in the United States: the Chevy Volt, though others were in the works. Plug-in hybrids are truer "hybrids" in the sense that they can use energy from electricity or from gasoline, but can get by on just one or the other. The Volt also deploys regenerative braking to save energy, and though its range is only 35 miles on electricity, that's enough to push its energy consumption per mile to only 1.5 times as much as a person walking at maximum occupancy (four passengers per Volt), and only five times a person walking at usual occupancy. Not bad, considering the Chevy Volt weighs in at almost two tons.

The most energy-efficient automobiles are, not surprisingly, electric. True electric cars eschew gasoline entirely and instead receive all their energy from a power plant or a wind farm stored in a battery and delivered via a plug. The most efficient electric car on the market when I was writing this book was the Nissan Leaf, which at full passenger capacity is actually more energy efficient than a person walking (!), and only three times more energy-consuming per person than biking. The Leaf is the latest in a small collection of electric cars sold by Ford, General Motors, and various foreign vendors over the last twenty years. Probably the best known American electric car was General Motors' EV1, the first and only one to carry the GM nameplate, which developed a small, incredibly devoted following in California at the turn of the twenty-first century. When GM canceled the three thousand leases on the EV1 in 2003, insisting all its owners return them, and then crushed the cars in the desert or disabled them for museum objects, stunned customers complained, picketed, and made a movie: *Who Killed the Electric Car?*

It turns out that many agents contributed to the demise of the EV1, not the least of which was the electric car's old nemesis: the rechargeable battery. The EV1 originally had a range of about sixty miles on a charge; battery upgrades, using nickel-hydride batteries, like the rechargeable ones in a toy car, eventually pushed the range up to 160 miles, but also upped the cost considerably. The 2012 Nissan Leaf has forty-eight lithium-ion battery modules, which weighs 660 pounds, affording the Leaf about a hundred-mile range between charges.

Transportation Space

We have choices about how we use the roads. During rush hour, cars turn out to be the least efficient use of the public domain, moving the fewest people in an hour; walking and biking move three to four times more folks; shared forms of transportation twelve to fifteen times more passengers. I prefer streetcars to buses because they run on electricity, are less compatible with cars, and are more genteel in general, but I'll take a bus in a pinch. My mom would never ride a public bus, but she's keen to try a streetcar.

Source: See assumptions below.

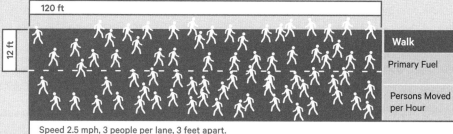

120 ft

12 ft

Walk	
Primary Fuel	Food
Persons Moved per Hour	17,600

Speed 2.5 mph, 3 people per lane, 3 feet apart.

Bike	
Primary Fuel	Food
Persons Moved per Hour	14,400

Speed 15 mph, 2 bikes per lane, 20 feet apart.

Car	
Primary Fuel	Oil
Persons Moved per Hour	4,400

Speed 20 mph, 1 car per lane, 60 feet apart, occupancy: 1.5.

Bus	
Primary Fuel	Oil
Persons Moved per Hour	52,800

Speed 20 mph, 1 bus per lane, 60 feet apart, occupancy: 25. Bus length: 40 feet

Streetcar	
Primary Fuel	Electricity
Persons Moved per Hour	66,000

Speed 20 mph, 1 streetcar per lane, 60 feet apart, occupancy: 35. Streetcar length: 52 feet.

Batteries, lest we forget, also need to be charged. Fast charging requires a dedicated charging station at high voltage (240 V; the usual household voltage is 110 V). Buying a Leaf doesn't include the purchase and installation of a garage-mounted charger for rejuvenation at home. Communal charging stations, the equivalent of gas stations, are doable, of course; we had plenty of them in electric truck garages of the 1920s. Perhaps they could be deployed again in take-out, drop-in battery exchanges such as the ones imagined back on Broad Street in 1895, if manufacturers adopted consistent standards for battery shape, size, and connection.

There is another automotive solution, though, suggested by the problems of the Leaf, which is to give up on range and speed expectations based on gasoline, and in-stead design electric cars that work well on their own terms, in town, at lower speeds. Mrs. Ford by all accounts was very happy with her electric car, which in fact was an early prototype of what we would call today a "neighborhood electric vehicle" (NEV), a kind of souped-up golf cart. These smaller, slower vehicles have conventional lead-acid batteries and an electric motor, they charge at a standard household outlet and can speed very happily up to 25 miles per hour while carrying 1,000 or more pounds of cargo. You have probably seen them zipping about in a gated community or amuse-ment park. The police, the military, and zookeepers use them, too. The government does not allow NEVs to play with gas cars on fast-moving boulevards or highways, restricting them to streets where the speed limit is under 35 miles per hour. (Thirty-five is not bad; it's the limit of many city streets.) Chrysler has a division that sells six models of NEVs under the brand name GEM for $8,000–$12,000 each, doors extra.

A Better Streetcar

I wish electric cars, small or large, could elegantly sweep in and replace gasoline cars and solve all our problems with a wave of the technological wand, but I can't see how it happens without a major breakthrough in automotive battery technology, which has eluded us for a century or more. The fact is that the only forms of powered trans-portation that give the kind of per-person bang-for-the-microwave-minute that we need are shared modes of transportation, particularly ones on rails: trains, light rail, and the streetcar.

Streetcars are the closest we know to the ideal motorized transportation. They roll with low resistance on steel wheels on steel rails, driven by efficient electric motors attached to the grid via overhead wires or underground cables, deploying regenerative braking for stopping. And they carry tens to a hundred passengers at a time, which gives more heads per cross-sectional area, thus dramatically dropping per-capita energy use. At full occupancy, streetcars best walking and rival biking in energy effi-ciency. Compared to a bus, they are more energy efficient, have more leg room, offer better views, and are more genteel; they are also more fun. Who doesn't like to ride a

Streetcars of the Future

Streetcars don't have to be a blast from the past. Contemporary ones often have a European feel, but that's because American car designers haven't had a chance to apply their skills. Future streetcars could have roof-to-ceiling windows and surfboard racks, and even cupholders, and will perform better for less energy than the car, especially if given their own space on the roads.

Source: First two sketches redrawn from images on Wikipedia; third is an original sketch by Yo-E Ryou.

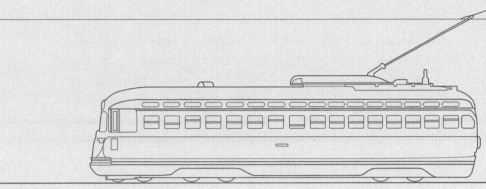

Old Fashioned Streetcar, Kenosha, WI

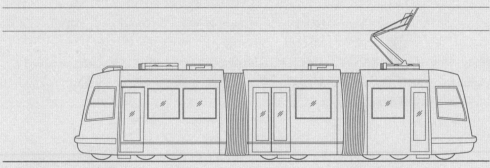

Modern Streetcar, Portland, OR

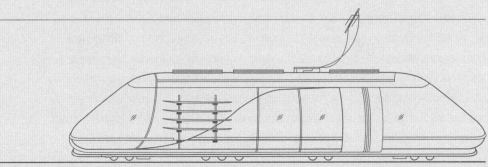

Concept Streetcar

streetcar? Once they are laid down, the rails reflect a tangible, significant investment in the city, something a bus stop can never hope to do. Some people don't like the overhead lines, but those can be buried so as not to interfere with the view of the phone and power lines that parallel so many American roadsides.

If streetcars ran on streets where they were the only vehicle, we could make them lighter, streamlined, and more stylish. They could also go faster because there would be no unpredictable cars to cross them. Twenty-first-century streetcars can be designed for contemporary times, to reflect a community's sense of itself. New York's can be sleek and elegant, Seattle's innovative and green. In Los Angeles streetcars can have sun roofs and surfboard racks. They could all provide free wifi, vending machines, and cup holders.

How viable is a nation of streetcar riders? Try this out: Sometimes I play a game with my son to pass the time while we wait for the bus. We count the cars going by and say: "One – two – three – four – five – streetcar!" We count to five because five cars use about the same amount of energy as one streetcar. On some residential suburban streets, you might need to wait ten minutes to get to five cars, but on City Island Avenue, our main thoroughfare, we could have a streetcar every other minute for most of the day for the same amount of energy we already lavish on cars. On busier city streets, they'd come in a constant stream. And whereas five cars might move five to eight people, each streetcar could handle seventy sitting or a hundred standing.

Try it next time you are stuck in traffic; if you can count four cars in addition to your own, then imagine yourself relaxing in a spacious, stylish streetcar, with a small number of your fellow citizens, quietly being transported by chauffeur toward your destination through clean, unpolluted air, unhindered by congestion, able to read the paper, text your friend, and admire the view. It could happen. It might be sorta thrilling: A streetcar to desire.

Here's the plan.

Roads to Rails

For short distances, it's clear we should do everything humanly possible to make walking and bicycling the preferred modes of transportation for as many people as possible. Currently, 49 percent of trips are already three miles or less, and 70 percent of them are taken by car, which suggests a huge potential. The ingredients are fairly simple: Pedestrians and bicycles need their own separate, pleasant spaces for movement— sidewalks and improved bicycle paths—and people need their everyday destinations within reach, whether they are for work, shopping, or school. Better, denser towns and cities designed for people are the means to the end of making walking and bicycling the cheapest, healthiest, fastest way to go for some 189 billion trips per year.

Walking and related modes, however, are not ideal when the weather is unpleasant or when we need to travel farther than a few miles. They also don't work for the very old, the very young, and the disabled, who need modes compatible with how they move; and businesses, emergency crews, and others need ways to move objects heavier than a person can conveniently carry. To obtain better trades of time for space, we still need vehicles powered by engines to apply greater energy than our bodies can. Small fleets of NEVs can help, streaming people and goods down to that paragon of motor propulsion: the streetcar.

When imagining the streetcar revolution, don't rely on your experience of public transit today, with long unpredictable waits, dingy subway tunnels, and motorbus diesel fumes. Instead, imagine what every city once had—lots and lots of streetcars running all the time (one for every five of today's cars) along every big street. Your wait won't be long, and it won't be uncertain, because thanks to GPS, wireless technologies, smartphone applications, countdown clocks, and a glance down the avenue you will know exactly when the next streetcar will arrive to whisk you away. As the transportation planner Jarrett Walker writes: Frequency is freedom.

Streetcars, NEVs, your bicycle, and your legs are the distributed beginnings of a new transportation network, reaching into New Town districts across America and bringing people to light rail trains running along major thoroughfares. Light rails are close cousins of the subway and elevated railway, except they run on the ground. They are heavier and faster than streetcars, able to race cars at 60–80 miles per hour. In the future, these local trains will shuttle between nearby cities, delivering people to high-speed rail systems that go cross-state, and eventually cross-country.

America already has a world-class freight rail system, moving 1.7 billion tons of goods each year. Today freight railways connect to trucks for the final delivery; in the future, they will connect to streetcars, and in the cities, the old subway tunnels. Subterranean movements will be set aside for inanimate things, rather than for people. At night specially designed flatbed streetcars will pull up to businesses or neighborhood receiving stations, the post offices of the future. Curb cutouts with loops of side track will provide lading sites out of the main flow. Small containers of standard size, and designed to fit within the large containers used by the shipping industry, will travel by rail and NEV. In the morning NEVs and folks with hand trucks will make deliveries to your door.

Instead of asking the car to do every transportation job for us, as we do today, transportation will be sorted by task. We will choose modes that work better and more efficiently for different distances and prioritize investment according to a formula that prefers human power over railways and railways over cars.

We make this happen by committing roads to rails, literally. Dedicating road space to rails resolves two problems simultaneously. First the roads turn out to be an excellent place to build railways at lower cost. The budgets of most rail projects today are based on an assumption that automobile traffic will continue ad infinitum. For street-

American Locomotion

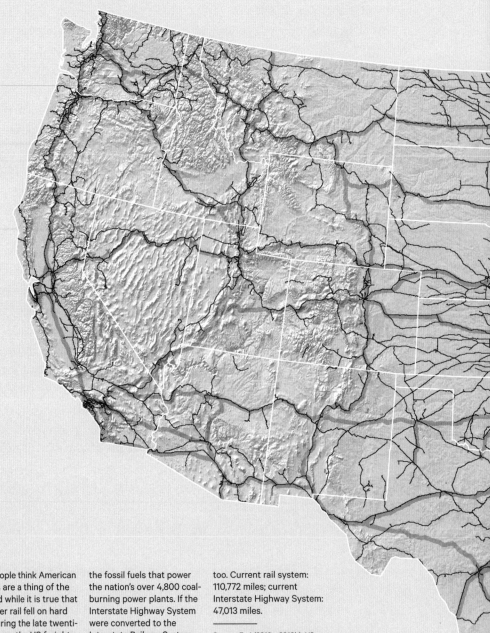

Many people think American railroads are a thing of the past, and while it is true that passenger rail fell on hard times during the late twentieth century, the US freight rail system moves 1.7 billion ton-miles of freight (as of 2011), including nearly all of the fossil fuels that power the nation's over 4,800 coal-burning power plants. If the Interstate Highway System were converted to the Interstate Railway System, then we could have fast and furious (and energy-efficient) passenger trains, too. Current rail system: 110,772 miles; current Interstate Highway System: 47,013 miles.

———

Source: Esri (2010a, 2010b); US Energy Information Administration (2010d).

cars, sharing the roads with cars necessitates extra staff to steer and see, extra weight for safety, limited choices about alignment (the technical term for where the rails will go), and extra expenses for switching and signaling. These problems are exacerbated for light rail and high speed (trans-region) rail systems that must have dedicated space to operate; they literally have nowhere to go in today's world because all our land is already given over to established public and private uses. (I shake my fist at you, John Locke!) What remains of the rail lines of the nation are mostly already spoken for by the freight industry (mixing freight trains and passenger trains is not recommended—different speeds, different agendas). As a result, the budgets for current railway plans, like the beleaguered high-speed rail plan for California, are swollen with funds to purchase right-of-ways and construct tunnels, overpasses, elevated lines, and other extraordinarily expensive acts of engineering necessary to find a route without disturbing the dominant car.

Making the counter-assumption of no cars provides extraordinary relief—now there is lots of space and reduced costs. Roadways are already engineered for transport, with bridges and tunnels in place. The electricity is already there in the power lines paralleling many roads. Dedicating roads to rail means that capital costs drop dramatically because land acquisition and grading expenses evaporate; it also means eventually we need less land dedicated to mechanized transportation, so we have more room for sidewalks, bike paths, parks, and garden cafés. Instead of dedicating a third of our city space to transportation, perhaps we can get by with only a quarter or a fifth, meaning that broad swaths of city land could become available for other uses. Think what we can do with all those parking lots!

Deploying railways down Main Street provides a second great advantage: It competes with the cars that remain. As streetcars on streetcar-only streets become more prevalent, they will force cars into a smaller number of crowded car-only streets. As congestion worsens for automobiles, and fuel costs rise, and free parking—and then all parking—vanishes, more people will see the wisdom of giving up on cars entirely and join the rest of the nation walking, biking, and on the rails. You can still get to work and your trip will be faster and more pleasant. Driving will persist in rural areas, where work necessitates infrequent trips over long distances, and on a recreational basis. (I'm particularly fond of the drive over the magnificent Million-Dollar Highway in Colorado.) Driving will become a hobby, not a burden.

Do you hear that jingling in your pocket? That's the 20 percent of your income now freed to be deployed elsewhere in the economy. Some of it will go back to transportation, but spent on an as-needed basis. Rather than writing out the insurance, registration, and car payments in lump sums each year, regardless of how much you drive, now you pay only when you ride. (Businesspeople call this process replacing fixed costs with variable ones.) Or we could establish a system where everyone makes a down payment—say, 50 percent of what we used to pay—and then all local transportation is free. You show a badge stating that you are a resident of New Oldtown USA, and climb on board. Exchange privileges give you free access in other towns, too.

To get the process started, we need to redirect funds from roads to rails. In 2008 government at all levels (local, state, and federal) spent a collective $182 billion of taxpayers' cash on capital and operating expenses related to roads and highways; the same year, we spent another $51 billion on transit projects. That's three dollars for cars for every one dollar for passenger trains and buses. Reversing this ratio would have enormous immediate effects on shared transportation in America without costing taxpayers a cent more than we are already paying.

Construction costs for new streetcar systems in the US over the last decade have run between $2 million and $20 million per track-mile. (Streetcars have grown in popularity over the last decade; as of summer 2012, at least thirty-five cities had streetcar or light rail lines.) If we assumed that we could achieve the lower end of this range through economies of scale and by building rails on roads without having to deal with car traffic, then a $150 billion investment could buy seventy-five thousand track-miles. If we assume track density and alignments so that everyone lived within a quarter-mile of a streetcar line, then those seventy-five thousand track-miles could serve 18,750 square miles of urban area. If those towns and cities were inhabited at a density of five thousand people per square mile, encouraged to move there by New Town districts, home-to-work rebates, and the new system of gate duties on fossil fuels, then those streetcars could serve ninety-four million people. If in a burst of enthusiasm and economic growth, the residential density pushed up to ten thousand people per square mile (remember that's only one-seventh of Manhattan density), then 188 million people could ride those streetcars, or 60 percent of the American populace. In other words, scratches on the back of an envelope suggest that after only a few years' worth of spending the money we already spend on roads, everyone in the country could have access to a streetcar, assuming that they inhabited happier, healthier, moderately denser locales than where most people currently live.

What Happened?

I know what you are thinking: If streetcars are so great, why didn't they succeed the first time around? And don't we need to know why they disappeared if we ever hope to rebuild them? It's like a beautiful forest eerily silent because all the animals have been hunted to extinction: We must understand why the forest is empty to fill it again. I don't think the answer to why the streetcar expired is as simple as some commentators have indicated—that there was a great conspiracy to replace it with automobiles, and that was that (though some unsavory things did happen, as we have seen in Chapter 5). Rather, the answer lies in the uneasy institutional relationships surrounding

Streets + Cars vs. Streetcars

Streetcars are a bargain compared to streets with cars if all the costs are accounted. Over thirty years, initial capital costs may be higher but opera-tion and maintenance costs are lower per mile, even at going rates. Because streetcars carry more people than cars do, per-passenger mile costs could work out to be 30 percent less to move 33 percent more folks. Economies of scale and gate duties will expand the profit margin further.

Source: Costs from Florida Department of Transportation (2012), American Automotive Association (2012), and District Department of Transportation (2010).

Components of Transportation System	Streets + Cars		Streetcars	
	Cost	Details	Cost	Details
One-time capital costs				
Vehicles	$21,900,000	New car every 5 years @ $20,000/car; 6,000 cars/day, moving 20 mph	$40,000,000	$40 million/mi, includes vehicles, tracks and power system
Road / Railway Construction	$4,800,000	Two lane city street with bike lanes @ $4.8 million/mi	$0	Included in above
On-going operations and maintenance costs				
Operators	$49,275,000	Driver's time @ $15/hr, 250 cars/hr @ 20 mph, 24 hrs/day for 30 years	$42,048,000	$144/hr (includes driver's wage @ $55/hr), a streetcar every 6 min., both directions @ 15 mph, 20 hrs/day for 30 years
Fuel	$9,309,690	14.17¢/mi	$0	Included in above
Vehicle Maintenance	$27,166,950	Tires and repairs @ 5.57¢/mi; Insurance et al. @ 35.78¢/mi ($4,293/yr for car driven 12,000 mi/yr)	$0	Included in above
Road / Rail Maintenance	$1,263,000	Roads resurfaced every 10 years @ $421,000/mi	$22,907,400	$5.23/service mile
Transportation				
Vehicles moved	54,750,000	6,000 cars/day for 30 years	4,380,000	400 streetcars/day for 30 years
People moved	82,125,000	Occupancy @ 1.5 persons/car (including driver)	109,500,000	Occupancy @ 25 persons/streetcar (excluding driver)
Cost				
Total cost	$113,714,640	More expensive	$104,955,400	Less expensive
Total cost per year	$3,790,488	More expensive per year	$3,498,513	Less expensive per year
Total cost per vehicle-mile	$2.08	Less expensive per vehicle	$23.96	More expensive per vehicle
Total cost per passenger-mile	$1.38	More expensive per person	$0.96	Less expensive per person

land, transportation, and money during the time of the first great streetcar blossoming at the turn of the last century.

The trouble started because city governments thought it was clever to give monopolies to the streetcar companies. In the heyday of the Standard Oil trust and the Selden patent, monopoly was considered good practice in transportation. Granting local, long-term exclusive franchises induced companies to make large upfront investments in infrastructure (the railways and the rolling stock), relieving the government of those costs. In return companies would recoup their expenses plus profits indefinitely through a captive ridership and real estate development.

To limit the monopoly power, however, local governments controlled the fare. At first, both sides agreed that five cents a ride was a fair deal. In the deflationary environment of the late nineteenth century, when the real value of every nickel was spiraling upward, each fare paid represented accelerating profits for the companies. For a time, they and their real estate subsidiaries made money hand over fist; a list of the richest men in America of 1900 included municipal transit magnates Peter A. B. Widener, Thomas Fortune Ryan, and Nicholas F. Brady.

We don't speak of Widener, Ryan, and Brady in the same reverential tones we do of the Rockefellers, Fords, and Carnegies because the streetcar kings' glory days faded fast. Inflation, labor strikes, World War I, and competition from electric and motor vehicles overtook the streetcar. Owners wanted to raise fares to keep up their lines, but government, subject to public pressure, refused. (New York Mayor James "Jimmy" Walker became famous by beating back a fare increase in 1928, for example.) Unionization was on the rise, demanding a greater proportion of profits, and during World War I, the War Labor Board instituted mandatory pay raises for railway workers, including on the streetcar lines, to compensate for wartime inflation.

With no way to raise revenues to cover their costs and with development along the lines already peaking, the companies had to make cuts to stay afloat, which meant deferring investment and reducing service. Even though ridership continued to increase through the 1920s, the trams and trolleys crowded with passengers were beginning to fall apart. Meanwhile, the automobile companies—producing vehicles that were newer, faster, and affordable, if relatively energy inefficient—had all the capital they needed. After 1931, the Texas Railroad Commission and interstate commerce legislation ensured everyone paid consistent, low prices for gas.

Gas and rubber rationing at home during World War II extended the streetcar's era for a few more years, but by midcentury, when General Motors, Firestone, and Standard Oil of California cobbled together their racket to replace the last streetcars with buses and then close the bus lines, the streetcar industry was economically crippled, the victim of deferred maintenance, high costs, and subsidized competition; the GM conspiracy was just the coup de grâce.

Democratic Transportation

Figure No. 61

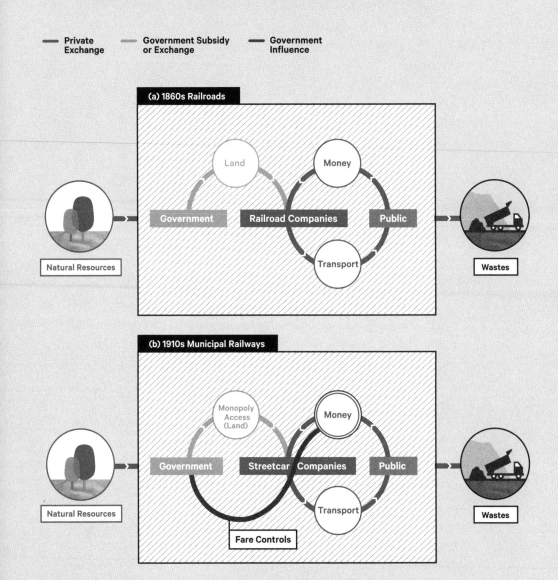

Private Exchange   **Government Subsidy or Exchange**   **Government Influence**

(a) 1860s Railroads

Land · Money · Government · Railroad Companies · Public · Transport

Natural Resources   Wastes

(b) 1910s Municipal Railways

Monopoly Access (Land) · Money · Government · Streetcar Companies · Public · Transport · Fare Controls

Natural Resources   Wastes

Transportation in a democracy needn't be complicated, but it does need to be clever. Here is what not to do: **(a)** Do not give away the land to railroad companies that then exploit the public; **(b)** Do not give monopoly access to companies and then limit the fares, thus ruining the companies; **(c)** Do not provide free roads and subsidies for cheap oil, damaging the economy, national security, and the environment; and please **(d)** Do not let the government run transportation companies, because then everyone loses. Here is a better way: **(e)** Let the government own and manage the public infrastructure in the public interest; let companies run the railways to make a profit and serve the people, subject to market competition; and let the citizens ride the rails to success, while speaking politely and specifically about necessary improvements.

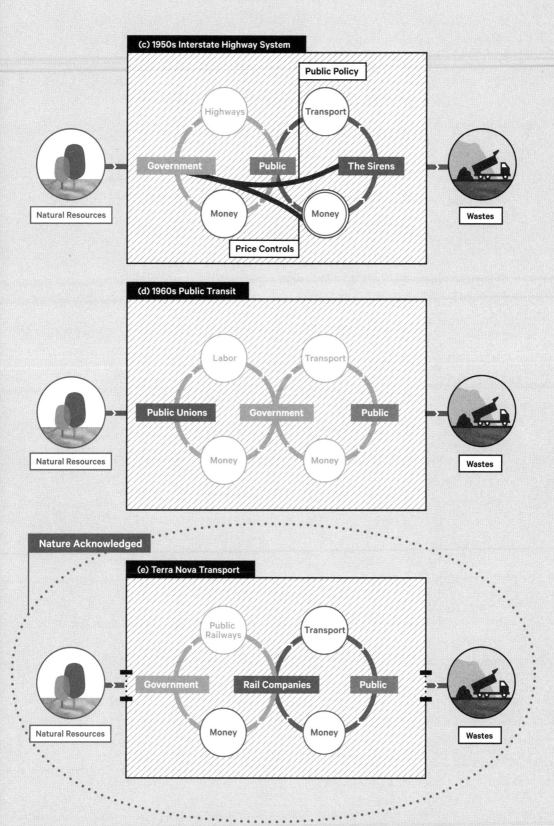

(c) 1950s Interstate Highway System

Public Policy

Highways

Transport

Government

Public

The Sirens

Money

Money

Price Controls

Natural Resources

Wastes

(d) 1960s Public Transit

Labor

Transport

Public Unions

Government

Public

Money

Money

Natural Resources

Wastes

Nature Acknowledged

(e) Terra Nova Transport

Public Railways

Transport

Government

Rail Companies

Public

Money

Money

Natural Resources

Wastes

Shared transportation in America is still haunted by the demise of the streetcar and its aftermath. In the late 1950s and early 1960s, government realized it had made a terrible mistake in its handling of the streetcar lines, and responded by making another terrible mistake: It took over transit. With a young president, John Fitzgerald Kennedy, in the White House in 1960, northeast politicians like Richardson Dilworth, mayor of Pennsylvania, and Senator Harrison "Pete" Williams of New Jersey, despairing of ever reversing the flight to the suburbs, saw an opportunity to win federal support to at least bring people back downtown for shopping. Thus began a subsidy war pitting us against us. With one hand, the government subsidized transit as a way of encouraging urban renewal, while with the other hand, it rolled out pavement for cars on a continental scale to help people flee town. In the epic battle of cars vs. transit, in the age of cheap oil, free roads, and low-density sprawl, transit couldn't win, no matter how big the subsidies. And many people questioned why we were writing checks for both in the first place. They still do. Like a gardener who planted two seeds that are now competing with each other to the detriment of both, we have to choose which will survive. One already seems to be failing.

What Business Does Best

Resolving the problems of public transportation means reforming the relationships between government, business, and the passenger once again. This time, we have to be realistic about the strengths and interests of each and play to them. Government owns the roads and looks out for the general welfare, for people today and in the future. Business is good at making a profit given a fair and competitive market with clear rules. Passengers know where they need to go and how much money they have.

Here's what I think we should do. Let's imagine that the government makes long-term investments in the necessary infrastructure for streetcar and other local rail systems. The public, via our self-instituted government, will own the tracks, signals, and maintenance yards and manage them in the public interest on the public land. The people will then rent out the rail lines to private companies to provide transportation services. The companies bring their knowledge of efficiency and the ability to flex and innovate; they also bring their own rolling stock and labor agreements. Passengers get a better bargain as a result.

Every few years, municipalities put out bids for contracts of limited duration, for example, three years. Short-term concession agreements ensure that companies are under the gun to provide excellent service, or the municipality will seek a different vendor next round. Companies are relieved of the capital costs of the rails and the real estate buys that have been the traditional argument for the necessity of long-term arrangements. The public runs the contracts on essentially a nonprofit basis, only asking for rent based on what is necessary to maintain the infrastructure, insure the rails, and keep up with inflation; no subsidies are involved, but no profits either to support other aspects of government. Contracts express the public interest: minimum levels of service, coordination across lines, bracketed fares, non-discrimination, electronic notifications, and bonuses for on-time service records and minimal passenger complaints. Within those bounds, companies are free to deploy service as they see best, including adding service to enhance profits. They can run more trolleys to accommodate the morning commute or the rush to the ball game.

In some cases, in coordination with the local authorities, companies might collect fares up front on an annual basis from residents, and then everyone could ride for free, with exchange privileges across connecting lines, facilitated by the same technologies that credit card companies use. Private service providers invest profits in advertising, better rolling stock, and transit-oriented development (e.g., shopping centers, housing stock) near the value-added transportation corridors, thus enhancing the market and bringing additional private funds into the towns and cities growing around them. New jobs will be created directly in service industries (steering and maintaining streetcars, local freight delivery, track maintenance), in manufacturing supply chains for streetcar construction, and through agglomeration economies generated by connected American neighborhoods, towns, and cities.

Once the streetcar is rooted within communities, then we will have the basis for a high-speed rail network between cities, not before. When streetcars and light rail systems bring people to the periphery, then high-speed rails can develop along the existing highway systems to connect cities across the vast expanses between. (In the meanwhile, temporary garages on the edge of town can store the cars reserved for rural travel.) Over time we transform long-distance travel from cars and trucks to trains, so that the Interstate Highway System morphs into the Interstate Railway System, with the federal government owning, maintaining, and coordinating regional rails, and private companies instead of government-owned corporations (like the hapless Amtrak), providing the service. Gate duties alter the economies of fuel and land, and higher functioning American towns and cities facilitate walking, biking, and public transport. The goal is to make American travel affordable, pleasurable, sustainable, and easy, a system to last for centuries, not just until the oil or the money runs out.

There is one final benefit to turning transportation over to the smooth whirr of electric motors: Those motors will use electricity. To produce it, we could continue to burn the black fossil fuel MacKays or build more radioactive nuclear power plants—or we can see the roads to rails program as a welcome opportunity to get our MacKays from warmer, breezier, brighter sources: the gifts of earth, wind, and the fire in the sky.

12

Invest in the Sun

Following the light of the sun,
we left the Old World.

Inscription on one of Christopher Columbus's ships (1492)

And so we return to the sun, from which everything begins. Treating the land and other natural resources with respect will make great towns and cities possible, and residents of those towns and cities will take advantage of a transportation infrastructure based on muscle power and electricity. Electricity of course can be garnered from many sources—whatever turns your dynamo—including those that spin with an elegant grace on a windy day, glint with azure solemnity under desert skies, and warm from sources deep within the earth.

Renewable energy sources have already proven their ability to generate power, but the knock on them is that they are intermittent and undependable, like the teenager who was supposed to mow your lawn but doesn't show up before the big party. Some critics sniff with a kind of Puritan disapproval: Renewables are weak and not to be relied on, unlike the sure, profitable fuels such as oil or coal. Never mind that the ecosystems that created the oil and coal in the first place got their energy from that unreliable light—our sun—and that it will be another hundred million years before nature replaces those deposits.

Deriving a greater proportion of our MacKays for transportation from electricity will create astonishing opportunities to rebuild the energy economy of the country, especially as gate duties nudge the markets, and towns rebound with creativity and streetcars. Electrified transportation will give the renewable markets the kick they need to catalyze economies of scale last seen with the railroads and the car, but now applied to the energetic foundations of our way of life.

The wind is already blowing in the right direction. But before we celebrate the good news about renewables, let's address the neo-Puritans' concern: Where are we going to store the energy that's harvested on a sunny, windy day?

Storage on the Mountain

Proposed large-scale electrical energy storage solutions sound like the inventory of a mad scientist's laboratory. Flywheels, supercapacitors, superconducting magnets, rechargeable batteries the size of a house, and compressed air captured in underground caverns have all been touted as possibilities. All have some potential, and perhaps in the future will make significant contributions, but today they are only so many ideas on the workbench. The solution I suggest, the one tried-and-true method, is prosaic in comparison, because it already works. In fact, it already has the capacity to store some seven percent (21 GW or 50 million MacKays) of US electricity, with facilities in eighteen states. It has an excellent track record here and abroad, with established safety measures and environmental safeguards. It's called pumped hydrologic storage (PHS), which I call the double-dam solution.

Most people are familiar with how conventional hydroelectric plants work as way stations on the hydrologic cycle. Every day the sun shining over the ocean evaporates

Figure No. 62 How Pumped Storage Works

Pumped hydrologic storage is a proven way of storing electricity. When electricity is cheap and abundant, water is pumped uphill into a reservoir. When electricity is needed, the water is released, regenerating the electricity. Paired with solar and wind power, pumped hydrologic storage can make otherwise "intermittent" sources as steady and reliable as coal and radioactivity. Like other kinds of power plants, pumped storage needs to be built in a safe, environmentally conscious way, toting its full costs.

Source: Redrawn from Mock (1972).

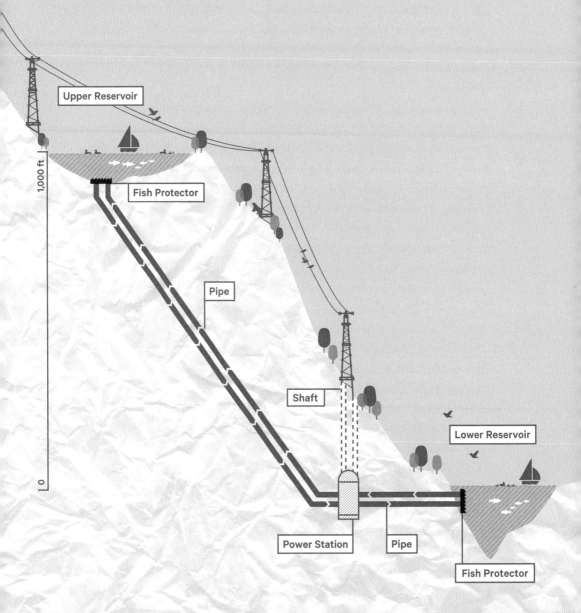

Upper Reservoir

1,000 ft

Fish Protector

Pipe

Shaft

Power Station

Pipe

Lower Reservoir

Fish Protector

0

water, which the wind blows overland in the form of clouds, which deposit their precious cargo as rain and snow. The water then runs downhill via brooks and streams into rivers, some portion of which—critical for the purpose here—is captured in a reservoir. The height of the reservoir over the valley below, known as the head, stores the energy of the water until it is needed. On demand, hydroelectric utilities release the water and it races down to spin a dynamo and generate electricity. Water power is renewed by the clouds and is widespread, much to the chagrin of the critters and communities that get flooded when the dams are created. (There is no free lunch when it comes to the environmental impacts of energy development—conventional fossil fuel and nuclear power plants pose different, less aquatic, social and environmental challenges—which is exactly why ecological use fees are important.)

Pumped hydrologic storage works on the same principle as hydroelectric power, except here windmills and solar collectors (or other sources) provide the power to transport the water from below to above. A reservoir is constructed 400–1,000 feet above a body of water (the sea, a river or lake, or another reservoir). When the wind is blowing and the sun is shining, the renewable power sources may generate more electricity than can be used directly, so the excess is diverted to a double-dam reversible turbine, which pumps the water uphill, from the lower source to the upper one. Later when energy is desired, water is released to run downhill through the turbine, to generate electricity again.

Some energy is lost in the process, of course, but less than you might think. Existing PHS power plants like Raccoon Mountain, Tennessee (capable of 1,714 MW or 41 million MacKays, 1,000 feet head); Ludington, Michigan (1,979 MW or 47 million MacKays, 363 feet head); and Bath County Pumped Storage in the Allegheny Mountains of Virginia (2,862 MW or 69 million MacKays, 1,262 feet head) show round-trip PHS efficiencies of 80–85 percent, another practical demonstration of the utility of Faraday's dynamos and regenerating motors.

For a site to qualify for PHS, it must have no outstanding environmental or safety concerns (PHS can be hard on fish and other aquatic creatures caught in the pipes), an abundant supply of water, space for two interconnected water bodies, and a slope steep enough to create a sufficient height difference between them. It struck me that North America has wet mountains in abundance, from the Appalachians of Georgia to the Catskills of New York to the White Mountains of New Hampshire, and out west in the Cascades, Sierras, and Rocky Mountains. Coastal cliffs in California, along the Great Lakes, and in New England could work, too.

To check my hunch, I used the computer to estimate the potential distribution of pumped-storage sites in the United States by analyzing elevation, water body, aquifer, and precipitation data, looking for sites that had at least a 300 m (984 feet) vertical elevation change and a source of water within a kilometer (0.62 mile). The water could

The Wet Mountain Country: US Pumped Storage Potential

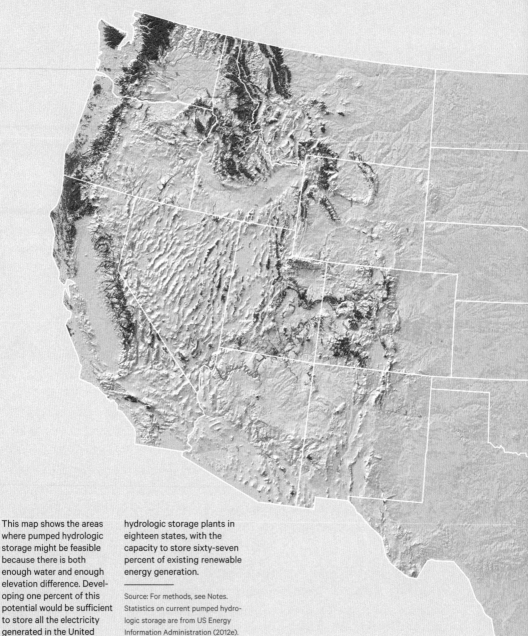

This map shows the areas where pumped hydrologic storage might be feasible because there is both enough water and enough elevation difference. Developing one percent of this potential would be sufficient to store all the electricity generated in the United States in 2010. The US already has thirty-nine pumped hydrologic storage plants in eighteen states, with the capacity to store sixty-seven percent of existing renewable energy generation.

Source: For methods, see Notes. Statistics on current pumped hydrologic storage are from US Energy Information Administration (2012e). Parks and reserves have not been excluded from analysis.

come from a major river, lake, or the seashore, an underground aquifer, or the sky, including sites with a mean annual precipitation of at least thirty inches. (These rain-fed sites would require two reservoirs.) I found 272,340 potential locations in the lower forty-eight states. If only one percent of those sites could be developed, and assuming an average 1,000 MW (24 million MacKays) capacity, then collectively those places could provide 2,723 GW (65 billion MacKays) of instantaneous power, 2.6 times the total amount of electricity that was generated nationwide from all sources (nuclear and fossil fuels included) in 2010. The existing PHS facilities, without building any new ones, can already regenerate almost 60 percent of all the current "intermittent" electricity generation from wind and solar (as of 2010); development of one percent of the new sites could multiply that capacity a hundred fold. In short, we can have as much pumped storage as we need. And that means that intermittency is not really the issue that critics have made it out to be.

Rules of the Grid

Interestingly most of the current pumped-storage facilities do not serve windmills and solar panels; rather, they were constructed to provide last-minute electricity during times of intense demand and, therefore, profit margin. Electricity is a squirrelly form of energy, impatient to be used and eager to flow. Once generated, the energy carried in electrons needs to reach the place where it will be used, which means that electrical utilities have the difficult task of balancing the load on a second-by-second basis. Utilities have no way of knowing when any one of us might choose to charge our smartphone or turn off a fan or run the dishwasher, so they have evolved a set of rules of thumb to operate the system.

The first rule of the grid is to generate the base load as cheaply as possible. The base load is the amount of electricity that is always required, day and night. Though the utilities don't know in detail who will be doing what, they do have a fair idea of what is minimally required at different times of the day and the year in aggregate. Nowadays the base load is supplied by nuclear power plants, which have very large upfront costs but relatively low operating costs (not counting, of course, spent radioactive wastes, which will attract extreme gate duties on exiting the economy of the future), and coal-fired power plants (which produce carbon dioxide and other atmospheric pollutants as they burn their fossil fuel, also currently uncosted), and in some areas, by conventional hydroelectricity. When the grid needs more power, the utilities fire up additional power plants. Each decision to bring power generation on line reflects a trade-off between dollars and electrons. The customers pay the difference.

Thus the second rule of the grid is to have additional capacity idling all the time, waiting for spikes in the load. These idlers are typically natural gas power plants, which can be controlled at the turn of a dial; before 1973, oil was also used, but after the oil embargo, President Nixon worked with industry to phase out most oil-fired electricity generation. Because peak loads can be more than twice the base loads, especially during summer heat waves when everyone is switching on the air conditioning, approximately half of the US power plants are idling, in various stages of expectant waiting, most of the year. (Another way to say this is on average the US uses just 47 percent of its capacity annually.) Since traditional power plants are very expensive to build and take a long time to construct, utilities seek ways to avoid those costs or at the least minimize them, which is exactly why relatively inexpensive, efficient measures like pumped storage were developed. A pumped-storage plant can be added right alongside traditional primary hydroelectric plants. At times of minimal profit, electricity is used to pump water uphill; then at times of peak demand the water runs back down to maximize dollars generated.

The third rule of the grid is an artifact of history: The grid is designed to carry electricity in one direction only, from large utilities to the many scattered, dependent consumers. When the country was first electrified, in the decades after Edison's discoveries, Big Brother was the only choice. Few people generated their own electric power. Windmills were known but hardly used for electricity; solar panels had yet to be invented. In the logic of the time, utilities were necessary evils, "natural" monopolies, which required restricted markets in order to raise the enormous capital required to build equally massive power plants. The government fixed prices for electricity, ensuring a fair price for consumers and a "fair profit" for the utilities. For decades this system of collaboration between government and industry worked—though under the surface there were always arguments about whether fair was actually fair.

Those debates percolated through the cheap oil window and beyond, finally culminating in a wave of deregulation that overtook the industry in the 1990s and early 2000s. Across the country, state governments forced vertically integrated electrical utilities to separate: From then on, power generation (i.e., the power plants), power transmission (i.e., the big power lines), and power distribution (i.e., the lines in the neighborhoods) would operate as individual units. Consumers could choose their generation company and utilities were required to transmit the power no matter who generated it; the notion was that competition, rather than government regulation, would set the fair price.

Consumers were thought to be mainly households and commercial enterprises, but as with oil and other commodities, speculators also wanted to play, consuming not power, but profit-taking opportunities. California circa 2000 is the poster child for the potential problems of deregulation. A Texas company called Enron and various others got into the action by buying electricity from generators and selling it to local utilities. They developed various ploys with names such as Death Star, Black Widow, Big Foot, Cong Catcher, Forney's Perpetual Loop, and Red Congo. Each scheme was designed to milk the system. For example, in Death Star mode, Enron and its intermediaries bought electricity from generators and sent it looping around the system to deliberately over-whelm the transmission capacity and thus trip automatic "penalty" fees payable to the speculators from the utilities who "failed" to provide enough transmission capacity. In April 2001 this kind of fraud forced the bankruptcy of Pacific Gas and Electric, the West Coast's largest utility, which was caught between having to buy wholesale electric-ity on the open market (sometimes generated from power plants it formerly owned) and selling it into a retail market at mandated prices. Consumers were bankrupt, too— of power. Prior to deregulation, blackouts had been rare, but in 2000–2001 California had more than twenty. Governor Gray Davis paid the ultimate political cost, when in 2003 he was recalled from office for his inability to control a system from which the controls had been removed, and Arnold Schwarzenegger was elected instead.

Despite such problems, deregulation is not a bad idea; well managed, with smart, transparent rules, the market does have the potential to provide better prices and more choices. The problem was that structural change wasn't taken far enough. Instead of focusing on the arrangement of institutions, reform should have targeted the grid itself. Edison's grid, designed in the 1920s to address problems of a pre-silicon chip society, is not really suited to answer modern needs. Computers, for example, turn out to be quite handy for many things, including managing electricity.

The three old rules of the grid (base load, idle capacity, unidirectional flows) need to be revised in light of current technology and demands. What we need is the "smart grid," touted by politicians, environmentalists, and economists alike. The smart grid is different from the old grid in two fundamental ways: (1) it transmits both power and information, and (2) it transmits those flows in two directions. The smart grid means everyone can be a producer as well as a consumer of electricity.

Modern grid engineering begins by recognizing that the wire doesn't care which way the electrons are flowing. The same wire that comes into your house with elec-tricity can take excess electricity from a solar panel on your roof back to the grid. In fact one recent government study notes that we already have more than twelve mil-lion distributed-generation facilities (solar panels, small-scale natural gas plants, even some hydrogen fuel cells) in the United States that could be connected to the grid but aren't. Two-way local distribution enables electricity to flow locally from where it's produced to where it's needed. Shorter transmission distances, like shorter commutes, are more efficient. (Seven percent of electrical energy nationwide is lost during trans-mission.) Better still, having many producers interacting with many consumers actu-ally creates a real marketplace, where it is much harder for any one utility (or any one speculator) to have an unfair advantage.

The key technical challenge is that at every moment supply must match demand. Here is where the double-dam set-ups and potentially other kinds of electric storage

A Smart Grid

A smart grid is a material manifestation of democracy. The current "dumb" grid transmits only electricity and only in one direction (from the large utilities to the dependent consumers). A smart grid allows electricity and information to travel in two directions, enabling everyone to be both a producer and a consumer of electricity, and thus make intelligent, real-time decisions about energy use.

Source: Redrawn from Fox-Penner (2010).

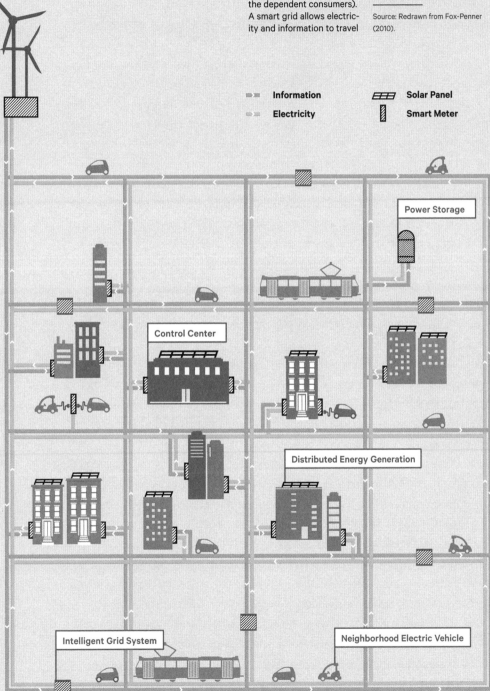

- ►► Information
- ►► Electricity
- 📱 Solar Panel
- 🔋 Smart Meter

Power Storage

Control Center

Distributed Energy Generation

Intelligent Grid System

Neighborhood Electric Vehicle

come in. Along with computer controls, the utilities can level excess supply and feed excess use, second by second, while communicating information back to end users.

Right now the only way most of us know how much electricity we are using, and how much it has cost, is by looking at the bill at the end of the month. At that point, it's too late to do anything about it. The smart grid will enable us to measure our usage and know its cost at any moment—whether it's peak energy or base load— and change our behaviors accordingly. We might be more inclined to turn off unneeded lights or choose to run the dishwasher off-peak if we knew what we were paying. Appliances can be fitted with internal switches to run now, later, or when power is cheapest. In studies, residences equipped with an indicator of the cost of their electricity (as simple as a light bulb that glows a different color depending on price) have reduced peak usage by as much as 25 percent. Manufacturers and office managers, always responsive to cost, can learn to accommodate activity to price. Magnified across the nation, smart electricity use means reducing the number of power plants by the hundreds and the capital costs by the billions of dollars; those savings will pay with interest a smart grid financed by government loans.

The smart grid is a framework for freedom, democracy, and an intelligent American way. Better still, it will enable us to use our native power.

Where the Wind Blows

What could be more beautiful than an elegantly engineered structure that turns for free as the wind blows? Windmills have a historical pedigree more than a thousand years old and picturesque precedents on Greek isles and Spanish hills; Dutch windmills turned in lower Manhattan long before skyscrapers were dreamt of. Windmills pumped water, the creaking sound of their mechanics accompanying independent-minded homesteaders staking claims on the Great Plains, and windmills sprouted like weeds on California hills when President Carter gave a favorable nudge through the tax code. In fact the only items besmirching the romantic resume of windmills are their relationship to Don Quixote (though he mistook them for monsters, not the lovely energy engines that they are), their propensity to knock down bats and birds on the wing (though cars do a lot more harm in that regard), the low buzz they emit up close (though have you walked down a city street recently?), and the fact that, from the American perspective, we were once very good at building them. America led the world in the 1970s and then we lost that advantage to the Danes, the Germans, and the Chinese; we became world leaders in sport utility vehicles instead.

Adding to the thrill is that America has a lot of wind, as has always been obvious to anyone who lives on the Great Plains or who sails off Nantucket. Some have called America the Saudi Arabia of wind power. The federal government has officially known what wind could bring since at least the Reagan Administration, when in

1981 the Pacific Northwest Laboratory of the Department of Energy published a series of regional atlases of wind-power potential in the United States. A revised national wind-power atlas published a few years later, reads like a Woody Guthrie song, noting that:

> Major areas of the United States that have a potentially suitable wind energy resource include: much of the Great Plains from northwestern Texas and eastern New Mexico northward to Montana, North Dakota, and western Minnesota; the Atlantic coast from North Carolina to Maine; the Pacific coast from Point Conception, California to Washington; the Texas Gulf coast; the Great Lakes; portions of Alaska, Hawaii, Puerto Rico, the Virgin Islands, and the Pacific Islands; exposed ridge crests and mountain summits throughout the Appalachians and the western United States; and specific wind corridors throughout the mountainous western states.

Recent studies estimate that if windmills operated at 30 percent of capacity and were excluded from unsuitable places (major cities, environmentally sensitive areas, etc.) full utilization of the onshore winds could generate 38.5 million GWh of electricity per year. (That's enough energy to provide 7.4 million minutes of microwaving for every person in the country.) For the sake of comparison, the total amount of electricity generated in the US for all purposes in 2009 was just 3.9 million GWh, or about one-tenth of the total American onshore wind-power potential. The offshore resource—the wind blowing over American seas—adds another 50 percent, meaning that just from wind the United States could potentially generate *fifteen times* our current electricity demand. European and Asian nations already have large offshore wind farms; as of this writing, America doesn't have a single one. (Though after years of debate, some are finally in the works off the northeast coast. For comparison, remember we have been drilling commercially for offshore oil since 1947.)

Although 2010 was a down year because of the recession, US wind power has been growing by 40 percent per year for most of the last decade, with a total installed capacity of over 40 GW (960 million MacKays) in 2010. In fact, the demand is so great that the cost of wind turbines has increased despite the recession. Gains have been made in many states including Illinois, California, South Dakota, Minnesota, Oklahoma, and Wyoming.

When the experts talk about wind power, they classify the quality of the wind, taking into account both how hard it blows and how often. Class 5 winds are excellent; they are winds that steadily blow at more than 16 miles per hour at ground level (and yet faster a hundred feet up in the air, where the blades turn). A 16 mile per hour wind is a moderate to strong breeze, enough to cause small waves to crest and small

The Windy Country: US Wind Power Potential

Other countries envy America's wind power potential. Developing wind power in Texas, Oklahoma, Kansas, Nebraska, and the two Dakotas would provide enough electricity to supply the entire country, even after excluding cities and environmentally sensitive areas. The wind power potential off the northeast coast alone is sufficient to provide the entire country and is conveniently close to major population centers. Little Denmark has giant windfarms. Why can't we?

Source: Redrawn from AWS Truepower (2011) and US National Renewable Energy Laboratory.

Wind Power Class	Resource Potential	Wind Power Density at 50m height (W/m²)	Typical Wind Speed at 50m height (mph)
3	Fair	300–400	14.3–15.7
4	Good	400–500	15.7–16.8
5	Excellent	500–600	16.8–17.9
6	Outstanding	600–800	17.9–19.7
7	Superb	800–1600	19.7–24.8

branches to swing back and forth. It's the kind of blow that makes you want to pull on a windbreaker and keep your head down.

Not surprisingly, places with unrelenting winds tend not to be places most people want to live, which means that the wind energy solution includes a discussion of transmission—how we will get the energy from where it is generated to where it will be used. On the positive side it means that most states with a lot of wind also have lots of wind power to export, because their populations are low. South Dakota, for example, already generates 23 percent of its electricity from the wind. Exporting that power could thus provide a substantive economic resource to windy rural places, largely compatible with other economic activities of wide-open spaces like farming, cattle ranching, and buffalo ranging. Offshore wind power is relatively closer to coastal markets but suffers from the added expense of building in the water and the political price of changing the view for wealthy seaside homeowners and their well-connected friends. Similarly, some ridgelines that provide excellent opportunities for wind power close to town have been shut down because people don't want their view spoiled, or because of concern about wildlife. Environmental, health, and aesthetic concerns about wind development, as with any kind of improvement, are appropriate, and should be dealt with through zoning and economics. But those decisions need to be taken in context. The proper comparison isn't between a windmill and unadulterated natural splendor; it's between a wind farm and a field of oil rigs, with all that both imply.

Many people have seen pictures of the San Gorgonio Pass Wind Farm area in the dry Coachella Valley, near Palm Springs, California, where four thousand windmills scattered across 70 square miles generate 893 MWh (54 million minutes microwaving) per year (in 2009). When San Gorgonio is added to the two other major wind energy zones in California (Altamont and Tehachapi Passes), California breezes already produce enough electricity to power a city the size of San Francisco.

Windmills, like wind farms, have been getting bigger over the years, with individual turbines rated up to 6 MW. Some machines have a wingspan wider than a football field is long. The Norwegians are testing a wind turbine with a rotor diameter of 160 meters (525 feet) that floats in the ocean, tethered to the bottom. The amount of energy that can be generated is related to the momentum with which the windmill turns; momentum is a function of mass multiplied by velocity (m × v), which means a large windmill turning slowly generates as much energy as a small windmill turning quickly. Slow windmills turn out to be easier for birds to avoid; but the sound they make causes trouble for bats and some people, so best to avoid placing windmills in sensitive bat and human habitats.

Windmills work by turning a coil of wire inside a magnet, thus inducing an electrical current. Nowadays these magnets are typically made with rare earth elements, like samarium and neodymium, which aren't really as uncommon as their name would

US Wind Power Cost + Production

American wind power grew 1500 percent between 2001 and 2011. In 2011 there were 781 recognized wind power facilities in thirty-eight states, and we have yet to build our first off-shore wind field. Wind power costs have fallen dramatically since the 1980s, and despite a small rise in prices because of the intense demand for wind turbines from 2004 to 2009, supply now appears to be catching up with demand. Wind and other renewables, however, have trouble competing with fossil fuels that don't bear their full environmental costs.

Source: US Energy Information Administration (2012a), Wiser and Bolinger (2012).

---- Nominal price ($/W)
—— Inflation-adjusted (2009$/W)

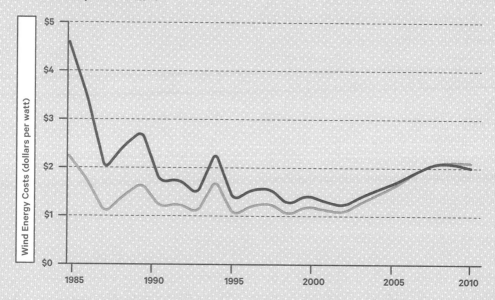

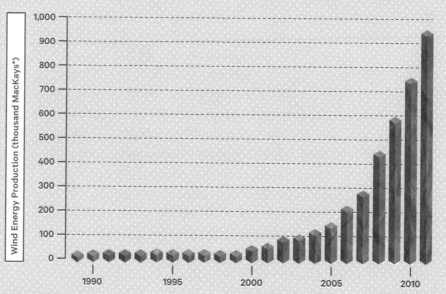

*MacKay = 1 kWh/day

suggest; neodymium is almost as common as copper, samarium as lead. One particularly rich bed of clay in southwest China currently supplies most of the rare earths for magnets found in coal-fired power plants, nuclear power plants, cell phones, laptops, televisions, and electric and hybrid cars. Since China began cutting back exports in 2011, it's fortunate that rare earths are found in the United States, too. In addition to magnetic minerals, windmills are also constructed from fiberglass, epoxy, and carbon fibers; they are another by-product of oil. And like that other by-product, the suburbs, windmills comprise a kind of "improvement" to the land; ecological use fees will apply.

Where the Sun Shines

And that's just the wind. The solar power potential of the country is extraordinarily massive and underexploited. Estimates of the potential are almost ridiculous in their magnitude: the Energy Information Administration put an initial estimate of the accessible resource of the sunlight striking the United States at 172 billion GWh, or some twenty-one million times (!) the wind-power potential we just considered. ("Accessible resource" is analogous to the "recoverable resource" estimates of the oil and gas industry; it means what can be had regardless of cost, as opposed to the economically accessible, or recoverable, resource.) The United States could power the entire world if the economics were right to harness our solar power potential. Fortunately each year for the past ten, solar panels have dropped in price as more people deploy them. And as the price drops, more people decide to invest in the sun.

In the sunniest climes, particularly in the Southwest, 600 square feet of the same photovoltaic panels green-minded suburbanites bolt to their roofs can produce enough electricity for a person for a year. Solar panels use photons from the sun to liberate electrons from silicon layers and generate a current. Flying over sun-drenched locales like Dallas, Phoenix, and Los Angeles, it is easy to imagine those millions of square feet of flat building roofs covered with panels. One estimate suggests 22 percent of residential and 64 percent of commercial rooftops in the United States could usefully accommodate solar panels rather near to where we use electricity, for example, downstairs. And on a roof, ecological use fees don't apply.

The authors of the 2008 *Solar Living Handbook,* a combination catalog and energy-efficiency manifesto, made some interesting calculations for the Nevada Test Site and Nellis Air Force Base complex, a sunny, mostly unpopulated area covering 5,740 square miles in southern Nevada that the military uses mainly to practice bombing. With the same kind of arithmetic that homeowners deploy to figure out how many solar panels they need to power their house, they showed that this parcel of land could, all by itself, produce 60 percent of the amount of electricity the country uses, if covered in photovoltaic solar panels. A square 97 miles on the side would be sufficient for all of us. The Air Force took notice. In twenty-six weeks in 2007, it installed a solar field comprised of 72,000 panels on 140 acres, which now generate 14 MW (equivalent to 336 million MacKays). In 2010 the base announced it wanted to double the size of its solar installation, which nicely demonstrates the scalability of photovoltaic panels. Solar panels can be laid a few at a time or can be massed together; they work at any scale.

Sending electrons out to the rest of the world from roofs and other scattered localities will ultimately require the smart grid discussed before. Because we don't have that form of the grid yet, most solar development heretofore has focused on massive capital-intensive facilities on the old model. There are a couple of workable methods: Some solar factories work by using mirrored troughs to concentrate sunlight onto blackened tubes filled with oil; the hot oil is collected to boil water that turns a steam turbine to generate electricity. (Natural gas is sometimes used to top off the steam on cloudy days.)

Another option is to aim the sunlight at a central tower (a "tower of power"), either to generate steam or melt pillars of salt. Molten salt holds its heat better than oil or water, which means that salt melted during the day can continue generating power overnight. In June 2011 the Spanish Torresol salt-solar plant in Fuentes de Andalucía became the first solar installation to produce electricity for twenty-four hours without interruption.

Another clever way of extracting power from the sun is with an external-, as opposed to an internal-, combustion engine. Recall that internal-combustion engines work by discharging an explosion inside a cylinder, which produces rapidly expanding hot gases that push a piston and thus generate motion. On the counterstroke, those same hot gases are expelled from the cylinder and then the process repeats with the injection of more fuel and air. Most of the energy goes out with the expelled gases, even in modern internal combustion engines, which contributes to low overall efficiencies. Robert Stirling, another inventive Scotsman of the Industrial Revolution, suggested in 1814 an alternative configuration, which he called a closed-cycle engine, that would be more efficient, safer, and as side benefits, quieter and more efficient.

One makes a closed-cycle engine by enclosing a gas in a sealed cylinder. (Sealing a container tight enough that no gases can escape is the reason Stirling engines didn't catch on until the twentieth century; they tended to leak or explode.) One end of the cylinder is heated (for example, by the sun, focused with mirrors) and as the end heats, the temperature of the gas inside also rises, causing the gas to expand and push a piston and generate motion and eventually electricity. The hot gas then passes into a regenerator (or what Stirling called an "economizer"), which pulls off the heat. The gas ends up in the cool end of the cylinder, where it compresses, and as it does so, another piston on the counterstroke forces the gas back through the regenerator, where it picks up some of the heat, and then back to the heated end of the cylinder, where the cycle is renewed. Because of the economizing regenerator, Stirling engines achieve

The Sunny Country:
US Solar Power Potential

The solar power potential of the US is, conservatively, at least 1,000 times greater than the wind power potential. For example, a single square of southern Nevada land, 97 miles on a side and covered with photovoltaic panels, could provide enough electricity for the entire country, not to mention some streetcars.

———————

Source: Redrawn from National Renewable Energy Laboratory (2008) for fixed panels tilted at latitude, US Energy Information Administration (2012a).

Solar Resource (MacKays/square meter)

< 4.5	5.5–6.0
4.5–5.0	6.0–6.5
5.0–5.5	> 6.5

US Solar Power Cost + Production

Solar power has taken off in recent years, growing 147 percent between 2001 and 2011. There are currently 326 commercial solar power installations in 20 states,

with lots of room to grow. One study estimated that 22 percent of residential roofs and 64 percent of commercial ones could usefully accommodate solar

panels nationwide. Solar power costs have dropped dramatically as economies of scale have come into play. Increasing demand for electricity for transport and

gate duties will tilt the economy toward the sun.

Source: US Energy Information Administration (2012a).

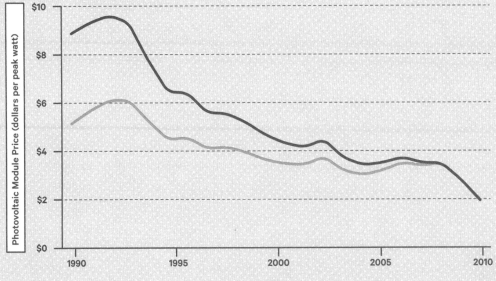

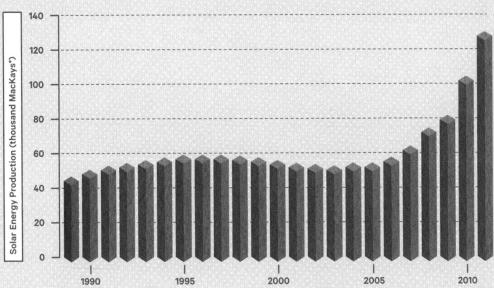

*Photovoltaic and solar thermal commercial electricity production. 1 MacKay = 1 kWh/day

theoretical efficiencies of around 35 percent. A demonstration project of six Stirling engines equipped with curved mirrors at Sandia National Laboratory in New Mexico set a new world record for solar-to-grid efficiency (31.25 percent), but technical problems remain; the largest commercial vendor of them filed for bankruptcy in 2011.

There is one more solar energy generator with an even longer pedigree than these others: plants. Not power plants, green plants. Our entire food supply is based on solar energy that plants collect through photosynthesis. You will recall that photosynthesis turns carbon dioxide from the air, water, and energy from the sun into carbohydrates. Later we harvest the leaves or fruits, or a cow, pig, or other animal eats them, and eventually the energy from the sun finds its way onto our dinner plates. The process is terribly inefficient, since photosynthesis only captures about two percent of the incident solar energy globally, and every subsequent transition loses more energy. That's why vegetarians can rightfully claim their dinner cost less energy than a more carnivorous one.

Because photosynthesis is inefficient and because we need to eat plants and animals that eat plants, biofuels like ethanol or biodiesel don't make a very convincing replacement for gasoline. Growing enough fuel to replace gasoline isn't possible given the current scale of motor fuel uses and the amount of cropland available; one estimate suggests we would need to convert every bit of American cropland to corn, and expand it at least another 50 percent, to produce enough ethanol to meet *half* the current transportation demand. Not only would we have no food to eat but we would also have to foot the cost of converting or purchasing new the current 246 million motor vehicle fleet running on pure ethanol or biodiesel. Biofuels are one of those things that sound good to the auto lobby and senators from farm states and terrible to everyone else; one writer calls them a "political" fuel as opposed to an actual one. Energy historians of the future will look at current government incentives for biofuels and wonder: What were we possibly thinking?

The Warm, Generous Earth

The dark horse of renewable energy, literally and figuratively, is the heat of the earth, an immense source of clean domestic power. The heat longs to escape to the surface, succeeding in spectacular form in Hawaii's volcanoes and Yellowstone's hot springs. Shallow geothermal sources like these are easier to access than oil, and have already shown great potential to generate electricity, but can be overexploited if the associated water and steam are pulled out faster than nature can replace them. People have used earth-heated waters for cooking and bathing for thousands of years; in 1904, Prince Piero Ginori Conti first produced electricity from nature's steam in Larderello, Italy. Today fifty-six geothermal plants generate 3,420 MW (82 million MacKays) in America, out of a potential resource of 30,000 MW (720 billion MacKays). The Philippines,

The Warm Country:
US Geothermal Power Potential

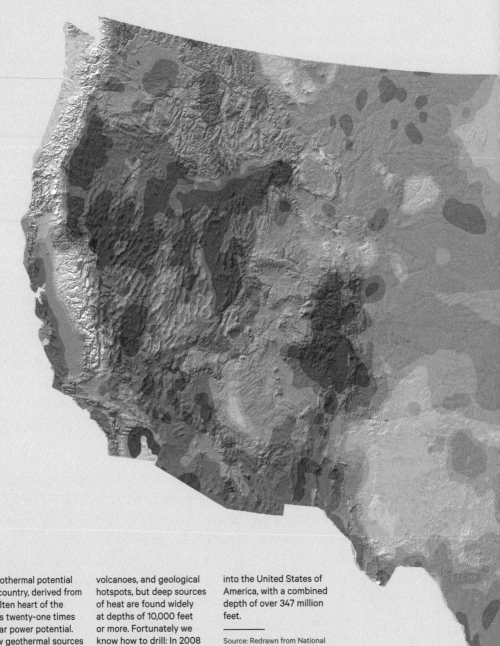

The geothermal potential of the country, derived from the molten heart of the earth, is twenty-one times our solar power potential. Shallow geothermal sources have a patchy distribution, associated with hot springs, volcanoes, and geological hotspots, but deep sources of heat are found widely at depths of 10,000 feet or more. Fortunately we know how to drill: In 2008 for oil and gas development, we put 55,735 new holes into the United States of America, with a combined depth of over 347 million feet.

———

Source: Redrawn from National Renewable Energy Laboratory (2008).

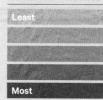

**Favorability of
Geothermal Resources**

Least

Most

Indonesia, Mexico, Japan, and New Zealand, other countries along the Pacific Ring of Fire, already generate significant electricity from geothermal sources, as do Italy and Iceland.

Shallow resources are only the beginning. Beneath them, at depths of ten thousand feet or more, the rock everywhere in the world is hot, heated by chemical reactions in the core of the earth ignited when the planet first formed. Thanks to more than a century of drilling practice, the oil and gas industry doesn't even blink when asked to dig that deep. (The Deepwater Horizon platform, for example, could reach depths of 30,000 feet, under a mile's depth of ocean water.) In fact, hot water extracted from oil wells is a significant nuisance and cost to the industry.

Geothermal engineers turn that nuisance into an advantage by drilling wells down to a depth with sufficient temperature, then cycling water through the rock to receive the heat. A geothermal plant on the surface is equipped with a turbine to generate electricity and the cooled liquids are recycled for another trip. Unlike hydrofracturing for natural gas, the liquids can be contained, reducing the chances of polluting groundwater supplies. If the environmental concerns are costed and avoided, then the potential resource is huge and available across large swaths of the United States. It does require a well and a pump and some capital to get started; economics as always apply.

When we make the economics work by costing all of nature's energy sources appropriately, geothermal will generate base load electricity as steadily as any coal or nuclear power plant. A recent study out of MIT estimates that "enhanced geothermal systems" (the pumping kind) have twenty-one times the solar power potential of the whole country. Or to put it another way, if only two percent of the potential geothermal energy were recovered, then there would be sufficient heat to provide you and every other American each year all the MacKays you could use for twenty-six hundred years.

Economies of Scale

Thinking about the renewable electricity potential of the United States puts me in mind of Thomas Jefferson, leaning back in his writing-chair at Monticello in 1803, fingering his contract with the Emperor Napoleon, wondering how in the world the American people would ever avail themselves of all of the new land, the Louisiana Purchase, that his administration had just brought into the nation. The answer history provided was economies of scale. Land sold and given away, railroads sponsored, homesteads improved, and huge, interconnected industries born from new scientific insights propelled Americans across a continent. Vast quantities of cheap resources created large incentives and significant rewards for innovation, and those innovations made fortunes for a few and raised the quality of life for many.

John Rockefeller, Thomas Edison, and Henry Ford, having learned the lessons of the earlier generation of economic titans, provided the basis for a second, even more

Figure
No. 70

US Geothermal Power Cost + Production

Geothermal electricity production has grown slowly but steadily for over a decade. If we could turn a few oil and gas rigs to geothermal instead, this supply could be the next great energy bonanza. Chevron, anyway, thinks so, having recently constructed geothermal plants...in the Philippines and Indonesia. Sinking the wells constitutes about 60 percent of the capital costs of most geothermal projects. Costs for drilling for heat compare favorably with oil and gas, and we already have over 360,000 oil and gas holes.

Source: US Energy Information Administration (2012a) and Bloomfield and Laney (2010).

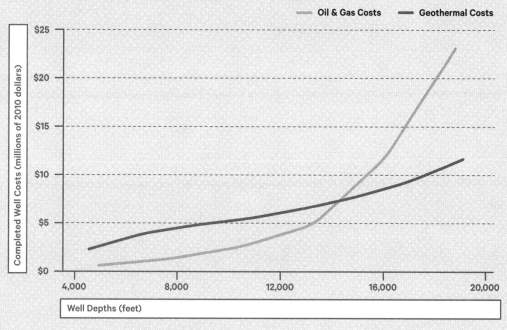

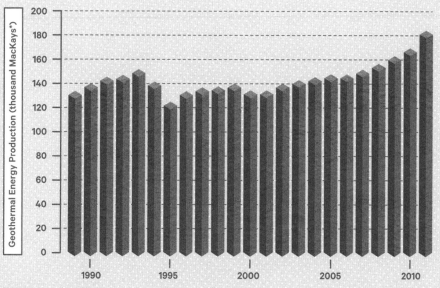

*1 MacKay = 1 kWh/day

Costs of Electricity Production

Even without gate duties, denser towns and cities, and widespread electrified transport, renewables are already becoming competitive with fossil-fuel sources.

Why? The true costs of fossil fuels—in terms of debt, wars, and the environment—are on the rise, a function of geological conditions fixed 100 million years ago, while

the costs of renewables continue to decline as economies of scale come into play. We can nudge the differential yet further through some extreme

measures of our own, Terra Nova style.

Source: U.S. Energy Information Administration (2010, 2012a).

Energy cost					
Fuel (Typical Plant Size and Actual Operating Capacity, 2011)	Construction Cost for New Power Plant	Fixed Operation and Maintenance Costs ($/Year)	Fixed Costs ($/MWh)	Variable Costs, including Fuel and Waste ($/MWh)	Current Costs ($/MWh)
Coal (650 MW, 63%)	$2,058,550,000	$23,380,500	$ 26	$ 4.25	$ 30
Natural Gas (540 MW, 29%)	$528,120,000	$7,770,600	$ 18	$ 3.43	$ 22
Nuclear (2,236 MW, 89%)	$11,929,060,000	$198,445,000	$ 34	$ 2.04	$ 36
Wood (20 MW, 71%)	$157,880,000	$6,775,800	$ 97	$ 16.64	$ 113
Municipal Solid Waste (50 MW, 59%)	$411,600,000	$18,688,000	$ 125	$ 8.33	$ 134
Onshore Wind (100 MW, 30%)	$243,800,000	$2,807,000	$ 42	$ 0.00	$ 42
Offshore Wind (400 MW, no US data)	$2,390,000,000	$21,332,000	$ 96	$ 0.00	$ 96
Solar Thermal (100 MW, 14%*)	$469,200,000	$6,400,000	$ 180	$ 0.00	$ 180
Photovoltaic (150 MW, 14%*)	$713,250,000	$2,505,000	$ 143	$ 0.00	$ 143
Geothermal (50 MW, 79%)	$278,900,000	$4,213,500	$ 39	$ 9.64	$ 49
Conventional Hydroelectric (500 MW, 30%)	$1,538,000,000	$6,720,000	$ 28	$ 0.00	$ 28
Pumped Storage (250 MW, no estimate)	$1,398,750,000	$3,257,500	$ 48	$ 0.00	$ 48

* Operating capacity for photovoltaic and solar thermal combined.

substantial reinvention of the economy in the early twentieth century, facilitated by vertical integration, consumer technologies, government policy, and plunging costs—in other words, economies of scale. Now we must learn from them. After leveling the playing field for energy development, we must place an order so massive for renewable energy that the unit costs drop and drop and drop.

The prerequisites are all in place. In 2010 the Energy Information Administration reported that the United States had 689 wind farms, 180 large-scale solar installations, and 225 geothermal plants feeding collectively 112 million MWh (6.72 trillion minutes microwaving) of electricity into the grid, meeting some 2.8 percent of the country's electricity demand. These facilities are our proof-of-concept installations, showing that electricity can indeed come from the wind, sun, and earth in significant quantities.

The next step is to figure out what we need. The current energy demand for transportation is 8,752 million MWh (525 trillion minutes microwaving; as of 2010). Neglecting that electrifying transportation will increase the energy efficiency of every trip taken in the future; forgetting that adopting bicycles and walking for short trips will lop off a significant portion of vehicle miles traveled; overlooking that moving closer to work will decrease the total distance traversed; and ignoring that having a robust, affordable, comfortable public transportation system will mean those distances will be shared, let's calculate how much we need to sustain our current levels of consumption: The 112 million MWh we already produce from sun, wind, and earth goes into the 8,752 million MWh we need approximately 78 times. Let's add another 22 times to cover transmission, storage losses, and other things we haven't thought of. In short, via the back of an envelope, increasing renewable electricity generation about 100 times over current levels could, with the changes already detailed, achieve national self-sufficiency in electric transportation.

Do you think it's possible? It depends on your estimation of the American people. Was it possible for the railroads to increase in mileage thirteen-hundred-fold between 1830 and 1860? Was it possible for the street railways to grow four hundred times between 1889 and 1912? Was it possible for the good roads of America to spread at an average rate of forty thousand miles per year between 1904 and 1954? If we can build sixty-nine million housing units, fight five wars overseas, endure eight recessions, and play 741,000 major-league baseball games between 1950 and 2000, I have to believe that twenty-first-century Americans can build enough windmills, solar panels, geothermal plants, smart grids, and double-dams to ensure the future of the economy, the environment, and the American Dream.

Energy is that which changes the physical state of a system. This time—in our time—the system that will change is America itself.

Part III
Ramifications

Figure No. 72

While the principal pleasures of Terra Nova are restored economic growth and improved national security, a number of collateral benefits will also accrue to the nation if we can break away from the old Siren song: a healthier populace, a greener continent, a cooler climate, happier communities, fresher food, a revitalized politics, and a better world.

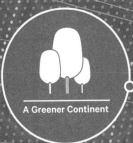

A Greener Continent

Fresher Food

Healthier People

13
A Future

The future ain't what
it used to be.

Yogi Berra

It's 2028, and after thirty years, I finally retire from the Wildlife Conservation Society. My wife and I decide to move back to California after a long, successful run in New York. With help from our friends on City Island and after a yard sale, we close up our little house and pack our things into six containers, sized to be picked up by the neighborhood electric delivery trucks that will come by later and take them to the nightly freight streetcar. We walk down to the avenue and say good-bye to the greengrocer, butcher, baker, pharmacist, bookseller, local Walmart retailer, school principal, cops on the beat, and diner owner. We've seen these people practically every day for years, and now they feel more like friends than "producers" selling to us "consumers." Streams of bikes speed by along dedicated paths beside the retro City Island streetcars, which rumble by every couple of minutes in the center lane.

Back in 2014 the local community board approved new zoning to phase out on-street parking and to allow higher density housing along the avenue (six-story apartment buildings with shops and offices on the ground floor), which meant that by 2017, we were awarded New Town district designation IIe (provided for densities >10,000 people per square mile; population size >10,000; where geographic boundaries constrain area). This paid for the streetcar tracks and improved sidewalks, bike lanes, and a new seaside community park with a small forest. It also created seven hundred new permanent jobs on the island. The green space caters to couples sitting on benches, families eating takeout, and kids swimming in the sparkling waters of Long Island Sound. Though there are more people and hustle and bustle on the streets, City Island is quieter without explosion cars and motorcycles; standing on the sidewalk, we hear the cries of oystercatchers from the shore.

We grab a streetcar to Pelham Bay Station, where we hop on the light rail that speeds us down to Manhattan along the old Cross Bronx Expressway (now covered over, with the neighborhoods rebuilt on top and high-speed tracks down below). Our ride is free to us as New York City taxpayers. Twenty minutes later, we change trains at the George Washington Bridge High Speed Rail Station, the "North Hub" in upper Manhattan, for Chicago. From here we could have run down to Times Square in another eighteen minutes on the Joe DiMaggio Westside Light Rail, or we could have taken another train out to JFK and caught a flight (one of the few modes of transportation still supplied by American petroleum, supplemented with refined biofuels), but since we have time and are now living on our savings and Social Security (bailed out thankfully in 2022 by increased tax revenues from the growing economy), we decide to take the cheaper high-speed train, which whisks us to the Windy City in just over seven hours. We stay a few days in Chicago, visiting friends and colleagues, going to a White Sox game, and taking a drive in an immaculate red-and-white 1957 Chevrolet Bel Air two-door hardtop along Lakeshore Drive, which has been set aside for recreational driving. We even splurge on a B&B that overlooks a community garden managed by the economics professors at the University of Chicago, founded by the largesse of John D. Rockefeller 130 years before.

After breakfast, we pack our bags again, grab the streetcar to Chicago's Union Station, and settle in for the eight-hour trip to Denver. I find myself drawn to the stunning landscape unfolding outside the train window, as we zip along the old I-88 in Illinois at twice the speed of cars from my dad's generation. In northern Illinois and Iowa, not only have the suburbs pulled themselves together and become separated by green-belts of forest and prairie, but in the long lovely stretches between the town centers, where there used to be endless fields of corn and soy, the landscape has returned to small crofter farming, woodlands, meadows, ponds, and streams.

It turned out the burden of ecological use fees on large-scale industrial agriculture tipped the tide of profits. Fewer cars meant less corn for ethanol to reduce air pollution—which was now practically nonexistent in any case—and the market turned to local, organic production, as consumers spent a portion of their newfound cash from lower transportation expenses on better food. The Environmental Protection Agency has been downsized too, its regulatory role less important. Now Iowa produces what Iowa is best at without anyone telling them how: fruits, veggies, dairy, flood control (an ecosystem service, provided by beavers and muskrats), and whip-smart kids.

We pull into high-tech, rebuilt Omaha around lunchtime. The development of wind power on the Great Plains in the '10s and '20s enabled a new industry of Internet servers and software companies to grow like weeds, taking advantage of the numerous small pumped-storage facilities in the hills along the Missouri River. Real estate developers, responding to new tax disincentives to build in wetlands, created a new city from parts of the old above the Missouri River floodplain. As a result, new Omaha (aka Silicon Prairie) doesn't have to worry about periodic inundation the way old Omaha did, and the Missouri River had reclaimed its age-old routine of laying soils and harboring gaggles of geese, sedges of cranes, and flights of doves, to everyone's delight.

Across the Great Plains of Nebraska and eastern Colorado we enjoy watching wind turbines in action. These enormous machines march across the landscape in astounding numbers, almost as amazing as the thousands of bison and elk grazing at their feet. All together, between 2012 and 2028, private companies in the United States had installed over half a million wind machines, offshore and on land—some eighty thousand on public and private land in Nebraska alone. Though some people farm under them, the majority of the ranchland has been given over to the bison, the national mammal, now grown to a herd two million strong, ranging across the plains and subject to sustainable hunts in fall. It turns out bison withstand the harsh winters and blazing summers here better than cows grown on feedlots, and they taste better too. We pass by prosperous Plains towns, encircled by fencing to keep the animals out of the green-belt edges, with sweeping views of the surrounding panorama.

Denver, too, saw surging revitalization as people moved back into town from the sprawling suburbs. Fortunately in the 1990s Denver began investing in its light rail, extending to Littleton and Lincoln; new lines were added in the 2010s and 2020s to reach northern and western suburbs. I long to go south on the high-speed line running along I-25 to see the mountain valleys of my grandparents, but now we have our own grandkids, so instead we take the overnight Zephyr train, which will arrive in San Francisco in just over nine hours. We sleep on a foldout bed while traversing the salt deserts of Utah at speeds in excess of 200 miles per hour, then slow to a more stately 180 miles per hour across Nevada's basin and range country, now known for its solar fields, geothermal plants, and pumped-storage facilities, which slurp and release, slurp and release water up and down the hills, feeding a smart grid that reaches from Everett, Washington to Key West, Florida.

In the morning light, we awake as the train emerges out of the Sierra Nevadas. Eating breakfast, we wave at the bicycles still wheeling around Davis and my wife claims she spots a grizzly bear ambling through an abandoned housing development on the edge of Suisun Marsh in redensified Fairfield. Finally, we cruise across the San Francisco–Oakland Bay Rail Bridge to disembark at the Embarcadero Station to meet my son and his family on time, at precisely 9:00. Total transit time, New York to San Francisco: 24 hours. Ticket price: $1,200 each, a bargain (equivalent to about $360 in 2012 dollars). Because of inflation, the dollar has dropped in value 400 percent since 2008 and everyone is a "millionaire," but it doesn't mean what it once did.

With my son's help, we had already leased a nearby apartment, so we hop on Sutro Tower bikes, available from the local bike share program, and our bags follow by hired NEV. Our furniture and other things arrived three days before, and my son and his family unpacked them for us. Our new home features expansive picture windows facing the Golden Gate, a small protected garden, and a study for my old-fashioned books and me. An Italian bakery nearby infiltrates the neighborhood with the aroma of fresh bread and coffee. The bell on the streetcar jangles as it passes by. The birds are singing, the prospect is good, and our grandkids are happy and healthy.

Standing sleepless by the windows late that night, facing west and reflecting on all that has happened, I hear Walt Whitman's lines reverberating among the waves:

> Long having wander'd since, round the earth having wander'd,
> Now I face home again, very pleas'd and joyous,
> (But where is what I started for so long ago?
> And why is it yet unfound?)

Whitman knows better than I do that no hope is without its risks, compromises, and disappointments. To these we turn next.

14

Cost, Sacrifice, and Evolution

Think of giving not as a duty but as a privilege.

John D. Rockefeller, *The Difficult Art of Giving* (1909)

The truth is: It's not going to be easy, whatever we do. Chasing the Sirens like we did through the twentieth century will lead only to more woe and despair, but the suggestions I've made are not without their own difficulties, especially over the next couple of decades. There will be cost and there will be sacrifice; evolution never occurs absent adversity. What might be gained, we will examine in the next chapter; in this section, we take on the downsides, direct and indirect, of adopting such sweeping rearrangements to our national modus operandi.

The good news is no one needs to die or to kill to bring a new form of American life to America. Earlier generations offered the ultimate sacrifice to create and protect the nation we live in today, at Bunker Hill, Gettysburg, Normandy, and Tet, and more recently, in Fallujah, Najaf, and Tora Bora. Many more have labored on remote homesteads or endured filth, disease, and crime in town, working long hours to give their children a glimmering hope of a better life. Now it's our turn to pitch in for the good of our country.

The war with the Sirens begins at home, on the domestic front, in our hearts and minds. Here's what we are up against.

Paying More for Gas

Yes, if oil and gas consumption is to decline, prices must go up. Economics has no surer principle than higher prices induce lower demand. As we have seen, the demand for gasoline is traditionally, even legendarily, inelastic, but it will stretch if we pull hard enough. That happened in the 1970s, when the oil shocks sent prices skyward and government and consumers worked to moderate oil consumption; subsequently we have seen a new flatter relationship between oil and GDP. We induced lower demand through surging unemployment during the course of the Great Recession, but that was less than optimal. It's better to choose not to drive as a matter of interest, not circumstance. Prices should be raised steadily, with plenty of advance notice, and they must rise inexorably beyond the demand destruction point, the price where we change our behavior. The bright spot, if there is one, is that you don't have to pay these higher prices; you know how.

For their part the oil companies are going to do everything they can to feed our addiction to their products; they will try to fill our heads with ranges of alternatives—all-hands-on-deck solutions, they will say; just wait until next year—while keeping us dependent on oil for transportation. They will promise us the moon in order to keep us driving on earth. And what they don't say to us in public, they will surely whisper to politicians in private via the lobbies that surround our government on all sides. That's exactly why gate duties are so important—to cut through the blather with economic common sense.

Giving Up the Car

I know, I can hardly face this, either. Though my Volvo has long since been taken to a metal recycling heap somewhere in the south Bronx, my family and I have grown quite attached to our five-door, forest green Subaru Outback with gardening stickers on the back. I won't want to give it up until there are other viable options, but it's difficult to imagine those options as long as we have an automobile. Before I moved to New York City, I never thought I could live without a car, but now I know it's possible, even preferable, given good bike paths and affordable, frequent, congenial public transportation. I have friends who don't drive anymore, and they are happier and richer for it. Ultimately, I know my bonds to the nation are stronger than my bonds to the car. The best choice—the only choice, the choice that makes all other choices possible—is, for most of us, to give up the car.

There. I said it. Let's keep moving.

Moving House

On top of saying good-bye to the beloved pickup or SUV in the garage, many Americans are going to have to relocate, especially folks in rural areas who don't need to be there for their work, and people who live in sparsely populated, economically failing, car-dependent environments. Moving is traumatic because many of us already live in settings we like, or at least in the places where we've grown up. Moving house is also painful financially, because some of us, perhaps many of us, may never see the investment we made in our homes recouped, unless we happen to live in a community with enough density, or a community with the will to build density, so that values might rise once again.

Moreover, our new places may be smaller and less private than what we are accustomed to, with shared walls and tinier yards; there will be less room for all our stuff and for the dog and the dog's stuff. We may end up spending more time outside of the home instead of glued to the television set or surfing the Internet. We will have to find other ways to keep up with the Joneses than owning a mansion on a hill.

It should be said that for some, moving might not be so bad. Many are already on the go in the latest economic dislocation. A lot of folks will see the benefits of higher-density living, jobs included. Having found work, they'll have a shorter commute, as companies move to them and/or promote neighborhood housing to cut their taxes. The best businesses will choose to be established in livable, creative cities, with excellent mixes of commercial, retail, and residential real estate. They will have the choice of the best employees and the lowest payroll costs; they may also see boosts in productivity from healthier, less stressed-out workers.

Towns and cities, in turn, will be competing for businesses and residents. Establishing New Town districts will bring funding to help achieve the density municipalities seek.

As cities become more successful, real estate prices will increase; appreciating property values in improving cities make other cities more cost competitive, so that they might improve. Thus the economic river will spill over into smaller economic streams as Americans actively seek affordable great places. Some small-to-medium sized cities like Chattanooga, Norfolk, and Boise have already invested in their downtowns; other cities will need to get on the bandwagon soon if they don't want to be left behind. And for those towns that choose not to play, it will be a long, excruciating, expensive decline. Look at what happened to Ford's Detroit.

All these changes will also require changes in the way we buy, sell, and finance real estate. Homeownership, as such, may decline, as fewer people will feel that locking into a thirty-year mortgage an hour's drive from the nearest job is in their best interest. Smaller apartments in town will enable smaller mortgages over shorter terms, and rental arrangements may offer more opportunity for many of us. Ecological use fees will reduce property taxes when shared among many. Families don't have to plan to live in a single structure their whole lives; they might move more freely, more flexibly. Real estate will come to seem like less an investment opportunity and more what it is first and foremost—shelter and home. Investment opportunities instead will be found in the marketplace, on Wall Street, and on Main Street.

Living with People

One challenge that will come with living closer together is that we will need to face up to a social and psychological fact clearly in evidence in our democracy: Other people can be really annoying. In the best of circumstances, density is fun and profitable; in the worst of conditions, it foments antipathy for humanity and certain individuals in particular.

It is also true, however, that people are annoying in less densely populated settings. Spreading out, sprawling among spacious lawns and hiding behind shrubby borders doesn't seem to help; in the suburbs, people nurse rivalries and find their neighbors exasperating, too. My grandparents from rural Colorado knew arguments that lasted for generations, even though the closest neighbors were a half-mile away. In the city, in the suburb, in the smallest rural community, it's always possible to find someone who gets your goat.

Excuse the digression, but think of the many ways we have to aggravate each other: People can be annoying because their behavior interferes with how we want to behave, like using loud motorcycles late at night or leaf blowers early in the morning. People can be annoying because they hold different views and perspectives—politics, religion, and culture all provide ample opportunity for difference, misunderstanding, and contention. We think we will be less annoyed by people more like us, so we tend to flock to neighborhoods of socioeconomic uniformity, and then our sons and daughters become annoyed by how similar everyone is. Other folks annoy when they endanger our lives, property, or freedom, like tailgating drivers or motorists racing through traffic. (I *really* hate that.) And people can be annoying when we are forced to interact at an inescapable disadvantage, like the beer vendor at the baseball game who uses his monopoly position to charge an outrageous price, or a pickpocket who uses her nimble fingers to grab wallets, or wealthy people who rewrite the laws in Washington to their own benefit. All very annoying.

If the past is any guide, people are not going to be any less irritating in the future, but we can blunt the effects of this phenomenon through the application of age-old principles: consideration, forgiveness, and, more than anything, communication. We will need to be more public-minded because we will all be spending more time in public. We will need to practice democracy by speaking up when our rights and customs are infringed upon, not just in the private voting booth or in anonymous comments on websites, but on the sidewalk and in the streetcar. Or if it's that bad, you can call the police.

Boundaries of various kinds can help when nothing else will. As the poet Robert Frost said, though "something there is that doesn't love a wall. . . . Good fences make good neighbors." Music via earbud and dark sunglasses provide personal seclusion. Physical barriers, like closed, locked doors and soundproof walls, provide angles of retreat in apartment buildings and offices. Laws give protection sanctioned by society when polite talking fails.

The most valuable deterrent to poor behavior in public turns out to be eyeballs, as Jane Jacobs famously noted. Recently I was downtown when a tourist asked me who told everyone on the subway to walk up the stairs on the right and to drop their bags on the floor when the carriages were crowded. I told him nobody told them; it was just the culture of the subway, enforced by riders watching other riders. We see each other and pay attention to what others are doing in the city. Human beings are social animals and it matters (for most of us anyway) what others think.

Small Moves

While practicing tolerance, we will also have to adapt to a finer-grained kind of economic life, characterized by a larger number of small exchanges rather than a few larger ones. For example, instead of going to the supermarket once a week in our SUV, we will stop at a corner market, and perhaps another store or two, each evening on the way home. Because walking, bicycling, and transit will limit how much one person can carry at a time, we will need to become accustomed to buying what we can conveniently carry or push in a cart or pack into a pannier bag on the bike. Bigger things will have to be delivered. All this will take more time, and the pace of life may seem to slow at first. We will bump into each other more, and spend more time chatting.

The smaller, oftener, closer economy will entail major changes for retail companies, some of which will see their profit models wrecked by the new alignment of transportation and land use. Others will find new profit models generated. The current big-box stores such as Walmart, Costco, Home Depot, and the like are dependent on customers providing the last link in their transportation chain. We drive to the edge of town where land is cheap and parking is free and mandated by zoning, buy what we will, and then deliver the goods to ourselves, at no cost to the retailer. It's all very convenient for them. Because we make those deliveries using motor vehicles, weight hasn't particularly mattered in the past, and so the retailers have adopted strategies of getting us to buy more quantity at once and storing it on our shelves, not theirs. We get lower unit prices, it's true, which the retailers conveniently provide by squeezing producers and labor and the local municipal government, therefore forcing down the economic chain the corrosive effects of economic strategy #3 (make something cheap). Ultimately, those pressures mean a lot of what we buy now comes from China and other countries with inexpensive labor, instead of our own. Even food has become a low-cost commodity as industry conveniently fills eatables with chemical preservatives and wraps them in plastic, so that what would naturally perish lasts practically indefinitely.

In the new world order, Walmart and Co. will have to downsize and propagate to stay in business. Instead of large retail locations on the outskirts, they will need to develop corner shops downtown, selling items in smaller numbers, or create specialty stores. The large companies will bring certain advantages to the neighborhood, like the efficiencies of supply-chain management, but now they will have to compete on a playing field where mom-and-pop–type retail outlets will have some advantages of their own, like personal relationships with their customers. Consumers will benefit from having multiple, smaller vendors to choose from; businesses will benefit from lower capital costs for each new location; and towns will benefit through reviving downtowns that mix employment, retail, and housing. Workers will benefit from having more work.

But we may pay more for the items in those stores. Gate duties on the raw materials used to make the stuff we buy may push up prices, though the absence of sales taxes will help mitigate these costs. Freight and shipping may cost more, too, especially in the early years as we build out the electrical transportation infrastructure and the renewable energy to feed it. Transportation costs will be going up anyway, as the recent past has shown, tar sands and Arctic Ocean oil notwithstanding.

What follows from these changes in price structure will be a cost laid at China's door: an increasing market for American quality over foreign quantity. Better stuff grown or manufactured in America, close to where it is sold, will have an economic advantage over more cheaper stuff dragged halfway around the planet. My hope and pleasure would be for these market forces to revive the dwindling American talent for the artisan trades—potters, tailors, carpenters, pastry-makers and the like—as the small-move economy rewards human creativity and skill, not only for making things more cheaply, but for making them well.

Industrial Changes

Artisans won't be the only ones changing; industrial manufacturers will also need to adapt to this strange new world. Oil companies may make less money than they would have if we maintained our addiction, but they will still do all right. As we have seen, oil is valuable for manufacturing a vast number of chemical commodities. These other uses of oil will not be going away anytime soon, and petroleum will still be needed for jet fuel, at least in the short run. Air travel will be more expensive, which is going to be a problem for global wildlife conservation and other transcontinental activities—but high-speed rail will become more competitive with the airlines, especially for regional trips. Gas stations on the corner will disappear. Oil companies will have to change refinery practices to slice and dice the contents of a barrel in other ways. Somehow, I'm sure they will manage.

More problematic, perhaps, will be adaptation in the domestic auto industry, which, once Ford and Olds faded from the scene, has not had a gift for flexibility. From their position as the most powerful industry in the world, American automobile companies lost market share for decades by insisting on building big, fuel-hungry cars long after the cheap oil window had closed and saturating Americans with messages about why we liked them. As recently as ten years ago, the Big Three (Ford, GM, and Chrysler) were arguing that Americans would never want to drive smaller, sleeker hybrid cars; then Toyota ate their lunch with subcompacts and the Prius, and now all the automakers are bringing out smaller, more efficient cars, even hybrids. GM jumped ahead in electrics, but then suffocated its own baby, the EV1. Despite car companies' intransigence, President Obama decided to save American automakers to save autoworker jobs in 2009; it worked in the sense that the companies didn't go bankrupt. But have they really changed?

Not enough, for sure. The car companies need to re-invent themselves as transportation companies, with capacities beyond the automobile. In the new world economic model, they can build streetcars, trains, or neighborhood electric vehicles while shutting down production of old explosion cars; some might even shift into bicycles. The American automobile industry will have to develop new expertise with electric motors faster than it has heretofore, and it needs to build innovative new designs out of lighter materials to achieve superior energy efficiency. Lighter NEV cars would help at lower speeds around town. And streetcars, if lighter and stronger, could really get a boost from weight reductions, which would also enhance safety and reliability, as Amory Lovins, the doyen of the American sustainability movement, has often argued. Right now the major streetcar companies are European—Skoda and Inekon in the Czech Republic and Siemens in Germany. United Streetcar in Clackamas, Oregon (the only American streetcar maker) has an arrangement to build Skoda designs in the United States. We need to do better: Ford streetcars, GM trains, and Chrysler NEVs need to dominate the world, fed by local steel, tool and die, woodworking, and plastic companies. In the meantime, the American steel industry must continue recycling iron for all the new railways we will need.

Government industries will also need to adjust. Wardens of the public safety—from the police to the fire departments to sanitation—have all adapted to the wide roads and combustion engines of the current era, and now enforce their adaptations on us all through regulations declaring how wide a street will be and what kinds of vehicles are permissible. In the future these public services will need to navigate on narrower access paths, through streetcar and bicycle traffic, and around more tightly spaced buildings. Police will spend less time in cars with radar guns, and more time on foot and on bikes (and so will the bad guys). Ambulances and fire trucks will need to scale down in size and become more numerous, and small clinics, rather than massive hospitals, will serve the hurt and needy. Doctors might even begin making house calls again. The military will downsize, having fewer overseas commitments, as will the military-industrial complex. These are large, painful changes. They will save us a ton of money.

Another industry due for reinvention is construction. The construction industry had quite a run during the Great American Expansion, building new homes out of new materials, but with gate duties in place builders will have to reuse existing materials as much as possible. A new business opportunity exists in tearing down old structures gracefully; expanded markets will open for once-used structural materials. The construction industry will also need to re-engage with architects skilled at designing multifamily homes, apartment buildings, and other structures necessary to bring New Town districts up to density. Some hardhats will find new jobs in ecological restoration, removing unneeded infrastructure that has become a tax burden on landowners (especially the largest of them all, the federal government) and restoring the lands to a semblance of ecological health. We will need men and women skilled with tools and terrain to give nature a helping hand by reviving streams, planting forests, and recovering species. People will have to adapt to new work and find new opportunities to learn how.

Intrusive Government

Earlier, I made various suggestions that might be interpreted as pushy, even intrusive—because they will nudge and poke, incentivizing changes that one might recognize as necessary in the abstract, but might not otherwise desire in the specific. It would certainly be irritating for my payroll taxes to increase because I choose to live a long way from work or not to be able to afford a driving vacation because of a Pigouvian tax. I might get angry, and blame the government for intruding on my life.

The fantasy would be for 314 million Americans to manage the transition to a new safe, sustainable, and surging economy without government having to do anything, but that is impractical. Individual action can't really create the changes necessary without some clues about which way to go. Large-scale interventions certainly weren't sufficient during the age of the Siren song and they won't be enough to build Terra Nova either. Remember history: the Texas Rangers riding in to clamp off the oil supply in Texas fields; the billions of dollars expended to pave roads across a continent; the hundreds of thousands of men and women who have fought in the oil wars— all organized by the government on your behalf, with or without your wholehearted consent, mostly before you were born.

Government exists because as our society grows larger than our individual social bonds can manage, we need formal ways of coming to agreement on joint action, even if we don't all individually agree. The alternative is anarchy. Checks and balances, the president and Congress, the local city council: American democracy as we know it, if decidedly imperfect, is still better than a lot of other ways people have tried (monarchy, tyranny, communism, plutocracy, theocracy, Bill Gates, etc.).

As we act democratically together, let's opt when we can for the economic prod over the regulatory mandate. Bureaucrats will sacrifice some power, but the rest of us will be free to see the true cost of our actions.

Shrinking Government

Finally, I predict that these measures will lead to government shrinking in size and scope, which depending on your perspective, could be a blessing or a curse. As land and natural resources are limited, so will government be, a result of decoupling the public purse from a bubbling and popping economy. Rather than cheerleading for more economic growth, we will look to our political leaders to fulfill an agenda we all share: the common defense, the domestic welfare, and long-term stewardship of the land community. And the costs of these tasks will decrease as the need for foreign interventions diminishes, economic growth of the new sun-train-town model picks up, and the land recovers from its long exploitation.

In this process, power will devolve to towns and cities. Most of the prescriptions we have been considering focus on local government, where decisions about transit, parking, zoning, New Town district applications, and ecological use fees will be made. Vesting more responsibility in our community leaders will require us all to pay more attention to our zoning boards and city councils, housed in the neighborhoods where we are, which will make it easier to keep an eye on them.

There is still a role for larger scale action, of course. State leaders will be called on to pass the enabling legislation for municipalities, establish intercity railways, coordinate among New Town districts, and adopt gate duties on the natural resources in their care. National political leaders will still set agendas, pass budgets, reform the tax code, consult with the Federal Reserve Bank, retract the money supply, foment the smart grid, and represent us overseas. Politicians at all levels will have to stop squabbling. The president and Congress need to lead us into the brave new world, or follow us otherwise.

And what might you do? I pray you encourage your government representatives to reform the tax system so that the economy and nature talk to—not past—each other. I hope you will allow your community to become denser (unless you live in the city already) and if you can't make that happen where you are now, you can vote for density by moving to it. I want you to back your officials in changing the zoning laws to allow apartment buildings and phase out parking* and provide for investments downtown, to ensure compact livable forms, rather than sprawling outward into the next farm. And please support that farm on the edge of town, too, so it stays a farm. Also, I wish that you might do everything you can to walk or bike to work, and if that is too difficult, that you will move closer to your workplace so walking and biking are easier. Support public transportation. Vote for sidewalks and bike lanes and streetcar streets, and then use them, use them, use them every day, with a calm, quiet, friendly, unshakeable American resolve, so that you inspire your neighbors to do the same.

All of these costs and sacrifices are made for a reason: To help us evolve. It will be tough for our generation and probably the next, but the changes we undertake today are providing to future generations a kind of American peace and prosperity that will last as long as the sun shines. And that will have some collateral benefits.

---

*Yes, parking. Parking is the ultimate lever communities have over automobile use in their precincts. No matter how much one loves one's car, you can't drive in it twenty-four hours a day; a car without a parking space is just an expensive hunk of metal with no place to rest, an albatross around the driver's neck. Take away parking and you will change America as profoundly as amending the Constitution or launching a revolution.

15

Collateral Benefits

What you are comes to you.

Ralph Waldo Emerson

I know what you're thinking: Sounds good but it can't happen; it's politically impossible. I know, I know, having heard it from no lesser authorities than my mother, who called to tell me, and the president of the United States, who offhandedly said in a press conference in March 2012: "I don't want the price of gas to go up in an election year, because I know how far some families have to drive." I couldn't have made the point better myself (except that I might have reversed the direction of cause and effect). Mom will love me in any event.

Some may find the unfeasibility of Terra Nova to be its largest potential benefit, because it would require reworking the political logic of the country along more rational and scientific lines for these suggestions to feel ordinary and realistic. Others, more pessimistically, may see the apparent impracticality of these ideas as a sacrifice laid at the altar of a once-great nation, now in irreversible decline. For my part, I think the time will come—sooner rather than later—when circumstances will force us to do something (as if the recent circumstances have been insufficient), and this something might be what we decide to do.

There will be some other benefits as well, and not just for the economy.

A Safer Country

The country will be safer once its transportation energy is provided by electrons instead of hydrocarbons. Oil seems to encourage a kind of narcissistic despotism unbefitting and harmful to the planet and its inhabitants. One does not have to look very far in the current geopolitics of oil to find bullies and seething discontent in Venezuela, Iran, Iraq, Libya, Angola, Nigeria, Bahrain, and Saudi Arabia. Even our friends in Canada, by no means despots, have demonstrated a remarkable voracity by ripping up a goodly part of Alberta to send tar oil south.

Over the decades, the US military has stepped up to the monumental task of protecting "our" oil and oil-producing governments, regardless of political form, in the Middle East. Since 1990 we have kept at least one aircraft carrier group continuously on station at all times in the Persian Gulf; that group requires another eight aircraft carrier groups to be in the cycle of preparation. The costs of this kind of insurance and related military protections have been estimated at over $150 billion per year. Another study, less inclusive in its definitions, suggests the United States from 1990 through 2007 spent $23–$73 billion per year on "peacetime and wartime expenditures" that wouldn't be necessary without our need for Persian Gulf oil. (These researchers also estimate that oil imports to the United States from the region were worth between $10 and $30 billion per year over the same period.) On top of these military investments, we give generously in foreign aid; US grants and credits to the Middle East have averaged over $10 billion per year since 2004. For context, the budget for the City Island K–8 public school is about $2.2 million per year; one year of

Persian Gulf defense and aid money could fund fifteen thousand comparable schools.

On top of the costs of defense, one adds the costs of actively prosecuting the War on Terror. Though the war may be simmering down, the bill continues to grow. Economists Joseph Stiglitz and Linda Bilmes estimated that the costs of fighting the Iraq War, circa 2003–2011, could eventually total $3 trillion dollars, when benefits to war veterans, replacement of military equipment, and loss of economic activity are taken into account. Another accounting puts the costs of war in Iraq, Afghanistan, and Pakistan over the last ten years in the $3.2–$4 trillion range. The Department of Homeland Security, separate from the Defense Department, spent $47 billion to keep us secure in 2011.

These monetary expenses, as massive as they are, pale in contrast to the lives lost and friends wounded in enforcing our peaceable desires in deserts and stony mountains on the other side of the world: 36,715 Americans killed and wounded in Iraq, 2003–2011; 19,639 more killed or wounded in Afghanistan, 2001–September 25, 2012; 2,753 dead on 9/11. For non-Americans, it is estimated that since 2001 over 200,000 people have lost their lives, while another 7.8 million have been displaced from their homes.

An America with windmills, streetcars, and lovely cities might fight wars, but not over oil.

A Healthier Populace

In this new world, we will be feeling better and living longer. The generally pro-motor, anti-exercise, drive-through-window lifestyle practiced in America today has made many of us overweight, and with too much body mass comes a host of other ailments. Stanford Hospital & Clinics, associated with the university, estimates that obesity-related conditions cost the American economy $150 billion in healthcare expenses and lost work time each year, and lead to 300,000 premature deaths annually. The list of related conditions is pretty terrifying: high blood pressure, diabetes, heart disease, stroke, joint problems, respiratory problems, cancer, metabolic syndrome, and psycho-social effects, including low self-esteem, depression, and suicide.

We need to lose weight. 34 percent of US adults aged 20 years and over are over-weight, 34 percent are obese, and 6 percent are extremely obese, according to the Centers for Disease Control and Prevention. Walking thirty minutes a day—about a mile and a half at a moderate pace—cuts the risk of heart disease, eases breathing, combats depression, bolsters the immune system, fights osteoporosis, helps prevent diabetes, and improves sex life. And in the future, it will get you to work. As Oprah Winfrey, the celebrity, tells it: "I've been through every diet under the sun, and I can tell you that getting up, getting out, and walking is always the first goal." Walking also helps us think more clearly and be more creative.

Walking, biking, and train travel are much less dangerous than driving. About three times as many Americans died in car accidents in 1900–2010 as have died in all the wars the United States has ever fought, the American Revolution through Afghanistan inclusive. Speed is the number-one reason why cars are dangerous. The second reason is that we are reckless. One study estimated that a 10 percent decrease in mean traffic speed would result in a 34 percent drop in mortality; at fifty miles per hour a person is twenty times more likely to die in a car crash than at eighteen miles per hour.

We compound the risk of dying in our jalopies by engaging in unsafe behaviors while driving: Fifty percent of fatalities are associated with people not wearing their seat belts. Sixty percent of accidents are associated with people not paying attention to the road because they were talking on their cell phone, texting a friend, reaching for the radio, or checking themselves out in the mirror. And then we make automobility more dangerous by driving when we are drunk or upset, running away from our problems in our chariots of freedom.

Most of these same bad behaviors can also be performed while walking, bicycling, or sitting on the train, but the consequences are less momentous. Hazards while walking include tripping or slipping on ice. Bicyclists need to be careful of their heads and genitals in a sudden stop. Trains do sometimes come into inadvertent contact or derail with devastating consequences; however, by any measure trains are safer than automobiles. According to the National Safety Council, the lifetime odds of dying in an automobile are 1 in 242, compared to 1 in 119,335 on a train. (Other lifetime odds: Death by plane: 1 in 4,608. Death by streetcar: 1 in 3,556,975.)

Another health benefit will be cleaner air, water, and land. Even though you can't see it, carbon monoxide, ozone, nitrogen oxides, and particulates all emerge from the tailpipe along with water and carbon dioxide. Air pollution from automobiles contributes to asthma and other respiratory ailments. Fine particulate matter, common in communities sliced up by expressways, can shorten average life expectancy one to two years. Oil leaking from cars on the street creates what's called "non-point source pollution." (It's not that the oil doesn't come from a point, it's that there are so many points that they are difficult to regulate.) Larger oil spills continue to blacken the news. The accident on the Deepwater Horizon cost BP more than $20 billion in 2010, about half of its profits that year, with unknown long-term health consequences. The US Coast Guard National Response Center received reports of 18,576 oil spills in 2011; 19,176 spills in 2010. Oil refineries are also major producers of industrial waste: 1.4 billion tons per year.

The list goes on and on, but the fact is that less oil in the environment means less interaction with the same, which will enable more of us, and other species, too, to live long and prosper.

A Greener Continent

Now we get to one of my favorites—a decisive move to more efficient land use would free up space for ecosystems to do their jobs better, rebuilding both the natural capital of the nation and the national capital of nature. Cars, it turns out, hit more than people and other cars; they knock into wildlife too. It's been estimated that more than a million vertebrates a day are killed on America's roads. I once murdered a horde of invertebrates—moths, grasshoppers, and other flying insects—on I-70 west of Glenwood Springs in Colorado; so many died in the Great Volvo Massacre that I had to pull over twice in thirty minutes to clean their corpses from my windshield before proceeding. Fortunately engineers and wildlife biologists have collaborated to devise underpasses and overpasses that allow larger animals like deer to safely cross our paths; the same can be easily adapted for trains, which, having a smaller cross-sectional area in aggregate will decrease the toll on butterflies and their kin.

Roads have other detrimental effects as fragmenters of habitat. Some animals, especially of the larger, more majestic kind—like grizzly bears, jaguars, and bighorn sheep—will deliberately avoid areas with too many roads; roaded lands become no-go zones, lost habitat. For smaller wildlife, roads can be outright barriers, or even traps, as when baby turtles tumble into storm drains or can't get up the curb. Fragmentation exacerbates the effects of outright habitat destruction, which also comes with roads. And roads lead to more access, which leads to more buildings, which leads to more development, increased fragmentation, and less nature: It's the familiar pattern we call the human footprint.

But with ecological use fees and urban concentration there will be more room for nature. Our planet does enjoy its open expanses, where wildlife populations rebuild, soil develops, and water cleans and percolates. Functioning nature also lessens loss of human life and property from catastrophic storm and disturbance. One study estimated that restoring 3 percent of the Upper Midwest to beaver habitat would be sufficient to mitigate the terrible floods of recent years on the Mississippi River. Salt marshes south of New Orleans would have reduced the impact of Hurricanes Katrina and Rita, and the same is true for Hurricane Sandy in the Northeast. Moving houses out of flood plains and fire-prone forests, off slippery hillsides, and back from the coast will all reduce predictable damage from natural disaster, while facilitating efforts to protect that infrastructure from the disturbances that nature needs but we could do without—flooding, fires, drought, hurricanes, tornadoes, and the like. In short, the less we are intermeshed in nature's business, the less it will bother us in ours.

But of course, for those of us who want to see what nature is doing, in the American future, green spaces will be closer at hand. Concentrated towns and cities and gate duties on wasteful land use mean that nature can be a walk rather than an airplane ride away. As the land goes through a long period of consolidation and repair, more of us will be able to enjoy its wonders near town, secure in the moral certainty that our demands for habitat do not deny a place to other members of the land community.

A Cooler Climate

Perhaps the best news on the environmental front is that in the absence of legions of internal combustion engines, there will be less greenhouse gases entering the atmosphere. It will be about time. Nothing is more important to the future of nature, and therefore humanity, than to limit these emissions.

As is obvious on a daily basis, climate is vital to the operation of every aspect of our lives and of the lives of everything else on earth. The climate delivers water, wind, and temperature; it provides cues to insects and to plants on when to come, when to grow, and when to leave; it predicates the kind of ecosystems that can exist in a place and sets the rhythm of the seasons. The climate turns the currents in the oceans and ensures we have water to drink and food to eat.

Climate change is deadlier than all other threats from the Siren song combined. It is the one threat with the potential, like nuclear weapons, to make the earth uninhabitable for people. We really don't understand the system well enough to know for sure, but the more dire predictions of climate scientists—such as the Gulf Stream shutting down, mega-hurricanes, and melting polar ice caps followed by surging sea levels—can keep you up at night. (They should keep you out of your car instead, but to date haven't been able to compete with the attractions of the morning commute.)

Meaningful global action on mitigating climate change has been held up almost entirely by the intransigence of one nation: ours. Other countries produce greenhouse gases of course, but until the US Senate ratifies a climate change treaty stipulating the rules for cooperation, it's going to be difficult for other nations, with so much more to lose, to join in. American leadership is required, but at the moment we don't lead, we prevaricate. Take away the dependence on oil-cars-suburbs and our intransigence will also fade. We will be climate change champions with our interests the same as everybody else's, at least when it comes to a temperate climate and a clear blue sky.

Happier Communities

Living closer to work, with less stress, and activated public spaces near home will translate into better community life and stronger neighborhoods. It is indeed possible that some of our neighbors are less annoying than we thought, observing them from the car window; bumping into them on the sidewalk or chatting with them on the front porch will engender conviviality and trust in our fellow human beings. This is what urban planner Jane Jacobs found most satisfying in her Greenwich Village neighborhood in New York in the 1960s—not the fine buildings, the agglomeration economies, the nightlife or symphony, but the small, quotidian experiences of companionship:

"People stopping by at the bar for a beer, getting advice from the grocer and giving advice to the newsstand man, comparing opinions with other customers at the bakery and nodding hello to the two boys drinking pop on the stoop, eyeing girls waiting to be called for dinner, admonishing the children, hearing about a job from the hardware man and borrowing a dollar from the druggist, admiring new babies and sympathizing about the way a coat has faded." Add in a bit of texting and a few tattoos, and it could be any community in America today, once we get out of our cars.

Greater Freedom

Adding to our sense of joy will be more personal freedom. For the dwindling driving public, we will need fewer traffic laws, fewer parking tickets, and less traffic enforcement from behind the shrubbery. On the environmental front, we will need less regulation as gate duties and the ecological use fees do their work; recovering species will make regulations, not critters, endangered. Less oil sloshing around the country, less oil refining, and less combustion means fewer opportunities for pollution, and therefore fewer rules required to curb pollution. Tax returns for businesses will become simpler as sales taxes are lifted and corporate business taxes removed. Special loopholes for the oil companies and other industries will be vanquished in the coming tax overhaul, making enforcement less burdensome and legalistic. Congress won't need to argue with the car industry about fuel economy standards any more. Etc.

You will also have more money in your pocket, which is a kind of freedom in itself. Not only will your transportation expenses be lower, but the economy will be more robust, employment more likely, the public health improved, and the amount of direct taxation reduced. Your consumption will include higher quality choices and greater rewards for every dollar spent. And if you have a good idea for a business, the smaller scale of enterprises will make it easier for you to get started, especially in New Town districts, where government small-business loans and concentrations of customers can be found.

Thirty years of mortgage debt will no longer be a rite of American life, as we adopt smaller, cozier, less expensive homes, with affordable choices in more locales. You can try more places, live in more kinds of situations, know more kinds of people. You can convey identity in ways other than putting stickers on bumpers. On the local scale, walking and bicycling mean you can go about your day at your own speed; and if you want to get to your destination sooner, then you can hop the next free streetcar trundling down the road. Where do you want to go? It's up to you!

Fresher Food

As you decide your course, perhaps you would like to have a piece of fruit from the orchard on the edge of town? Having more compact communities will free up nearby productive fields for growing food again. Buying directly from producers allows fruit, vegetables, and dairy to be harvested later and brought to sale sooner over shorter distances. Fresher food is healthier, tastes better, and makes more money for the farmer, rather than feeding the chains of grasping middlemen and -women. If you have a farmers market near you, you already know about this collateral benefit of wiser land use. It's one part of our new world that we can already taste and smell.

A Revitalized Politics

In these ideas lies the groundwork for a new kind of political dynamic that is neither right nor left, but rather perpendicular to the traditional political axis, and focused on the future rather than rehashing quarrels of the past. I hope liberals will love Terra Nova for its openness and rationality, its emphasis on an assertive public, its equilibrating effects with regard to power and income, and its relief for the planet. Conservatives will support it for its emphasis on sustainable economic growth, its mechanisms for using markets to advantage, its opportunities for independence and personal choice, and because it respects the oldest traditions of all, the traditions of nature.

If we pull this off, then generations of Americans—indeed generations of the world, billions of lives yet to be lived—will come to thank us, Americans of the early twenty-first century, for our costs and sacrifices, for our willingness to evolve, for our ability to dream. It will provide something very large to live for.

A Better World

Where America leads, other nations may follow, or as Abraham Lincoln is said to have said: "My dream is of a place and a time where America will once again be seen as the last, best hope of earth." Though mostly I have written for my American kin, all nations can take part in building a better world.

The good news is that the physical bases are already in place at the global scale for the worldwide energy revolution. The earth's total wind-power potential has been estimated at 72 TW (or 1.7 trillion MacKays); just a quarter of this potential is sufficient to provide as much energy as the entire world used in 2008, nine times the global electricity demand. The solar power potential is even more massive and less exploited. A recent study estimated that concentrated solar power plants connected by high voltage direct current transmission lines could provide 1,386 TW (or 33.2 trillion MacKays) from sunny places to the not so sunny. The geothermal resource base worldwide has been estimated at 1.2 million TW (or 28 quadrillion MacKays; assuming an extraction efficiency of 33 percent). These numbers are so large as to be practically meaningless, except for the lesson: The world has more than enough energy in renewable sources to supply the human population for a very long time.

The bad news is that not all countries, not even all continents, share the natural endowments of renewable energy equally. As with oil and other mineral resources, the boundaries of countries are drawn for reasons that have little to do with wind or sun, yet each country's ecological heritage—whatever its form, potential, and current state—is as essential to success as its history, culture, and politics.

Over time countries have varied in how they have used the blessings of nature. Some nations have been first to a good idea and exploited it to its bitter end; other nations have just been the suppliers of natural resources and received, relatively speaking, little in return for what they gave. The colonization by fiat of the Americas was just an interlude in the colonization of Africa, Asia, and Oceania; once in place, the colonizing powers from Europe were not shy about extracting the colonies' resources and sending them home.

The decolonization of the world in the nineteenth and twentieth centuries has been a story of uneven and wobbly success. Economically, the lingering effects of historical appropriation and empire mean that some former colonial citizens have hundreds of MacKays at their beck and call (for example, those of us in America), while others have none beyond what they provide by dint of their own muscle and bone (most of sub-Saharan Africa). Energy poverty looms large in the lives of over a billion people worldwide. And the cities of the developing world face terrible density-dependent problems. Billions of people worldwide suffer daily in traffic jams, polluted streets, and dismal shanties on the verge of town, yet migrate to cities anyway, longing for a chance to build better lives for themselves and for their families.

Though the details, contexts, and provisions will vary from country to country, the general recipe outlined here can help nearly all, as long as political and economic institutions are tailored to the ecological particulars of each place. The first and last lesson is to balance land, labor, and capital, just as Adam Smith said, in a just and transparent economy. With equanimity, everyone shares a chance for a better world.

I can see a future where ten billion people live as part of the community of nature, where the traditional fighting points over natural resources, ethnicity, and religion are quaint leftovers from another time, themes for history books rather than everyday realities. Nations will continue to compete over who builds the best wind turbine or who designs the most fantastic streetcar, while contesting monumental soccer tournaments around the globe. I even envision a day when other countries take up baseball and give the Yankees a run for their money.

That would make me smile in my northern California retirement, twenty years hence. One Saturday, I have a date for a bicycle ride with my grandson. After grabbing some sandwiches from the weekend farmers market close by, he and I flag down a streetcar to carry us to the edge of town. We hook our bikes on the front rack of the trolley, and my boy's boy gets to ring the bell as we cruise along, leaning out into the sweet-smelling air. At the end of the line, he and I retrieve our two-wheeled steeds, wave good-bye to the driver, and then pedal past strawberry fields, over grassy hills, and through oak woods, until at last we come to blue Pacific waters. We walk hand in hand over sun-warmed sand. Playing at castles, he asks me to tell him a story of heroes. Looking at his expectant face, already grown so much, I think of Odysseus, who avoided the Sirens' grasp and returned safely home. But then I decide on a legend of the new world.

> Now Nanapush, the Strong Pure One,
> The Grandfather of Beings and the Real People,
> Made a great ceremony, the first thanksgiving.
> He called for a helper who would receive and carry the new Earth.
> Taxkwâx, the Turtle, stepped forward and was chosen.
> Nanapush placed the mud obtained by Tamask'was, the Muskrat,
> On the back of the Turtle, and blew his life-giving breath into it.
> Immediately the mud, part land and part water, began to grow;
> It grew and grew until it became the great island
> Where all of us are living today.
> Because Turtle carried the Earth on his back,
> This is why this land on which we live is called
> Taxkwâx Mènâ'te, Turtle Island.

People who came after would call it America.

Notes, Sources, and Elaborations

Chapter 1
How the Sirens Sing

We begin in myth and we end in myth, or as Joseph Campbell (1949) puts it in *The Hero with a Thousand Faces*: "Furthermore, we have not even to risk the adventure alone; for the heroes of all time have come before us, the labyrinth is fully known; we have only to follow the thread of the hero-path. And where we had thought to find an abomination, we shall find a god; where we had thought to slay another, we shall slay ourselves; where we had thought to travel outward, we shall come to the center of our own existence; where we had thought to be alone, we shall be with all the world."

The story of Odysseus's experience with the Sirens is from Johnston's (2000) translation of Homer's *Iliad*. I wish I could say I was the first to use the metaphor of the Siren song, but H. K. Hudson (1963) beat me to the punch.

If you wish to dive right into the problems that oil, cars, and suburbs pose to American society, primary references are Frumkin et al. (2004), Hamilton (2011), Financial Crisis Inquiry Commission (2011), Intergovernmental Panel on Climate Change (2007), MacKay (2009), Millennium Ecosystem Assessment (2005), National Commission on Terrorist Attacks Upon the United States (2004), Putnam (2001), Yergin (1991), and the nightly news.

Not Listening

I grew up in Walnut Creek, California, which was established in the late nineteenth century as an American farming community. Prior to 1849 the land was conveyed by a Spanish grant, recognized by the Mexican authorities, to Juana Sanchez de Pacheco, and prior to that the area was inhabited by Bay Miwok Indians. Our backyard was probably once a riparian forest in the flood plain of the creek, surrounded by oak woods and perennial grasslands; later a walnut orchard; then a suburban housing tract. For more on the historical ecology of the San Francisco Bay Area, see Grossinger et al. (2007a,b) and related publications.

My parents rode the California version of the housing price escalator to success, buying and selling at the right time. To adjust for inflation in the price of their house and elsewhere in the book, I converted the prices to their equivalent value in a given year using the consumer price index (CPI), which is available from the Bureau of Labor Statistics (2012a). The time series for the CPI is indexed so that the average prices from 1982 to 1984 are set to 100, though it can be easily re-indexed for prices in any other year. In certain cases, when I needed a longer time series, I use the indices compiled by Oregon State University political economist, Robert Sahr (2012). As an example, for my parents' former house, I divide the nominal price in 1969 by the annual CPI for 1969 divided by 100, and then do the same for the nominal price in 2003, when my Dad sold, to find the inflation-adjusted "real" prices in constant-dollar units (averaged for 1982–84).

While converting prices for inflation and doing other things, I work for the Wildlife Conservation Society (WCS), founded in 1895 in New York City. Our mission is to save wildlife and wild places around the globe and connect people to nature. Once known as the New York Zoological Society, we currently manage over 500 conservation projects in more than 60 countries, from our headquarters at the Bronx Zoo. At the Bronx Zoo, the New York Aquarium, and three other parks, we educate and inspire millions of visitors each year on the importance of wildlife and wild places to humanity and nature. You can read more about WCS at www.wcs.org.

The human footprint mapping is described in detail in Sanderson et al. (2002) and Sanderson et al. (2006). The human footprint map and last of the wild analysis have become widely used tools in conservation planning: See Leu et al. (2008), Redford et al. (2008), and Thoisy et al. (2010) for examples. You can download the data for your own analysis at www.ciesin.org/wild_areas.

Against the Mast

There have been many books about the reasons behind and consequences of the global war on terrorist extremism, aimed particularly against groups like Al Qaeda based in the Middle East. The National Commission on Terrorist Attacks Upon the United States (2004), Filkins (2009), Coll (2004), and the PBS documentary series *Frontline* (Kirk, 2003; Roberts, 2005) provide useful and compelling perspectives. Bilmes and Stiglitz (2008) provide an accounting of the money spent to fight the Iraq War. I was living on City Island at the time of the attacks. From the beach, we used to mark the southern end of Manhattan by the tall towers standing on the distant island's edge. In my mind's eye I can still see the smoke plumes rising against a clear blue sky. During the writing of this book a new tower has arisen in lower Manhattan.

The 9/11 Commission (2004) inquiries after the attack record that Osama Bin Laden found the deployment of US soldiers in Saudi Arabia objectionable. The 9/11 commissioners write: "In August 1996, Bin Laden had issued his own self-styled fatwa calling on Muslims to drive American soldiers out of Saudi Arabia. The long, disjointed document condemned the Saudi monarchy for allowing the presence of an army of infidels in a land with the sites most sacred to Islam, and celebrated recent suicide bombings of American military facilities in the Kingdom." Bin Laden's (1996) fatwa led to the reprehensible attacks on American embassies in East Africa in 1988, the attack on the USS *Cole* in 2000, and eventually to the 9/11 attacks.

Chapter 2
An Ode to Oil

Elvis's commentary on the sun and truth is ubiquitously quoted on the Internet, though I haven't been able to track down the original source—see BrainyQuote.com.

The concept of nature is more fraught than it needs to be, so to avoid the tangled academic conversations about "what is nature," I define what I mean right up front. If you care to be fraught, try out Cronon (1995).

The Celestial Campfire

Berman (2011) provides a lively current account of the sun and its many virtues; also see fun facts from NASA (nssdc.gsfc.nasa.gov/planetary/factsheet/sunfact.html.) For additional background, turn to textbooks like Zeilik (2002). The tendency of the universe is for energy to spread out and decrease in order, but various "hang-ups" keep that process from happening all at once, which is why we are still here—see Dyson (1971). Woolfson (2000) reviews theories of the origin and evolution of the solar system.

Closer to home, the evolution of life on earth is described in popular form by Zimmer (2009). Blankenship (2010) reviews the evolution of photosynthesis, which started when eukaryotic organisms managed to ingest but not digest cyanobacteria; Taiz and Zeiger (2010) describe how photosynthesis is practiced today by plants. Ecosystem ecology is rooted in the groundbreaking work of the Odum brothers, Eugene and Howard, in the 1950s and '60s; see an up-to-date account of how energy moves through ecosystems, among other things, in Odum and Barrett (2004).

How Nature Makes Oil

Any basic biological oceanography text will tell you about the importance and beauty of plankton—for example, Mann and Lazier (2005). Ironically the plankton that fueled the production of oil so many years ago now appear to be in decline, with lost productivity of approximately one percent per year over the last century, perhaps because of a changing climate, itself caused by the release of carbon once sequestered by fossil plankton and other lifeforms—see Boyce et al. (2010).

The origins of oil were fiercely debated once upon a time. Old theories of abiogenic origins of oil, from methane trapped in the planet's crust, are reviewed in Glasby (2006). Famously Edward Teller, the nuclear scientist, claimed in 1979 that no one understood how oil forms. Teller also thought it might be useful to use nuclear bombs to dig a deep-water harbor for oil shipping in Alaska, a program considered by the US government under Operation Plowshare; see US Department of Energy (1997). Science since Teller's day strongly supports organic origins for oil; see Head et al. (2003), Seewald (2003), and Selby and Creaser (2005) for details. The one-tenth of one percent figure for oil deposits on the coastal shelf is from Deffeyes (2006) citing Carmalt and St. John (1986). We probably won't know the full distribution of oil until all oil is gone.

The End of Oil?

Much of this book depends on energy statistics published by the US Energy Information Administration through its ever-evolving website: www.eia.gov. Particularly important is the Annual Energy Review (AER), which compiles national-level statistics in time series and is published in September each year (e.g. US Energy Information Administration 2012a). Also important is the International Energy Agency's World Energy Outlook, published annually (www.worldenergyoutlook.org). Recent book-length treatments of the state of world oil include Huber and Mills (2006), Roberts (2004), Simmons (2005), and Yergin (2011). Robelius (2007) and Simmons (2006) provide detailed summaries of the twenty-first-century state of giant oil fields; Höök et al. (2009) show how those fields are broadly in decline.

Hubbert makes his case in papers from 1956 and 1982 and many others in between; you can hear him speak in a widely circulated YouTube clip from 1976, from a show originally produced by the American Hospital Association—check out: www.youtube.com/watch?v=ImV1voi41YY. Kenneth Deffeyes, professor emeritus from Princeton University, expands and elaborates on Hubbert's legacy in easier-to-read books from 2006 and 2008. It's important to note that Hubbert defined "discovery" as the cumulative amount of oil produced, plus the known "proved reserves" (also referred to as proven reserves) for a given year; it's more like "oil discovered" at a given point in time rather than "new discoveries" per se.

The main argument against Hubbert's peak is that human cleverness and technological prowess will overcome any shortages, spurred on by rising prices (e.g., Yergin 2011, or earlier, Simon 1996). Hubbert addressed this argument by noting that each generation brings new technologies to bear, so the effect of technological progress is built into the observed discovery rate on which the analysis is built. Hubbert, Yergin, and Simon would all agree, I believe, that though the amount of oil we started with does not change, the economically recoverable fraction does, particularly as the price goes up. As the price goes up, we innovate; this book is one such innovation.

The US Energy Information Agency (2012b) defines "proved reserves" this way: Proved reserves are those volumes of oil and natural gas that geologic and engineering data demonstrate with reasonable certainty to be recoverable in future years from known reservoirs under existing economic and operating conditions. You can see the problem of telling the fortunes of oil given definitions like this: Not only is geological and engineering data necessary to make projections of future reserves, but so are economic and operational conditions, leaving ample room for disagreement.

Although large oil spills like the Deepwater Horizon event are rare, smaller oil spills are relatively common. The US Coast Guard National Response Center received reports of 18,576 oil spills in 2011; 19,176 spills in 2010—see the data at www.nrc.uscg.mil/download.html (filter "Materials Involved" tab by all the codes starting with "O" except "ONG"). The comprehensive report by the National Commission on the BP Deepwater Horizon Oil Spill and Offshore Drilling (2011) describes the spill in mind-numbing detail; a more personal account comes from oil journalist Antonia Juhasz (2011). PBS *Frontline* also has a fascinating video documentary—see Smith and Gaviria (2010). Although the well was deep, as was the water, it's interesting that the Deepwater Horizon platform could go deeper still, capable of drilling wells up to 35,000 feet deep, while standing in waters over 8,000 feet in depth. The platform was an example of extraordinary technical achievement until April 20, 2010, when it blew up.

You can read more about the Canadian tar/oil sands in IHS Cambridge Energy Research Associates (2009) or watch the *60 Minutes* television treatment (Schorn 2006). My WCS colleague J. Michael Fay, also explorer-in-residence at the National Geographic Society, plans a megatransect across the Canadian oil regions in 2013. The Marcellus Shale formation is a source of natural gas that has become economically extractable only recently—see Sageman et al. (2003) for the natural history and Arthur et al. (2008) for a description of rock fracturing ("fracking") and horizontal drilling. Andrews et al. (2009) summarize the technical and policy issues for Congress and the rest of us. For an assessment of Arctic oil, see Gautier et al. (2009); for the politics, see Shoumatoff (2008).

Black Gold

My account of early uses of oil and development of oil in the United States leans heavily on Yergin (1991). McCollough and Check (2010) describe how baleen whales were saved by petroleum-based lighting fuels during the nineteenth century. Black (1998) provides a mordant account of the use of Oil Creek as an improvised oil transportation mechanism. Ida Tarbell, the muckracking journalist who wrote so movingly of the Standard Oil Trust, grew up in the Pennsylvania oil fields and describes them grimly in her 1904 account. Chernow (1998) provides a fascinating blow-by-blow record of John D. Rockefeller's efforts to bring order to the chaos of business in Petrolia. Canadian Abraham Gesner first described kerosene from coal in 1854, and described the description in 1861; Silliman's report is from 1855.

The rule of capture and other legalities of the oil extraction business are but one string in the tangled web describing American conceptions of private property. Freyfogle (2003) helps navigate the tangle; Platt (2004) reveals the interesting intersections between land use, geography, and law; Lowe (2009) provides specifics about current oil and gas law. How the legal system reflects the physical reality of the resource and the ways people want to use it can result in tragedies of the commons if we are not careful, as described by Hardin (1968). Elinor Ostrom, 2009 Nobel Prize–winner for economics, updates the many clever ways that people have invented to avoid the tragedy, or at least limit it. See book-length treatments like Ostrom et al. (1994) or easier-to-digest video explanations such as www.env-econ.net/2009/10/ostroms-take-on-the-tragedy-of-the-commons-.html. I hope this book makes a contribution to that literature.

As one of the many examples of a common tragedy, in the early days of petroleum refining, the gasoline fraction was just poured into the ground or local waterways, creating pollution problems that we still face today, in unreformed "brownfields." Rockefeller recalled of gasoline during the kerosene era (cited in Chernow 1998): "We used to burn it for fuel in distilling the oil and thousands and hundreds of thousands of barrels of it floated down the creeks and rivers, and the ground was saturated with it, in the constant effort to get rid of it." Chernow (1998) also writes that the runoff from refineries made the Cuyahoga River in Cleveland so flammable that if a steamboat captain shoveled burning coals overboard, the river erupted in flames; it happened many times over, from the 1880s up to June 22, 1969, a fire that ignited the modern environmental movement. As a result of that movement, the river is doing much better nowadays—see Maag (2009)—and Cleveland is a better place to live.

Chapter 3
Flexible Power

I borrowed the Ickes quote about the age of oil from Yergin (1991).

A Brief Chemistry Lesson

To learn more about the chemical soup from which your ancestors emerged and how such a remarkable event might have happened, see Miller and Urey (1959), Jortner (2006), and Theobold (2010).

Bubble baths are made of detergents, which are lipids with charged and uncharged ends (which is why they make bubbles in water, circling the wagons as it were, with the charged ends out and the uncharged ends in); these lipids can be derived from crude oil. Plastic lawn chairs and thousands of other useful and not-so-useful products are made from the many varieties of plastic, including vinyl, polyethylene, polypropylene, polystyrene, polyester, nylon, and the like, all of which have antecedents in oil. Candles also are derived from waxes usually derived from petroleum if not bees. Pharmaceuticals and insecticides and other bioactive compounds have organic structures that mimic or trick bodies into doing things, like relieving symptoms (drugs) or falling over dead (pesticides). Bombs are made of trinitrotoluene (TNT), and the toluene comes from oil. Oil can be made into a large number of useful gadgets and materials, including clothes, food, building materials, and your favorite computerized tool. A good background book on materials science, with a fascinating chapter on the uses and abuses of oil and other hydrocarbons, is Geiser (2001); Downey (2009) provides a useful molecular breakdown of a barrel of crude.

Crude Refinement

Downey (2009) goes on to describe the refining process in detail, including a discussion of refinery gain. For a textbook treatment, see Gary et al. (2007). Wang et al. (2004) provide an interesting analysis of the energy inputs to various parts of the refining process, quantifying how much energy is supplied by the fuel stream itself and how much from outside sources (like electricity). The amount of petroleum refined products each year is reported by the US Energy Information Administration (2012a,b). David Blume explains how Model T's ran on alcohol in a YouTube video from 2009 (see www.youtube.com/watch?v=5qDYoEupl28).

Unintended Consequences

Downey (2009) explains more about distillation modes at oil refineries; also see Gary et al. (2007). To go inside the refinery, try Margonelli's (2008) lively account.

If too abundant oil leads to too many discarded plastic bags, perhaps the solution is to make discarding the bags more expensive—see Roach (2003) and Skumatz and Freeman (2006) for discussion of options to price garbage. US Environmental Protection Agency (US EPA) (2010a) reports on the municipal solid waste, but that garbage is only a fraction of the total waste stream; on a mass basis, the industrial waste, which is not reported as frequently, may represent up to 94 percent of the total waste stream—see Dernbach (1990) and US EPA (1987). The EPA estimates the industrial waste, much of which comes from the petroleum refining and chemical manufacturing industry, at some 6.7 billion tons per year (US EPA 2010b).

The story of the saturation of the environment with cheap chemicals is still told most movingly by Rachael Carson's book of 1962. Numerous books have updated the story, for example Marco et al. (1987). How much these chemicals circulating through the environment affect human health is unclear, but what we do know is troubling; see recent reviews by Perera and Herbstman (2011) on prenatal exposures to chemicals including diseases and epigenetics; Soto and Sonnenschein (2010) on the role of endocrine disruptors and health; and Wallace et al. (2011) on the role of genetics and environment in the incidence of cancer internationally. Steingraber (2010), Smith and Lourie (2011), and Williams (2012) all provide more personal, scientific accounts.

All chemicals have a material safety data sheet (MSDS) produced by the chemical industry. Here are a few choice quotes from the Amerada Hess Corporation (2004) MSDS for gasoline:

"Eyes: Moderate irritant. Contact with liquid or vapor may cause irritation.

"Skin: Practically non-toxic if absorbed following acute (single) exposure. May cause skin irritation with prolonged or repeated contact. Liquid may be absorbed through the skin in toxic amounts if large areas of skin are exposed repeatedly.

"Ingestion: The major health threat of ingestion occurs from the danger of aspiration (breathing) of liquid drops into the lungs, particularly from vomiting. Aspiration may result in chemical pneumonia (fluid in the lungs), severe lung damage, respiratory failure, and even death. Ingestion may cause gastrointestinal disturbances, including irritation, nausea, vomiting and diarrhea, and central nervous system (brain) effects similar to alcohol intoxication. In severe cases, tremors, convulsions, loss of consciousness, coma, respiratory arrest, and death may occur.

"Inhalation: Excessive exposure may cause irritations to the nose, throat, lungs, and respiratory tract. Central nervous system (brain) effects may include headache, dizziness, loss of balance and coordination, unconsciousness, coma, respiratory failure, and death. WARNING: the burning of any hydrocarbon as a fuel in an area without adequate ventilation may result in hazardous levels of combustion products, including carbon monoxide, and inadequate oxygen levels, which may cause unconsciousness, suffocation, and death.

"Chronic Effects and Carcinogenicity: Contains benzene, a regulated human carcinogen. Benzene has the potential to cause anemia and other blood diseases, including leukemia, after repeated and prolonged exposure. Exposure to light hydrocarbons in the same boiling range as this product has been associated in animal studies with systemic toxicity. See also Section 11–Toxicological Information.

"Medical Conditions Aggravated by Exposure: Irritation from skin exposure may aggravate existing open wounds, skin disorders, and dermatitis (rash). Chronic respiratory disease, liver or kidney dysfunction, or pre-existing central nervous system disorders may be aggravated by exposure."

Joules, Watts, and MacKays

Energy densities for different fuels can be found in Smil (2007) and the annually published Transportation Energy Databook—see Davis et al. (2011). For more details on measuring energy, see MacKay (2009) and Forinash (2010) or any introductory physics textbook like Halliday et al. (2007). Read in MacLean and Lave (2003) about the energy efficiency of internal combustion engines; Lindeman (1942), Odum and Barrett (2004), and McNab (2002) describe the energetics of ecosystems and organisms.

The Industrial Revolution is a lot more fun than one might think learning about it in school; try Rosen (2010), Weightman

(2010), or the first few chapters of Black (2007). They compare the pitiful efficiencies of early steam engines, which account in significant part for the debt of greenhouse gas emissions in the atmosphere. On a total emissions basis, small England compares unfavorably with large America because of its creative, smoky, early contributions to the Industrial Revolution (MacKay 2009).

The National Institute of Standards and Technology (NIST 2000) website explains about the *Système international d'unités* (international system of units, or SI). The minutes microwaving and the MacKay are not official SI units; I made them up for your convenience; see a list of conversions following these notes. Energy utilization rates of various gadgets can be found listed on the device or by Googling.

The Energy that Drives America

To feel powerful, read the energy statistics reported in the Annual Energy Review. Statistics on US military energy use are from the Defense Energy Support Center Fact Book (US Department of Defense 2009) and a fascinating article by journalist Robert Bryce (2005). Fuel consumption for military planes is from Hoy (2008). I estimated the energy consumption for the Lockheed Hercules C Mk.1P (C-130K) from the website of the Austrian Armed Forces (see www.doppeladler.com/oebh/luftfahrzeuge/c130.htm). Although information in the public domain is sparse, rest assured that the military is very cognizant of its energy usage and working hard to decrease it; for them it can be a matter of life and death as well as money. See Kanigher (2010) and Fein (2011) for Department of Defense leadership on this issue, as well as critics, like Karbuz (2006) and Eggers (2008).

The MacKay-wise summary of American energy consumption is calculated from the US Energy Administration (2012a) Annual Energy Review using data from 2011—see Tables 2.1a–f. and 2.2. Petroleum manufacturing data include coal products, so might be slightly exaggerated; source data from 2006. Population figures are from US Census Bureau (2012a).

Chapter 4
The Cheap Oil Window

Captain Anthony F. Lucas, of the chapter opening quote, was the engineer who brought in the first gusher on Spindletop—the quote is cited in Yergin (1991). This chapter is deeply in debt to Daniel Yergin's scholarship. Less eloquent but considerably briefer is Morgan Downey's "just the facts, ma'am" summary of oil history in *Oil 101* (2009). Isser (1996) is a more academic treatment. Sampson (1975) valuably fills in the personal details of the "seven sisters," the principal oil companies of the twentieth century, from the breakup of the Standard Oil trust to the mid-1970s. Yergin (2011) has a new book *(The Quest)* to bring the story up-to-date, though as often happens with history, the deeper past runs clear, while recent times seem a persistent muddle.

"Drill baby drill" was originally used at the 2008 Republican Party Nominating Convention by Michael Steele, who then became the party leader (Carnevale 2008). Republican vice presidential candidate Sarah Palin picked up the slogan in her debate with her Democratic adversary, Joe Biden, in the fall of 2008. "Drill baby drill," in the ears of some listeners, may echo the chant, "Burn baby burn," called out by rioters in Watts, California, in 1965. Unfortunately the call was heard again during 2012 presidential election campaign (Cappiello 2011).

Where Oil Used to Come From

Yergin (1991) covers the various early-twentieth-century oil discoveries in detail, including the dramatic story at Spindletop. Prior to discoveries in Ohio and Indiana in the 1880s, western Pennsylvania was the only source of oil in the United States, leading to an early prediction by the state geologist of Pennsylvania that oil would run out (Chernow 2004). Then oil was discovered a few valleys over, near Bradford, Pennsylvania, and on the other side of the planet, near Baku, on the Caspian Sea in the Russian empire. The longest time series of oil (and other fuels) production in the US and internationally is from Etemad and Luciani (1991), covering 1860–1985. The US Energy Information Administration (2012a,b,c) and BP (2011) bring the data up to date. The number of oil wells in the United States for 2009 is from US EIA (2012b). Approximately 95 percent of US wells require artificial lift, which means a liquid or gas has to be pumped into the well to get the oil out. Average productivity of American wells peaked in 1972 at 18.6 barrels per well per day; average productivity in 2009 had dropped 46 percent to 10.1 barrels per well per day (US Energy Information Administration 2012a). Recall that a barrel is 42 gallons.

Ida Tarbell's 1904 history of the Standard Oil trust still rings with indignation. Standard Oil's problems had started long before her book, with investigations of its curious "cooperative" relations (for example, Lloyd 1894), but it took a long time to bring down the giant, which then, like the hydra, sprouted thirty-four heads—for details, see Chernow (1998) on Rockefeller. Yergin (1991) and Sampson (1975) provide additional context. Grover Cleveland lost his presidential bid in 1888 describing the dominant political climate of the time: "the Government, instead of being the embodiment of equality, is but an instrumentality through which especial and individual advantages are to be gained. . . . The arrogance of this assumption is unconcealed. It appears in the sordid disregard of all but personal interests, in the refusal to abate for the benefit of others one iota of selfish advantage, and in combination to perpetuate such advantages through efforts to control legislation and improperly influence the suffrages of the people. . . . Communism is a hateful thing and a menace to peace and organized governments, but the communism of combined wealth and capital . . . is not less dangerous than the communism of oppressed poverty and toil." (The quote is cited in Richardson 2008.) Richardson (2008) and Chernow (1998) describe the conditions that led to the Sherman Antitrust Act, which is named after the "Ohio Icicle," Senator John Sherman, brother to Civil War General William Tecumseh Sherman.

In recent times some of the heads separated forcibly in 1911 have rejoined, creating huge concentrations of power and renewed concerns about anti-competitive tendencies, now entrenched on a global scale (e.g. Aune et al. 2010). Chevron assembled a fascinating tree diagram showing the companies forming and re-forming; find it on-line as part of the history of Standard Oil by Droz (2004). Anti-competitive accusations have dogged the industry since 1911. The US Senate investigated in 1952 and firmly lodged the terms "international oil cartel" and "seven sisters" in the vernacular of the industry and the nation. See US Department of Justice (1952) and Kaufman (1978).

Yergin (1991) and Sampson (1975) both described the anti-competitive "as-is" or Achnacarry agreement.

Texas Steps In

Yergin (1991) writes about World War I, the Teapot Dome, Calvin Coolidge, the Red Line Agreement, the Texas Railroad Commission, and the events surrounding the Black Giant field in 1932—see also Smith (2010) on the last. McCartney (2009) provides a recent retelling of the Teapot Dome scandal. For more on the pedantic operations of the Texas Railroad Commission than you probably want to know, pick up Childs (2005). Burrough (2009) provides an entertaining account of the colorful characters, which big oil amplified in the oversized state of Texas.

"A War of Engines and Octanes"

Yergin (1991) also steers us masterfully through oil and World War II; the seven billion barrel figure for the war is from his account, as are the discussions of the technological advances, TNT, bombing raids, dinners between world leaders, and quotes from Ickes and Speer. Yergin is actually quoting Ickes (1943) quoting Stalin on American industry. Hitler was keen on oil from early on; he told two men from I.G. Farben in June 1932, "Today an economy without oil is inconceivable in a Germany which wishes to remain politically independent. Therefore German motor fuel must become a reality, even if it entails sacrifices." For more to chill your soul, see Miller (2002), Goralski and Freeburg (1987), Shirer (1960), or any comprehensive history of World War II such as Roberts (2011).

The Oil of Arabia

Yergin (1991) takes us through the postwar history of oil in the Middle East, including FDR and Ibn Saud meeting on the USS *Quincy,* serial discoveries of oil in the desert, the scramble to get production moving, and the evolving negotiations on prices. Sampson (1975) covers much of the same ground, but with more attention to the corporate players involved; Brown (1999) fills in details on the oil company consortium, Aramco (now Saudi Aramco—see www.saudiaramco.com/en/home.html), where Sampson leaves off; and then Simmons (2005) picks up the story, with a worried account of how Saudi oil has been used and abused. The Federal Trade Commission report on the International Petroleum Cartel released in 1952 is available online at www.mtholyoke.edu/acad/intrel/Petroleum/ftc.htm. Congress would investigate again in 1973–74 to little avail. I coined the term "cheap oil window" after plotting oil and gasoline prices available from the US Energy Information Agency (2011a,b); see sources like Pogue (1921) to extend the time series backward.

50–50

Yergin (1991) again is a great help in deciphering the deal-making around oil up to the 1970s. Sampson (1975) describes the tax deal which made the companies and the Saudis rich at the expense of the American tax payer. *Time* magazine also covered the arrangement (Editors of *Time* magazine 1957).

For those wondering, a placer deposit is "an alluvial deposit of valuable minerals usually in sand or gravel; a lode or vein deposit is of a valuable mineral consisting of quartz or other rock in place with definite boundaries" (Humphries 2002). The Oil Placer Act of 1897 stated that petroleum or other mineral oils were to be governed by the laws relating to other placer claims (Humphries 2004). The Pickett Act of 1910 approved the right of the executive branch to withdraw some public lands for public purposes, which President Taft used to reserve a few oil lands for military use. The Pickett Act was repealed and replaced by the Federal Land Policy and Management Act of 1976, which affirmed that some public lands could be set aside for "a particular public purpose or program" (Humphries 2002). To this day, there exists a Strategic Petroleum Reserve under the ground in northwest Alaska and other localities around the US. The Mineral Leasing Act of 1920 determined how deposits on public lands would be leased. See Humphries (2004, 2006).

In more recent times, to encourage the highly risky and dangerous (and also expensive) work of deep water offshore oil extraction, the American oil industry has applied for various forms of royalty relief and tax incentives, summarized by Congressional Budget Office (2000), Humphries (2007), and Sherlock (2011). These include code words that stand in for billions of dollars of special tax treatment, or "subsidies" if you prefer. Current subsidies include expensing for intangible drilling costs (like the wages, fuel, and hauling associated with oil and gas drilling), reduced amortization periods (as short as two years), excess of percentage over cost depletion (explained below), temporary expensing for equipment used in oil refining, and credit for enhanced oil recovery costs (which allows a 15 percent income tax credit for extracting domestic oil using methods to inject fluids or gases into the ground to force the oil out because the natural reservoir pressure has been depleted, i.e., fracking). In other words if saving elephants was treated the same as oil drilling, I could deduct the costs of the wages, transportation, and food for my conservation work; amortize the cost of my project over two years even if it takes thirty years to complete; write off any elephants lost or killed during the course of the project; and get a 15 percent deduction for any novel conservation techniques I developed in the meantime.

Sherlock (2010) estimates that between 1968 and 2000, these various deductions and credits for the oil and gas industry cost the US Treasury $137 billion (constant 2000 dollars) and another $19.2 billion between 2001 and 2009 (constant 2009 dollars), on top of the extraordinarily low royalty fees the oil companies pay to extract public oil on public lands.

I have to say my favorite tax deduction for the oil industry (and probably theirs, too) is the cost depletion allowance, whereby oil and gas companies are allowed a tax deduction based on how much oil they have extracted. The more oil and gas taken away, the greater the deduction, up until all the oil and gas is gone and the entire capital expense is deducted. This deduction is meant to be analogous to depreciation; for example, as a company uses a truck or a boat its value declines over time through wear and tear, making it worth less to the company. The difference is that, whereas a truck or a boat can be replaced, the nonrenewable oil resource can never be replaced. And the nonrenewable resources in this case are an asset that ultimately belongs to the

work of millions of years of natural action, but is exploited in a matter of a few years by a private owner.

Here is an explanation of how the cost depletion allowance works in Texas from Bryce (2004): "An oilman drills a well that costs $100,000. He finds a reservoir containing $10,000,000 worth of oil. The well produces $1 million worth of oil per year for ten years. In the very first year, thanks to the depletion allowance, the oilman could deduct 27.5 per cent, or $275,000, of that $1 million in income from his taxable income. Thus, in just one year, he's deducted nearly three times his initial investment. But the depletion allowance continues to pay off. For each of the next nine years, he gets to continue taking the $275,000 depletion deduction. By the end of the tenth year, the oilman has deducted $2.75 million from his taxable income, even though his initial investment was only $100,000."

Yergin (1991) describes in detail the nationalization of oil in Mexico, Iran, Venezuela, Libya, and eventually Saudi Arabia. Today most oil resources in other countries are owned and sold into the market by national oil companies. Oil price data are from US Energy Information Administration (2011a).

The Cheap Oil Window

College textbooks cover the basic macroeconomics of prices, inflation, and the money supply developed during the neoclassical synthesis in economics—for example, see Case et al. (2008) and Samuelson and Nordhaus (2009). The classic history of the American money supply is the detailed and provocative work by Friedman and Schwartz (1963). It shows just how tightly the US money supply was controlled after the American Civil War until the early 1970s, despite all the hoopla about the monetary standard in the 1890s and the mistakes made in the Great Depression. Nelson (2007) describes Friedman's influence on US monetary policy in 1961–2006.

Long series of the consumer price index put inflation of the past four decades in context, like a mountain rising above a plain.

Here is the full quote from James Truslow Adams in 1931 on the American Dream: "But there has been also the American dream, that dream of a land in which life should be better and richer and fuller for every man, with opportunity for each according to his ability or achievement. It is a difficult dream for the European upper classes to interpret adequately, and too many of us ourselves have grown weary and mistrustful of it. It is not a dream of motor cars and high wages merely, but a dream of social order in which each man and each woman shall be able to attain to the fullest stature of which they are innately capable, and be recognized by others for what they are, regardless of the fortuitous circumstances of birth or position."

Oil consumption data are from the US Energy Information Administration (2012a). Growth in the gross domestic product is charted by US Bureau of Economic Analysis (2012a), Schurr (1977), Heilbroner and Singer (1998), and Gordon (2005) among others describe the remarkable economic history of the postwar era. Sampson (1975) describes in some detail the inquiries into oil price-fixing in the 1950s, and Yergin (1991) describes the factors that led to the formation of OPEC; also see Skeet (1991).

The Window Slams Shut

Again see Yergin (1991) and Isser (1996) on the Texas Railroad Commission and its changing policies. Yergin (1991) and Sampson (1975) describe the 1973 oil embargo; supplement with Boyne (2002) and Bailey and Farber (2004). See congressional testimony about the '73 oil shock in Committee on Foreign Affairs (1973) and see pictures from that time in Maniaque and Russell's (2008) brilliant *Sorry, Out of Gas*. The dismantling of the Bretton Woods system in 1971 under Nixon followed a run on US gold supplies; arguably Nixon did the only thing he could at the time. However that decision, made in the midst of one crisis, only set up a series of other crises that we still have not resolved. Eichengreen (2008) provides a history of Bretton Woods and its unraveling.

Monetary policy is complicated but fundamental; like the effects of heating up the atmosphere on ecosystems, altering the money supply changes many aspects of economic behavior, sometimes in ways that are difficult to predict.

Economists of Milton Friedman's generation decided that a pro-inflationary monetary policy was good for investment and economic growth, which is arguable, but the political aspects of monetary policy are more problematic. For the politician whom the public unfairly holds responsible for the state of the economy, the temptation is overwhelming during an economic recession to use monetary policy to stimulate the economy; in the short term it's politically cheaper than fiscal stimulus (that is, the government spending money), it shows the government is doing something and it can work, temporarily. See Friedman and Schwartz (1963), Friedman (1970), Bernanke (2006), and Billi and Kahn (2008). The recent rounds of quantitative easing, discussed in Chapter 7, demonstrate the temptations. Back in the 1970s, the Fed increased the monetary base by $4.5 billion between September 1973 and April 1974, an increase of 5.65 percent. The inflation rate in 1974 was 11 percent.

Barsky and Kilian (2004) and Hamilton (2009) review various theories of how oil shocks might be transmitted to the macroeconomy, causing productivity declines, recessions, inflation, and "stagflation," in which high unemployment and inflation occur simultaneously. The relationship in time between oil prices (or shocks) and macroeconomic activity in the United States seem to be strongly connected; higher

oil prices increase inflation and typically result in recessions, as defined by the National Bureau of Economic Research (2010)—see www.nber.org/cycles.html. Economists, including some of the best economic minds of the last thirty years, have spilled considerable ink over how and why oil shocks deliver recession every time.

Here are some theories: increased costs of oil reduce profits when oil is a key component of total costs, decreasing investment (Rotemberg and Woodford 1996); wealth transfers between nations as more money flows from oil-consuming to oil-producing countries, leaving less capital available in consuming countries like the US for other economic activity (Olson 1998); increasing oil prices cause sectoral shifts, where higher prices decrease consumption of related products, like automobiles, propagating declines to other sectors as oil prices increase (Hamilton 1988); a "waiting effect" model where firms wait for oil prices to subside or to stabilize before investing, causing productivity declines (Bernanke 1983); a monetary effects theory, where the Federal Reserve Bank adjusts monetary policy to mitigate the effects of the oil shock but overcompensates, tipping the economy into recession (Bohi 1989; Bernanke et al. 1997); a "wage-price spiral" based on the theory that firms set wages in response to prices in the marketplace (gasoline prices are considered highly visible in this context), higher wages cause inflation because more money is available for expenditures, and thus wages and prices feed off each other, causing an upward spiral on both (Bruno and Sachs, 1985); and/or a capital loss of energy-intensive (or inefficient) equipment, which is no longer viable at the higher fuel price, resulting in loss of invested capital for production in the short run, and therefore declines in production until new capital replaces the old equipment (Barsky and Killian 2004).

What seems most compelling, however, is the theory argued

in the text and borrowed from Hamilton (2009, 2011, and earlier papers): As demand increases over a tightening supply, prices go up; because the demand for oil is relatively inelastic, the amount of money available in the economy for other activities decreases, perhaps through the mechanisms described above, resulting in lower overall production, recession, and inflation. See also: Edmonson (2005) and Radchenko (2005).

Of course all these effects can and will be exacerbated by financial speculation on the price of oil (Yergin 2011; Hamilton 2009). Following the weakening and then removal of federal government price controls in the early 1980s, a strong and perhaps all-too-creative commodities market has developed in oil, especially on the New York Mercantile Exchange. Investors buy oil futures contracts as a financial asset, betting that the price will go up before delivery is due; their activities drive spikes in price as they hold oil off the market and take profits (Masters 2008; Downey 2009). During the 2007–08 oil spike, these effects, in some combination, appear to have been sufficient to tilt the global economy into recession, particularly as the real estate fundamentals, as discussed in Chapters 6 and 7, appear to have been simultaneously exhausted.

Oil prices in 1979 are from Hamilton (2009) for West Texas Intermediate; oil consumption and trade deficits in the '70s are calculated from US Energy Information Administration (2012a). Klare (2005) lays out the political connections between oil and national security, including the Carter Doctrine. The Carter Doctrine built directly off the Eisenhower Doctrine, stated in a 1957 special message to Congress on the Middle East:

"It [Eisenhower's proposal on the Middle East] would, first of all, authorize the United States to cooperate with and assist any nation or group of nations in the general area of the Middle East in the development of economic strength dedicated to the maintenance of national independence.

"It would, in the second place, authorize the Executive to undertake in the same region programs of military assistance and cooperation with any nation or group of nations which desires such aid.

"It would, in the third place, authorize such assistance and cooperation to include the employment of the armed forces of the United States to secure and protect the territorial integrity and political independence of such nations, requesting such aid, against overt armed aggression from any nation controlled by International Communism."

President George W. Bush put it this way in 2002, in what is now known as the Bush Doctrine (cited in National Security Council 2002): "The security environment confronting the United States today is radically different from what we have faced before. Yet the first duty of the United States government remains what it always has been: to protect the American people and American interests. It is an enduring American principle that this duty obligates the government to anticipate and counter threats, using all elements of national power, before the threats can do grave damage. The greater the threat, the greater is the risk of inaction—and the more compelling the case for taking anticipatory action to defend ourselves, even if uncertainty remains as to the time and place of the enemy's attack. There are few greater threats than a terrorist attack with WMD.

"To forestall or prevent such hostile acts by our adversaries, the United States will, if necessary, act preemptively in exercising our inherent right of self-defense. The United States will not resort to force in all cases to preempt emerging threats. Our preference is that nonmilitary actions succeed. And no country should ever use preemption as a pretext for aggression."

These doctrines, repeated by multiple presidents over the last sixty years, are the rationale for massive US public investments through military and foreign aid programs in oil-rich Persian Gulf regions since World War II, including the first Gulf War of 1991 (Atkinson 1994) and the Iraq War of 2003–11 (Filkins 2009). See notes to Chapter 15 for the figures. To be fair, Donald Rumsfeld, as Secretary of Defense at the initiation of the Iraq War, called suggestions that the US is really after Iraq's oil "utter nonsense" during an interview in February 26, 2003 (cited in Rhem 2003). He went on to say: "We don't take our forces and go around the world and try to take other people's real estate or other people's resources, their oil. That's just not what the United States does. We never have, and we never will. That's not how democracies behave."

That's not how democracies should behave, anyway.

Nine Percent

The US Energy Information Administration (2012a) Annual Energy Review provides summary data on petroleum imports; details are available on-line: www.eia.gov/dnav/pet/pet_move_impcus_a2_nus_ep00_im0_mbbl_m.htm. The figures cited in the text are for crude oil imports as a percentage of total consumption in 2009. Gasoline prices are also available on-line from the Energy Information Administration at www.eia.gov/petroleum/gasdiesel. According to Belasco (2011) the fiscal year 2009 appropriations were $95.5 billion for Iraq and $59.5 billion for Afghanistan. The US population in 2009 was estimated at 305 million people.

Oil company profits are reported as part of the National Income and Product Accounts Tables from the US Bureau of Economic Analysis (2011a). The data series runs from 1929 to 2010. US oil production figures are also given in the 2011 AER and the data in Etemad and Luciani (1991). For those of you optimistic about the recent surges

in shale oil and gas production, they are a function of higher prices enabling more expensive extraction techniques, i.e. just another verse in the Siren song, cursing your kids as our parents cursed us.

Chapter 5
Time for Space

The opening quote is cited in most biographies of Henry Ford—for example, Brinkley (2004). Ford had a compelling way with words as well as cars.

Oil was cheap when I drove across the country in 1998 because the markets were confused how much oil there was. As Simmons (2005) recounts, at an OPEC meeting in Jakarta, Indonesia, in November 1997, members approved a production quota increase as part of their continuing campaign to meet mounting oil demand; however the oil markets perceived the quota increase as evidence that there was too much oil in the market, so people sold and the price of oil began to drop. Dropping prices fed further speculation of an oil glut and a search for the "missing barrels," with various accusations made of countries hoarding oil around the world. *The Economist* magazine even got into the act, running a cover story in March 1999 entitled "Drowning in Oil." Finally Saudi Arabia, Venezuela, and Mexico, losing money over the low prices of oil, instituted a production cut, and the price of oil surged threefold over the next eighteen months. Oil has not been so cheap since.

To calculate the energy demands of my trip east, remember that one gallon of gas holds 34 kWh of energy, and a 1000 W microwave uses 1 kWh every 60 minutes. The calculation in the text is based on driving 4,052 miles with an average fuel economy of 18 miles per gallon (about what my old 1977 Volvo got in combined city and highway driving), which implies that an estimated 225 gallons of gasoline was combusted, releasing energy equivalent to 7,643 hours of microwaving (or 0.87 years of microwaving). 225 gallons of gasoline is the same as

5.4 barrels of gasoline (one barrel = 42 gallons). Those 5.4 barrels of gasoline would be refined from 11.9 barrels of crude oil assuming that American refining yields 45 percent gasoline from each barrel of crude oil (US Energy Information Administration 2010a). Adjusting for refinery gain of 107 percent (Downey 2009) and assuming that the refinery is 88 percent energy efficient (that is, some crude oil is consumed for energy in the process of refining—see Wang et al. 2004), then approximately 12.7 barrels of crude oil were refined to produce 225 gallons of gasoline to drive me 4,052 miles across the country.

In 1998 US crude oil extraction in the field was 8,011 thousand barrels per day and imports of crude were 9,764 thousand barrels per day (US Energy Information Administration 2012a).

To estimate the flying energy costs, I assumed that I might have flown in a Boeing 737 direct from Sacramento to New York–La Guardia Airport, traveling 2,506 miles with 117 other passengers. According to figures available from Boeing (2010), a 737 with that occupancy rate would expend 0.59 kWh per passenger per mile. Naphtha-based jet fuel holds 37.77 kWh/gal according to Davis (2011), which means my personal share of jet fuel for that trip was 39.2 gallons, or 0.93 barrels. Only 0.088 of each barrel of crude yields jet fuel, which means that 10.6 barrels of crude had to be refined, which after adjusting for refinery gain and efficiency, as above, results in 11.3 barrels of crude oil to make my trip by airplane from California to New York. Those barrels could be the same ones that yield gasoline for the cross-country driver, of course.

Finally my stuff moved by truck in a standard 20-foot dry container weighing 4,850 lbs with a capacity to hold 62,350 lbs of stuff. Assuming I shipped 1,500 lbs of things and the container was only 80 percent full when it made the trip, then my share of the total load was about 3 percent on a weight basis. Total distance traveled by truck from Davis to Los Angeles, and from LA to New York is approximately 3,194 miles according to Google Maps. I assumed that my stuff traveled by diesel truck, and that truck made 5.8 miles per gallon, estimated from data in the Vehicle Inventory and Use Survey (US Census Bureau 2002). In total, the truck used 550 gallons of diesel, but my stuff's share of its journey was only 3 percent, or 16.5 gallons, containing 678 kWh of energy. (Diesel fuel has 40.95 kWh per gallon according to Davis 2011.) Twenty-two percent of the average barrel of crude becomes diesel fuel, so those 16.5 gallons required refining 1.9 barrels of crude, after adjusting for refinery gain and refinery energy efficiency.

Vehicle miles traveled and other transportation statistics are compiled annually by the Bureau of Transportation Statistics (or, US BTS 2012). See recent reports on leveling trends in VMT by Puentes and Tomer (2008) and Young (2010). The National Household Travel Survey data for 2001 and 2009 is available on-line at nhts. ornl.gov. Freight statistics are from US BTS (2006, 2012) and Dennis (2007); see summary by US General Accounting Office (2011).

Trading Time for Space

Einstein would say that timespace is the fundamental dimension (singular, not plural) of the universe, but for the purposes of this book, let's assume we live in a Newtonian universe where and when we can think of time and space separately (for description, see Dolnick 2011 on Newton's view and other aspects of sixteenth-century science).

Weightman (2010) provides an overview of the industrial revolution and its economic consequences. Railroad statistics are from US Census and Social Science Research Council (1949). The remarkable ease by which steel wheels roll over steel rails is reported in MacKay (2009), backed up by Lindgreen and Spencer (2005). White (2011) and Ambrose (2001) provide histories of the transcontinental railroad. Draffan (1998) provides a blow-by-blow history of US railroad land grants; see additional details in Gates (1979). We forgot how contentious and wildly munificent land grant policies in the second half of the nineteenth century were. Draffan (1998) quotes Senator Howell in 1870 arguing against further subsidies to the Northern Pacific Railroad: "I now wish to prevent a perpetual monopoly of over 50,000,000 acres of lands by an immense railroad company… I hope that the American Senate … will not by their action here to-day cause their posterity to curse their memories for thus building up such an immense monopoly to the detriment of the country, to the oppression and injury of all who may settle in that region." The generous railroad land grants set off remarkable booms in land values—see for example, Coffman and Gregson (1998) and the histories cited above.

The quote from Henry Adams about a boy in 1854 is from his own biography of 1907. Adams, the descendent of two presidents, was ringside for the enormous change, exultations, and disappointments of the second half of the nineteenth-century. For a shorter, but highly useful, gloss on the Gilded Age and its effects on society, see Putnam (2001). For the economics of deflation, see Friedman and Schwartz (1963); for a nineteenth-century perspective, see Wells (1889). Deflation and inflation of prices both have their costs; recent generations of economists have tended to prefer the risk and experience of inflation over deflationary pressures, and we have reaped what they sowed, but deflation can be unpleasant as well—see Krugman (2010) for a succinct summary why. For more on the silver standard, populism, and William Jennings Bryan, see Goodwyn (1976). Measurements of personal wealth of the Gilded Age plutocrats are from Phillips (2003). The quote from William James about the bitch-goddess is from a letter to W. Lutoslawski in 1920 (see Hardwick 1993).

New Ways to Move

The activation energy required to ignite a chemical reaction is a staple of undergraduate chemistry class (cf. Gonick and Criddle 2005), and provides a useful analogy for getting things started in other areas of endeavor. For your information, the spark to set in motion combustion of iso-octane (a surrogate for gasoline) is 175 kJ per mole (a mole is ~6 x 10^{23} molecules); it results in enthalpy of combustion of 5460 kJ per mole. Not bad! Chemical reactions are governed by the Arrhenius equation; I wish I knew the analogous formula for affirmative social change.

My brief history of internal combustion transportation technologies is cobbled together from Yergin (1991), Wolf (1996), and Black (2007). Black (2007) follows the history of the Selden patent in all its outrageous turns with more original research than his provocative book's title would suggest lies within—also see Greenleaf (1961). Grayson (2001) provides a pictorial history of early internal combustion engines. Ahlfeldt and Wendland (2009) highlight the economic importance of transportation innovations.

A Spark

For basics on the physics of electricity, see Gonick (1992) and Forinash (2010). For the history of science of electricity see Baigrie (2006); Ben Franklin's part is remembered in Schiffer (2006). Baldwin (2001) and Israel (2000) contribute excellent biographies of Thomas Edison. John D. Rockefeller's empire started in Cleveland, which was conveniently close to the oil regions of Pennsylvania, but access to money, seaports, and consumers (especially overseas markets) led Standard Oil to shift the center of its operations to New York City. Rockefeller didn't move to New York himself until 1883, a year after Edison's experiments that would fundamentally rock his world—see Chernow (1998).

Black (2007) details the development of the electric battery; see also treatments by Schiffer (1994) and Kirsch (2000). The quote from Edison about the unreliable qualities of the secondary battery is cited by Black (2007) from the February 17, 1883 volume of *The Electrician* magazine. Edison would later develop his own secondary battery when converted to the cause of electric transportation after the turn of the twentieth century. There is no shortage of dramatic, and in retrospect, quite forward looking, exclamations about electric transportation from that era. Another trade journal, *Western Electrician*, exulted in January 1898: "Electricity is the natural medium for the application of motive power. Its supply is unlimited. It is everywhere. It is to movement what the sun is to growth."

The First American Cars

The history of the development of streetcars is told in Miller (1941), Schiffer (1994), and Jones (2010), and remembered in many nostalgic picture books, including Diers and Isaacs (2007) and Meyers (2005). The development of electric cars is told in Black (2007), Kirsch (2000) and Schiffer (1994), which nicely makes the connection between bicycles, batteries, cars, and streetcars. The problems with urban horse-based transportation are memorably described by Tarr and McShane (1997). The quotes from Pedro Salom are cited in Black (2007) from an essay Salom wrote for the *Journal of the Franklin Institute* in 1896. Shukla et al. 2001 provide a recent appraisal of electric vehicle propulsion prospects, and the commercialization of neighborhood electric vehicles, hybrid cars, the Chevy Volt, and the Nissan Leaf show that even in the twenty-first century, electricity can propel vehicles.

Ford's Gift

For histories of Henry Ford, Ford Motor Company, and the early days of the American motor industry see biographies by Brinkley (2004) and Lacey (1988). Bryan and Ford (2002) provide lots of fun details and pictures of the Fords and friends, including Thomas Edison and William Burroughs. See also Black (2007) and Kirsch (2000) for ladies' preferences in terms of vehicles. The quote from Ford about a car for multitudes comes from the launch of the Model T in 1908—Brinkley (2004) cites Garrett (1952). The quotes from an essayist, from future president Wilson, and from the *Breeder's Gazette* are also from Brinkley (2004). The quote from Burroughs is from 1913 and is found in Bryan and Ford (2002). Burroughs, Ford, Edison, and Harvey Firestone became fast friends, and took family car camping trips together in 1918–20.

Black (2007) and other chroniclers (cf. Paine 2006) have made a big deal about the failed electric car collaboration between Edison and Ford, citing a mysterious fire at Edison's research facility during the summer of 1914, but the evidence for foul play seems equivocal to me; it could have been the same technical difficulties that still give electric car advocates headaches. See Strohl (2010) for another perspective.

Parallel Infrastructure

Statistics on American road development, registered vehicles, and licensed drivers are provided in a series of Highway Statistics publications available at www.fhwa.dot.gov/policy/ohpi/hss/hsspubs.cfm (US Public Roads Administration 1947; US Federal Highway Administration 1967, 1977, 1987, 1997, 2011a). To see the current state of affairs, download data from US Census (2010). The Good Roads Movement and other bureaucratic, engineering and political developments critical for creating the modern American road network are covered academically by Gutfreund (2005) and delightfully by Swift (2011), including the rise and fall and train riding habits of the Good Roads Movement. Herlihy (2006) covers the role of bicycles, with many pictures and illustrations that catch turn-of-the-century spirit; the Editors of *Bicycle Magazine* (2003) provide a loving photo-documentary history of the bicycle. Some are still trying to extend the Interstate Highway System—see Dellinger's (2010) account of the effort to build I-69 from Canada to Mexico. Black (2007) is back with details of the conspiracy to rid the country of streetcars; here he is clearly on the mark since General Motors was convicted in a court of law. See Klein (1996) for the film documentary version of the story in *Taken for a Ride*—clips are available on YouTube. US Forest Service roads are described in Coghlan and Sowa (2000); the distance from road statistic is from an analysis by Forman (2000).

Driving America

For more on the social psychology of cars, see academic treatments like Seiler (2008) and popular ones like Kay (1998). In various essays and books, Lewis Mumford was particularly vociferous on the pernicious aspects of car culture—see volumes from 1961, 1968, and 1979. In more recent times, James Howard Kunstler has taken up the call: *The Geography of Nowhere* (1994) and *The Long Emergency* (2006) are notable for their eloquent vitriol. For a more positive view of the influence of cars on American culture, see Hinckley (2005) and Wollen and Kerr (2004), or talk with any of your neighbors. According to the American Time Use Survey conducted by the Bureau of Labor Statistics (2011d), the average American civilian spends 1.15 hours per day traveling on weekdays; 1.10 hours per day traveling on weekends. Over the course of a year, that adds up to approximately 415 hours of travel per person, or 2.47 weeks. Linguist George Lakoff (2006) addresses the words and metaphors of freedom; for mental models revealed through language, see Lakoff and Johnson (2003). There is much to be learned from the metaphors we use and how we use them.

Not Driving

The philosopher Arthur Schopenhauer expressed the conundrum proposed by free will this way in his essay on the *Wisdom of Life* (1901): "Everyone believes himself *a priori* to be perfectly free, even in his individual actions, and thinks that at every moment he can commence another manner of life. . . . But *a posteriori*, through experience, he finds to his astonishment that he is not free, but subjected to necessity, that in spite of all his resolutions and reflections he does not change his conduct, and that from the beginning of his life to the end of it, he must carry out the very character which he himself condemns." How true!

The differential in who drives and who doesn't is perhaps starkest in the commuting data—see Pisarki (2006) and McGuckin and Srinivasan (2005). *The Economist* (2012a) summarizes how the young and the rich are eschewing the car. The American Automotive Association (AAA) publishes costs of driving surveys each year—see the latest at www.aaaexchange.com. For example, AAA (2012), as cited in text.

Transportation planners have spilled a lot of ink and research dollars over the question of what induces us not to drive, considering all the problems posed by the driving lifestyle—see Taylor et al. (2009), McDonald (2007), Levine et al., and references therein. Handy et al. (2005) discuss why Americans choose to drive excessively. Shrank et al. (2011) provide data on urban congestion trends. Economists Glaeser et al. (2008) make the case for the poor moving to the city because of economic and transportation concerns; Glaeser enlarges the argument in his 2011 book on the triumph of the city. Salon (2009) discusses the choices made by New Yorkers as a case study. The percentage of car owners can be found in the Census for New York County—see Kazis (2011).

Chapter 6
The Great American Expansion

This chapter draws heavily from the pathbreaking work of Kenneth Jackson's (1987) book on the crabgrass frontier, including the opening quote. Other important histories about the suburban experience over time include Duany et al. (2000), Hayden (2004), Gillham (2002), and the second chapter of Newman and Kenworthy (1999), which provides an international perspective.

What People Want

Defining a suburb and estimating how many people live there is surprisingly difficult. The estimate of the number of people in the suburbs cited in the text is derived from the 2010 Census, using the method suggested by Altshuler et al. (1999). See further details below.

The cited paper about residential choice is Fernandez et al. (2005). Related academic studies of where people want to live and why, of which there are many, include Brown and Robinson (2006), Cho et al. (2008), Dietz (1998), Ioannides and Zabel (2008), and Pinjari et al. (2007); you can also ask your local realtor. Ironically, even after all the re-sorting into the suburbs, more than half of Americans wished they lived somewhere else—see Taylor et al. (2009).

The point about durability of urban infrastructure is made by Glaeser and Gyourko (2005). Students of urban form document city structures that have lasted over thousands of years (cf. Kostof 1993), which underscores why building cities with flexibility and livability in the first place is so important. See advice in Lynch (1984), Gehl (2010), and Alexander et al. (1977) for examples and wisdom. Knowing how to build livable cities isn't the problem; creating the conditions to actually build them is.

For historical analysis of the conditions surrounding home building, see Jackson (1987) and Hayden (2004), and for a more contemporary example, Rybczynski's (2007) entertaining account of the last harvest of a farm in Pennsylvania. Despite the harrowing bureaucracy of land use control, it is remarkable how much influence, and therefore, responsibility, real estate developers have in shaping the human landscapes of the future. As Rybczynski shows, interests in the development community are multiple, sometimes contradictory, and can include considerations beyond the pecuniary, but don't usually.

It is instructive to remember just how difficult life can be for new immigrants, especially in the nineteenth century, as I was reminded recently reading a biography of Jacob Riis (Buk-Swienty 2008). We don't want to go back. Let's go forward.

Density Dependence

Population data are from the US Census (2012a), available on-line at www.census.gov/popest/estimates.html and through the Census Bureau's factfinder website: factfinder2.census.gov, including for City Island which is in the Bronx, New York, Census Tract 516, in zip code 10464. Historical population figures for New York City are from Jackson (1995).

There is a vast literature on the economic effects of urbanization. They are reviewed in more detail in the notes for Chapter 10, but to whet your appetite, try out the effusive Glaeser (2011) or the more convincing Jacobs (1985). That literature should be read alongside the literature on density dependence: the classic ecological reference is Andrewartha and Birch (1961). Also see Fowler (1981) and Brown (1995). Livi-Bacci (2006) provides a useful review of past human population trends, and the UN (2011) previews the future. The quote from Riis is from 1901; see his book republished with additional photographs in Riis and MCNY (1971). Unsafe, crime-ridden density is not what we want either. Interesting in this context are recent observations of how New York has become one of the safest big cities in America; see Zimring (2011).

The First Suburbs

George (1879) appreciated the dynamics of real estate markets earlier than most and proposed a remedy. Hayden (2004) and Jackson (1987) describe the development of the streetcar suburbs, building off the work of Warner (1962) in Boston. Statistics on the expansion of the streetcars into the suburbs are from Social Science Research Council (1949) and Steuart (1905). Hayden (2004) makes the connection between Sears and Roebuck and prebuilt houses, and the development of the early savings and loan associations offerings of thirty-year amortized mortgages; see other references therein. The etymology of mortgage suggests it is a pledge made to the death; the length of these loans certainly makes it feel that way. During the Great Depression, President Franklin Delano Roosevelt pushed legislation to extend mortgages to larger swaths of the population and to stem foreclosures through the Home Owners Loan Corporation, and later, the Federal Housing Association—see Jackson (1987) for details, or for a television version, Jakabovics (2010).

"Build and Build"

Jackson (1987), Hayden (2004), Rybczynski (2007), and Kushner (2009) will all introduce you to the Levitts and their creations. The "buy all you can" quote is from Kushner (2009). Hales (2009) provides additional details and some wonderful images on-line. You can read more about the Hempstead Plains in Harper (1911). The number of housing units constructed over time is tracked in the statistical abstracts collected by the US Census Bureau 2012a,b).
I probably underplay the role of the home mortgage interest tax deduction in my account of the expansion of suburban housing after World War II. It was a critical incentive, not only making more capital available affordably to homeowners (but not renters), while also expanding the market for banks, who reaped the benefits of all those 30-year debts ordinary

people signed up to. Jackson (1987) and Gillham (2002) do a better job.

The Great American Expansion

I'm sorry to say: Defining suburbia is much harder than it needs to be, because of the various definitions of urban and rural (but never suburban) over the years employed by the Census Bureau. It seems almost criminal in retrospect that we don't have a more concrete quantitative description of the most important land use trend of the twentieth century. See the current definitions in the Federal Register (US Census 2011a), and US Census (1995).

The suburban figures are from Altshuler et al. (1999). Modern computer mapping however allows one to map census data across density, and therefore circumvent definitional problems partially. For example, Hammer et al. (2004) used the census data to show people moving out of the cities and clustering in what most people would call suburbs between 1940 and 1990. For another approach, see Pozzi and Small (2001). Radeloff (2010) took this data further by creating a high resolution map of housing units across the nation; his interest was to study housing growth near parks and other protected areas; my colleague Kim Fisher and I used that same data (provided graciously by Volker Radeloff) to demonstrate the growth of housing in rings expanding out from the city for all the metropolitan regions in the country. For comparison, McDonald et al. (2010) estimated that 3.5 million acres of open space was lost in US metropolitan areas between 1990 and 2000, with per capita land consumption rates of 15,650 square feet per person over the decade. See also Acevedo et al. (2006).

The practice of redlining was set in motion by the federal government during the Depression. For example, the US Federal Housing Administration's underwriting manual (1938) included provisions about race: "Prohibition of the occupancy of properties except by the race for which they are intended. . . . Schools should be appropriate to the needs of the new community and they should not be attended in large numbers by inharmonious racial groups."

Redlining was outlawed by the Fair Housing Act of 1968 and the Community Reinvestment Act of 1977; the latter required that banks use the same lending criteria in all communities. For more on redlining, see discussion in Jackson (1987).

The near collapse of American cities in the late twentieth century manifested in many pernicious ways, from increased crime, to fiscal mismanagement, to landlord abandonment, to urban blight. I know the literature best for its terrible consequences in the Bronx, but by no means are the phenomena limited to my borough: See Mahler (2006), Smith (1999), and Jonnes (2002). The simultaneous decline of small-holder agriculture is told with poignance in Pollan (2006) and Conkin (2008). Also see Linder and Zacharias (1999) on Brooklyn. Berry (e.g. 1996) provides essays on the cultural changes this economic transition has wrought. Cultural critics of the suburbs include Kunstler (1994) and Mumford (1968) as noted above.

The good news is we seemed to have learned from our mistakes; we know how to make better cities and better agriculture, and even to combine them (e.g. Donahue 2001).

The Price Escalator

Historical housing prices are studied through indexes, like the Case-Shiller Index (Shiller 2006). Shiller's data can be downloaded through the website for his book: www.irrationalexuberance. com. Additional data can be had from commercial companies, e.g. Standard and Poors (2012). See also studies by Davis (2009) and Davis and Palumbo (2008). The value of total housing stock nationwide as measured over time is from US Bureau of Economic Analysis

(2011b); average values by state are available from the US Census Bureau (2011c). Housing as a percentage of the consumer price index is from US Bureau of Labor Statistics (2012b).

California Proposition 13 is more formally known as the "People's Initiative to Limit Property Taxation" and was an amendment to the State of California's constitution (www.leginfo.ca.gov/cgi-bin/waisgate). Article 13 now reads in part: "(a) The maximum amount of any ad valorem tax on real property shall not exceed One percent (1%) of the full cash value of such property. The one percent (1%) tax to be collected by the counties and apportioned according to law to the districts within the counties."

And: "SEC. 2. (a) The "full cash value" means the county assessor's valuation of real property as shown on the 1975–76 tax bill under "full cash value" or, thereafter, the appraised value of real property when purchased, newly constructed, or a change in ownership has occurred after the 1975 assessment."

Fixing tax rates to a historical time period is an interesting way to finance a government, and not one that has worked out very well for California. See for example Martin (2009) or in shorter form, Lewis (2011).

It has become a truism that housing leads the US economy out of recession, as *The Economist* put it in 2011: "There are two things everyone knows about American economic recoveries. The first is that the housing sector traditionally leads the economy out of recession. The second is there is no chance of the housing sector leading the present economy anywhere, except deeper into the mire. In the two years after the recession of the early 1980s housing investment rose 56%; it is down 6.3% in the present recovery. America is saddled with a debilitating overhang of excess housing, the thinking goes, and as a result is doomed to years of slow growth and underemployment."

Chapter 7
The Crescendo and the Crash

I haven't been able to track down the original source of Jack Welch's quote, despite its frequent repetition on the Internet; search BrainyQuote.com. Welch's injunction to change seems the flipside of Gandhi's oft-cited: "We need to be the change we wish to see in the world" (also on BrainyQuote.com).

Clinton's Song

Population and housing statistics are from US Census Bureau (2012a). Katz (2010) details the drive for homeownership in the 1990s and early 2000s, and its consequences for many American homeowners, including predatory subprime loans, house flipping schemes, and mortgage fraud. She led me to decisions made in the Clinton White House (see US Department of Housing and Urban Development 1995) and carried forward by Bush II (see Becker et al. 2008). John Snow, US Treasury Secretary under President Bush, said in late 2008: "The Bush administration took a lot of pride that home ownership had reached historic highs. But what we forgot in the process was that it has to be done in the context of people being able to afford their house. We now realize there was a high cost." The standard interpretation of rising house prices before the housing bubble can be found in the second chapter of Shiller (2006) or in Dokko et al. (2011). Note the link to money supply.

President Clinton's comments on homeownership are available on-line at www. presidency.ucsb.edu/ws/index. php?pid=51448#axzz1nVfxfCEO and the resulting National Homeownership Strategy adopted in 1995 (US Housing and Urban Development 1995). Interesting in light of later events, Clinton started out his briefing on homeownership with sharp barbs at Congress, then considering a counter-terrorism bill promoted by the Administration. He predicted: "You can be sure that terrorists around the world are not delaying their plans while we delay the passage of this bill. It is within our reach now to dramatically strengthen our law enforcement capabilities and to enhance the ability of people in law enforcement to protect all kinds of Americans. We have an obligation to do that." Unfortunately the bill died in committee six years before 9/11; some of its recommendations would be taken up in a more extreme form with the Patriot Act of 2001.

The Hidden Cost of Housing

The account of the increasing relative value of land is from real estate economist Morris Davis. His incisive analysis is described in a 2009 paper; the data itself is available from morris.marginalq.com. See supporting analysis by Davis and Palumbo (2008), which found that the value of residential land accounted for about 50 percent of the total market value of housing in 2004, up from 32 percent in 1984. Other analyses of land prices include Nichols et al. (2010) and Kocherlakota (2010). All agree that land prices have appreciated considerably over time, peaking during the early 2000s, and crashing dramatically since 2008. Rybczynski (2007) provides an anecdotal but detailed account of the difficulty of finding greenfields to build on in the early 2000s, and therefore the impact on prices; be sure to read the two afterwords. Willis's (2006) advice from before the crash reads with a certain kind of pathos now.

Rodkin (2005) contributed the review of Chicago real estate, working off the 2005 report of the Office of Federal Housing Enterprise Oversight to Congress.

Land to Build On

The Sanderson and Fisher analysis of buildable land takes advantage of the National Land Cover Databases, which can be downloaded from www.mrlc.gov; see Fry et al. (2011) and Homer et al. (2004) for details. Topographic information is available from the National Elevation Dataset available at ned.usgs.

gov; see Gesch et al. (2009). The distribution of government and other privately protected lands is from databasin.org/protected-center/features/PAD-US-CBI; see Conservation Biology Institute (2010).

Describing legal, as opposed to geographic, land use restrictions is a challenge given the large number of jurisdictions involved, but see the work by Gyourko et al. (2008) and Pollakowsi and Wachter (1990).

The Practical Commuting Horizon

The definition of the practical commuting horizon is mine and should be considered a rule of thumb. Pisarski (2006) is the conventional authority on commuting in America. The costs of commuting are measured annually by Texas Transportation Institute—see Shrank et al. (2011). Vehicle occupancy statistics are from the National Household Travel Survey (Federal Highway Administration 2011b); household vehicle ownership from the Transportation Energy Databook (Davis et al. 2011). Commuting times are from Pisarski (2006) and suburb density from Altshuler et al. (1999). Gas prices in 2009 constant dollars are from US Energy Information Agency (2011b, 2012b).

Sexton et al. (2012) bring together oil, cars, suburbs and the financial crisis in a way not dissimilar to my analysis.

The Crash

Detailed conventional explanations of the Great Recession include Faiola et al. (2008), Sorkin (2009), and Lowenstein (2010). For the inside perspective, see Paulson (2010). The official account is from the Financial Crisis Inquiry Commission (2011). Bezemer (2009) pulls together accounts of economists who predicted the collapse before it happened, but no one paid attention. Data presented in Davis (2010) show how the twin triggers of falling house prices and increasing unemployment triggered foreclosures across the country. Maps produced by the Center for Housing Policy and

partners show a disproportionate number of foreclosures on the outer peripheries of the suburbs (CHP et al. 2011).

Employment statistics are regularly reported by the US Bureau of Labor Statistics (2012c) and homeownership and poverty data by the US Census Bureau (2012a,d). Quantitative easing is a process by which the Federal Reserve Bank buys government bonds from private banks and pays for them with new money created electronically. (They don't actually print new banknotes, but electronic money spends just like any other kind.) Data on the monetary base is provided by the Federal Reserve Bank of St. Louis (2012c). For brief, mordantly amusing explanations of how these open market operations are made, read or listen to Davidson and Blumberg (2010) and Kestenbaum and Joffe-Walt (2010). Thornton (2010) describes potential downsides of quantitative easing including the risk of long-term inflation and extension of high unemployment. Although inflation has been relatively low in 2008–2012 because of the recession, the Fed needs to be mindful of positive feedback loops that might trigger hyperinflation and the social unrest that comes from everyone's money becoming valueless (see Shirer 1960 for comparative example in Germany in the 1920s). Theoretically the Fed can pull the money back and destroy it through open-market operations with the small group of large banks that they foist the money on in the first place. Ironically many of these institutions are the same ones that played such important roles in the 2008 financial collapse; see Federal Reserve Bank of New York (2007, 2012).

Economic vital signs (Figure No. 39) from the following sources: The media regularly reports the Dow Jones Industrial Average (Federal Reserve Bank of St. Louis, 2012a); employment status for US civilian non-institutional population (US Bureau of Labor Statistics, 2012c); and the Federal debt (US Office of Management and Budget, 2012a). Although less regularly noted, but just as important for our democracy, we should monitor the money supply, as indicated by the monetary base (Federal Reserve Bank of St. Louis, 2012b); income inequity, now at quite extreme levels historically (Alvaredo et al. 2012); and the concentration of carbon dioxide in the atmosphere (Tans and Keeling, 2012). Note that the Dow Jones is not adjusted for inflation; prior to 1947, employment statistics are for workers over thirteen years of age and after, for workers over sixteen; and the government debt is not adjusted for inflation. Also note that the monetary base is not seasonally adjusted or for inflation (obviously), but is adjusted to minimize discontinuities associated with regulatory changes in reserve requirements. Income inequity is indicated by the percentage of all income earned by the top one percent of US earners (that's right—one percent of the people rake in over one-fifth of total earnings) and includes capital gains. Atmospheric trends are the same on Hawaii (Mauna Loa) as they are in the Antarctic and show a definitive upward trend since the beginning of the Industrial Revolution. Seasonal greening and dieback in the northern hemisphere accounts for annual variation. (For more, see Intergovernmental Panel on Climate Change, 2007.)

The American Presumption

Nowadays you can learn how land is created and shaped in any textbook of geomorphology, for example, Goudie and Viles (2010). Ecosystem services are described in any standard ecology textbook such as Odum and Barrett (2004). Although there is more yet to learn in ecology, we know enough to act responsibly.

Many historians tell the tale of American discovery and settlement; in many ways, it is *the* story of America, at least the one that underlies all the others. Classic references for the exploration and discovery of America include Sauer (1971), Sauer (1980), and Morison (1971). Weber (1992) fills in some of the Spanish history often not told in more Anglo-American–focused accounts. For how the

natural abundance of the continent shaped the perceptions of the founders, see Drake (2011). For how the nation shaped nature, see Opie (1998).

For an eclectic history of drawing political lines over ecological ones in America, read Linklater (2003); also see Neunzert (2010). Henry David Thoreau, the nature writer, made some money as a surveyor; it's fascinating to read Chura's (2011) account of the contradictions and complementarities of a surveyor-naturalist.

The story of the Native Americans who met the Europeans has also been reclaimed, though it's not clear we have learned the lessons thoroughly enough. To refresh, see Crosby (1973), Cronon (1983), and Mann (2005, 2011). To understand the conflict of perspectives firsthand, read William Penn (of Pennsylvania) try to make sense of the Lenape he found living in his new colony, available in Myers's edition (1970). To hear accounts of contact and beyond from the Native American perspective, read Nabokov (1999) and Blaisdell (2000).

Land preemption (the practical application of the American presumption) once took up much attention by American political writers and theorists (see Johnson 1951 and references therein); a sign of the times, the term today is mostly used with respect to federal law trumping local and state. In any case, having obviated any prior claims, the principle of preemption set in motion the dramatic land, economic, and social history of the American state. Summaries are available from Anderson and Martin (1987), Draffan (1998), US Bureau of Land Management (2012), and Linklater (2003). Dennen (1977) takes you into the economic implications.

Dewey (1918) provides the data to show how exceedingly cheaply the US government sold the American land during the nineteenth century, yet it ran on proceeds of land sales for nearly 120 years. (Tariffs and duties on imports were also

important revenue sources.) The government, as much as the citizens, benefited from the vast territory of the country, territory which is now entirely parceled off and settled, the practical result three hundred years later of a philosopher's penned thoughts on divine and natural rights. For the modern situation, read Banner (2011) and Freyfogle (2003).

Read John Locke's essays in a critical edition with an excellent introductory essay by Laslett (1960). Although Locke's thinking was critical for how private property was formulated in America, his immediate problem in 1690 was replacing the Divine Right of Kings. Before the English chopped off Charles I's head in 1649, property flowed from God, through kings, to men. If men could replace the king, then how could property, given by God through nature (or nature through God) be justified? Locke's solution was to give people natural rights, inherent in their existence. That idea found a ready audience in America, recited in a more dramatic form back to the British crown in 1776: "We hold these truths to be self-evident, that all men are created equal, that they are endowed by their Creator with certain unalienable Rights."

The problem with Locke is not that men and women have rights, but that those rights carry with them concomitant responsibilities. Expressing those responsibilities through the economic system is the topic of Part II.

Part II: Terra Nova

Chapter 8
Holding Council

The opening quote is one of many retellings of the myth of Turtle Island; in this case, adapted from Hitakonanu'laxk (2005). See also Bierhorst (1995), Converse (1908), and Snyder (1974) for others.

Chapter 9
Gate Duties

Aldo Leopold is arguably the greatest American conservation

thinker, and the concept of the land community, alluded to in the chapter opening quote, is his greatest idea. Read recent biographies of him by Newton (2008) and Meine (2010), or read him in his own easy, elegant prose (Leopold 1949, 1953, 1999).

Muir webs are described in my book on Mannahatta (Sanderson 2009). The idea of networks is one of the most powerful and influential in science today—see popular accounts of the science of networks by Barabasi (2003) and Buchanan (2002). The human footprint map is cited in the notes for Chapter 1.

Adam Smith's Pins

Though Adam Smith's *The Wealth of Nations* crackles with wit in places, it can be plodding in others—Heilbroner's edited essential version from 1987 provides the most important bits, with helpful, witty commentary. Heilbroner (1999) also gives a fascinating comparative biography of major economic thinkers, the worldly philosophers, including Smith. Phillipson (2010) provides a recent biography, which cites a newspaper description of the Professor in his latter years: "a cathetic [that is, debilitated] habit, his appearance was ungracious, and his address awkward. His frequent absence of mind gave him an air of vacancy, and even of stupidity." Nevertheless his writing cast a spell on the world that has never been broken.

Smith, by the way, can be interpreted as an advocate of living close to work, which he sees useful to minimize time otherwise lost to "sauntering." He writes in one of his *Letters in Jurisprudence* (1766), reprinted by Heilbroner (1987): "There is always some time lost in passing from one species of labour to another, even when they are pretty much connected. . . . This is still more the case with the country weaver, who is possessed of a little farm; he must saunter a little when he goes from one to the other. This in general is the case with the country labourers, they

are always the greatest saunterers; the country employments of sowing, reaping, threshing being so different, they naturally require a habit of indolence, and are seldom very dexterous. By fixing every man to his own operation, and preventing the shifting from one piece of labour to another, the quantity of work must be greatly increased." He later makes the same point in Book 1 of *The Wealth of Nations*.

Robert Solow's Nobel Prize–winning work on economic growth stems from two papers in 1956 and 1957. He begins his 1956 essay this way: "All theory depends on assumptions which are not quite true. That is what makes it theory. The art of successful theorizing is to make the inevitable simplifying assumptions in such a way that the final results are not very sensitive. A 'crucial' assumption is one on which the conclusions do depend sensitively, and it is important that crucial assumptions be reasonably realistic. When the results of a theory seem to flow specifically from a special crucial assumption, then if the assumption is dubious, the results are suspect." I quite agree.

The 1957 paper delivers a quantitative analysis of American economic growth, according to the theory laid out the year before; the 1957 analysis focuses on the "aggregate production function," written as: $Q = f(K,L;t)$, where Q is output, K is capital, and L is labor, measured over t, time. Where is N, natural resources? Nowhere to be found. In a paper from 1974, in the midst of the controversy stirred up by the *The Limits to Growth* (Meadows et al. 1972), Solow takes up the question of natural resources in the economy and states that "If it is very easy to substitute other factors for natural resources, then there is, in principle, no problem. The world can, in effect, get along without natural resources." (And if wishful thinking can't be substituted for oil? Yeah, anyway. . .) Solow dissects this assumption, considers how technology will affect natural resource extraction, and ends by worrying whether society can ever effectively estimate the correct

discount rate for the future value of natural resources. The discount rate is the amount of profit one can make at some point in the future divided by the rate of profit one can make today. Solow finally, wearily, concludes that "Maybe the safest course is to favor specific policies—like graduated severance taxes—rather than blanket institutional solutions." That is, even Solow eventually comes down on the side of gate duties.

For another perspective, open textbooks in ecological economics by Daly and Farley (2010) or Common and Stagl (2005). Popular treatments of ideas in ecological economics include Hawken et al. (2000), Braungart (2002), and McKibben (2008). Older but valuable contributions in the traditionalist line include Dasgupta and Heal (1974), Stiglitz (1974), and Hotelling (1931). Natural resource economics writers make similar points, though from a less ecologically oriented, and more consumption based, view; they worry that resources are going to run out if used at unsustainable rates, therefore cutting into profits, therefore we should be efficient in use of natural resources and take into account the full costs: See textbooks by Conrad (2010) and Field (2008).

The Household and the House

The circular flow of the economy is explicated at the beginning of many introductory economic textbooks—see for example Figure 1 in Mankiw (2008) or Figure 3.1 in Case et al. (2008). These diagrams typically show exchanges between households and firms, which I have relabeled as consumers and producers, but the idea is the same. Compare the economic circular flows to ecosystem flows in texts like Aber and Melillo (2002). A beautiful paper of economical and ecological analogy by Bloom et al. (1985) compares a plant and a business.

The problem of how to establish a price measured in money, given conditions of supply and demand, is a key contribution of economic thinking. After reviewing Adam

Smith in Heilbroner (1987), other key texts include Stigler (1987), Friedman (2007), and other sources cited in the aforementioned economic textbooks, including the texts on ecological economics. Appadurai (1988) provides a contrasting perspective, drawing from cultures around the world. Also see Graeber (2011) on debt.

When it comes to natural resources, ecologists are not concerned about use per se, but rather use in excess of the rate of renewal. In the 1970s, these concerns were expressed forcibly by Meadows et al. (1972), which sparked a bitter, but ultimately useful, discussion of how much human innovation (the "ultimate resource" sensu Simon 1996) can compensate for ever accelerating rates of finite natural resources.

The argument continues, for example, in the firestorm lit by Lomborg's (2001) *Skeptical Environmentalist*. For me the bridging point between the practically infinite capacity for human creativity and the unmistakably finite physical world we live in, is prices; the prices of natural resources will go through the roof before the actual amount of them literally runs out. But as explained in the next section, if prices include the full costs of use, they will signal a more sustainable balance between supply and demand.

Gatekeeping

We take a lot from nature. Crude oil consumption is from US Energy Information Administration's Annual Energy Review (2012a). Carbon dioxide emissions from petroleum combustion were estimated in the 2011 US Greenhouse Gas Inventory (US Environmental Protection Agency 2011).

Wastes from oil exploration, extraction, and refining are unfortunately not tracked on a regular basis; most are considered exempt wastes under Subtitle D of the Resource Conservation and Recovery Act of 1976 (amended in 1984). The most recent assessment of the amount of these wastes was made in the late 1980s (US Office of Technology Assessment or OTA 1992), which estimates that 1.4 billion tons of industrial waste were produced in 1985 by crude oil and natural gas exploration and processing activities, not including "produced waters" that were recycled for enhanced oil recovery (i.e. pumping waters from wells back underground to push more oil to the wellhead). Additional details are provided in US Environmental Protection Agency or EPA (1987). Total industrial nonhazardous wastes amounted to 11 billion tons in 1985 (OTA 1992). Municipal solid wastes (i.e. garbage) are much better characterized than industrial wastes. US EPA (2010a) estimated that 243 million tons of municipal solid wastes were produced in 2009 (compare to 11 billion tons from industrial processes). Approximately 12.3 percent of those 243 million tons were discarded plastics, or approximately 29.9 million tons, in 2009.

In 2007 the United States applied 1,133 million pounds of active ingredients of insecticides, herbicides, fungicides, and other pesticides on agricultural lands, lawns, and gardens, approximately 22 percent of global pesticide usage (Grube et al. 2011). See also Smith and Lourie (2009).

Pharmaceutical wastes are not as closely tracked as they should be, especially given the rising number of reports of frogs with too many legs and other unnatural mutations (see review book edited by Kümmerer 2008). An Associated Press investigation (2008) estimated that 250 million pounds of pharmaceuticals and contaminated packaging are discarded by hospitals each year. That estimate does not include industrial wastes from making pharmaceuticals or from doctor offices and pharmacies, or the pharmaceutical by-products excreted after human consumption. It does not include veterinary or illegal uses of drugs or personal care products such as creams or lotions. For further details see Barber et al. (2005), Brooks et al. (2009), and Celiz et al. (2009).

A Better Way to Tax

Externalities were first identified by the great British economist of the turn of the twentieth century, Alfred Marshall. Marshall (1890) wrote:

"Not only does a person's happiness often depend more on his own physical, mental and moral health than on his external conditions: but even among these conditions many that are of chief importance for his real happiness are apt to be omitted from an inventory of his wealth. *Some are free gifts of nature; and these might indeed be neglected without great harm if they were always the same for everybody; but in fact they vary much from place to place.* [Italics mine.] More of them however are elements of collective wealth which are often omitted from the reckoning of individual wealth; but which become important when we compare different parts of the modern civilized world, and even more important when we compare our own age with earlier times."

Arthur Pigou took Marshall's place at Cambridge University, picked up his idea of externalities and expanded it through his book *The Economics of Welfare* (1932). I haven't found (oddly) a good authoritative biography of Pigou; however a concise one is available on-line (see Henderson 2008.) You can also read Mankiw's "Pigou Club Manifesto" on-line and see the membership list (gregmankiw. blogspot.com/2006/10/pigou-club-manifesto.html). Caplan's (2008) brief on externalities provides rapid entry into the standard economic treatment of the subject; Cornes and Sandler (1996) provide the textbook treatment.

Most economists—Pigou among them—will tell you that ideally the costs of externalities should be equal to the costs borne by others, whether those people live in the future or represent squirrels in the tree. But how to assess those costs is difficult. One cannot ask a squirrel to speak any more than one can ask the future. Some economists worry that if you can't develop a way to exactly value the cost of the externality, the taxes may create other economic distortions, which could be worse than the original problem. Fair enough. The traditional neoclassical response to the problem of externalities comes from Ronald Coase of the University of Chicago in 1960, in what has since become known as the "Coase Theorem."

Coase theorized that externalities would not be a problem if the transaction costs were zero. Transaction costs are expenses incurred from making an economic exchange—for example, the costs of searching for the cheapest product in the market, negotiating with the other party to acquire it, and then making sure it is safely yours. At the time of his study, Coase was thinking about the problem of radio frequencies then being auctioned off by the government. Conceivably after buying a frequency, one radio station could broadcast a signal strong enough to swamp the signal of another station with a nearby frequency on the radio dial. Coase used some mathematical reasoning to show that as long as the two stations could communicate with each other, then they should strike a mutually advantageous deal, regardless of who owned what radio frequency first or how much they paid for it. Coase's theory is often mentioned as justification to extend private property rights because if everything is owned by someone, then someone will have an interest in its efficient use. According to this line of reasoning, the best way to avoid the tragedy of the commons is to delete the commons entirely through privatization.

Unfortunately Coase's theory, while mathematically correct, even clever, is founded on two bad assumptions. The first dubious assumption is the obvious one: Transaction costs are never zero, a point Coase readily admits. These costs can be minimized, but no one has yet found a way for two humans to transact without some cost.

More importantly, Coase makes the social assumption that the two parties in conflict can talk to each other—that they have the capacity to make an economic transaction, or at least, have a negotiation. But the hard part of the problem of externalities is of parties unable to communicate, whether we mean the externalities foisted by people today onto people in the future, who may curse or applaud their ancestors, but cannot meaningfully negotiate with them; or we are discussing the land community, wonderfully still amid our unending chatter. Trees, squirrels, and the water cycle do not talk in our terms; they have no legal standing; so if we value their participation in the land community, then we will have to speak for them. Economist Steven Cheung (1973) attempted a demonstration of the Coase Theorem by paying attention to how a beekeeper and an apple farmer negotiate over the positive externality of bee pollination; the bees didn't get a say and so their welfare was neglected, even though they are the ones doing the pollinating. Because externalities are more often between parties that cannot negotiate, negative externalities are passed down the generations, passed out of America to other parts of the world, and passed from humanity to long-suffering nature.

Another attempt to wiggle out of the inevitability that resources are not infinite comes from Paul Samuelson, Robert Solow's close colleague at MIT, who defined public goods as those that are not owned privately and which do not diminish as the result of use (Samuelson 1954). Think of air. As you can imagine, real public goods, formulated this way, are quite rare. More common are excludable goods, like oil and land, which lead straight away to negative externalities under the current economic framework.

Gate Duties

Gate duties as proposed are a general form of externality payment, also known as a severance tax or a natural resource consumption tax. For some other examples of costing externalities, particularly negative ones, read Russell (2011), Skumatz and Freeman (2006), Stoft (2008), Tietenberg and Lewis (2008), and Lewin (1982). George (1879) had a similar remedy for poverty in the midst of progress. The greatest successes have been with the negative externalities of wastes, in particular pollution and garbage. Cap-and-trade proposals are a kind of soft form of externality pricing, though without the prices, it becomes necessary for government to pre-set allowable emissions, a kind of regulation (US EPA 2010d; Houser et al. 2008). Carbon taxes are a limited form of gate duty, focusing on only one waste from combustion. A related line of study on the positive side is the economic evaluation of ecosystem services, like flood protection or provision of clean water: See reviews by Wilgen et al. (1996), de Groot et al. (2002), and Sutton et al. (2012). Valuations like these are inexact, complicated, and, as suggested in the text, never "correct." Instead we will have to adaptively manage them through the political process. (Don't roll your eyes at me! This is what it means to live in a democracy.) Adaptive management is a structured, repetitive process to make decisions in the face of uncertainty and changing conditions; it's what we do in conservation all the time—see Wagner et al. (2007). For those wondering, the amount of carbon dioxide in a gallon of gasoline is explained in US Environmental Protection Agency and Department of Energy (2012).

Taxes and the Law

A readable account of the current US tax system is Slemrod and Bakija (2008). The US Bureau of Economic Analysis (2012b) provides regular summaries of tax receipts at local, state, and federal levels, and from the websites of individual states, counties, and municipalities. Johnston (2003) provides a stunning account of how the rules are made. One of the side benefits of adopting a system of gate duties would be the ability to wipe away the spider's nest we have made of our tax code. Search, read, and weep over the federal code itself at US Internal Revenue

Service (2011). In 2010, local and state governments collected $438.3 billion in sales tax, and the federal government collected $72.9 billion in excise taxes. State and local government collected $57.9 billion in corporate income tax, and the federal government collected $329.6 billion. The federal government does not collect property tax, but state and local governments collected $430.6 billion from property owners. These estimates give an idea of the magnitude of revenues that would need to be generated from gate duties.

Entry, Exit, and Use

The proposed rules of thumb for setting gate duties arise out of basic principles of conservation biology, which in turn arise out of good sense. See Groom et al. (2005) for the basics.

Costanza et al. (1997) is the standard reference for the value of the world's ecosystem services, but it has its many critics, which Costanza and colleagues try to address in Costanza et al. (1998). Related work estimated that the value of the US Fish and Wildlife Refuge System in the contiguous US at approximately $26.9 billion per year (Ingraham and Foster 2008). Loomis et al. (2000) found that some people in the Platte River Basin would be willing to pay to enhance ecosystem services, and that the potential payments exceed the potential costs by a margin of 10:1. Balmford et al. (2002) in a review article estimated that an effective program of global conservation would pay benefits of 100:1, at a cost of approximately $45 billion per year. For comparison purposes (and not to suggest we could get by without these things, but . . .), the US population in 2011 spent $63.6 billion on "games, toys, and hobbies"; $94.9 billion on "tobacco" products; and $51.6 billion on "hair, dental, shaving, and miscellaneous personal care products" according to the US Bureau of Economic Analysis (2012c). See Jones (2008) and West and Williams (2004) for discussions of an optimal gas tax.

Ecological Use Fees

The scientific literature brims over with examples of how human land conversion for buildings, agriculture, mining, transportation, and other purposes influences the ecology of the site. The effects, the scale, and the extent of our understanding are all quite astonishing—see McNeill (2001), Millennium Ecosystem Assessment (2005), and Mackenzie (2010). To know what to do about it, you might feel better to read Groom et al. (2005).

The technology for map overlay analysis suggested in the text is conveniently available via geographic information systems, or GIS; this technology is commonly used for military, real estate, natural resource extraction, and natural resource conservation activities: See Ormsby et al. (2010) or Maantay and Ziegler (2006) for introductions.

Historical ecology has produced a large number of studies of past environmental conditions, suitable for analysis using GIS, for example, Grossinger (2012), Beller et al. (2011), and Sanderson (2009). Over larger areas, potential vegetation mapping may be suitable, for example, Brown et al. (2007), Ricketts et al. (1999), and Slaats (2000). Such maps even exist at the global scale, for example, in the work of Olson et al. (1997). For an introduction to the precious heritage of the United States, see Stein et al. (2000).
We also already have a standardized national system of land cover classification, with well-developed methods for satellite observation: See Fry et al. (2011) and Homer et al. (2004). These national land cover databases, with vintages from 1992, 2001, and 2006, would be appropriate for county or state level assessments; probably higher spatial resolution data would be required for more local assessment. However with Google Maps–type aerial photography, taxing agencies could easily do it and neighbors can verify.

The debate about whether several small patches of nature or a single large one would be better, all other things being equal, exhausted a lot of energy among conservation biologists in the 1980s: See Simberloff and Abele (1982) and Wilcox and Murphy (1985).

The American Society of Civil Engineers is keeping an eye on the country's infrastructure and giving it generally low marks—see its report from 2009 for the $2.2 trillion price tag. In cities like Detroit, society is already making some hard choices about which parts of the infrastructure to retain and which to let go.

Tax Shifting

Fox (2003) provides a brief history and assessment of the sales tax. Gregory Mankiw lays out his ideas for moving taxes in an essay from 2009. The idea of reducing or removing the corporate income tax is a direct bid to all of business to support the notion of gate duties.

Chapter 10
Moving to Town

The opening is from Mumford's 1979 memoir.

Cities Are Our Streams

To read about how and why cities develop, find various opinions among Mumford (1961), Jacobs (1961), and Weber (1966). Kostof (1993) provides a photographic review of the actual development of cities. Mumford (1961) states the difficulty of "city" definition this way: "What is the city? How did it come into existence? What processes does it further: what functions does it perform: what purposes does it fulfill? No single definition will cover all its transformations, from embryonic social nucleus to the complex forms of maturity and the corporeal disintegration of old age. The origins of the city are obscure, a large part of its past buried or effaced beyond recovery, and its further prospects are difficult to weigh." The same could be said of any patch of forest.

The economic justification for the city is taken up by O'Flaherty (2005) and in triumphal popular form by Glaeser (2011). Agglomeration economics were first identified by Marshall (1890); Rigby and Essletzbichler (2002) update the story for modern US cities. Jacobs's now canonized view of the city as a series of overlapping, small towns is supported by Kevin Lynch's (1960) studies of the mental geographies of urban dwellers. Jacobs also wrote two small books on the economics of cities: In *Cities and the Wealth of Nations* (1985) she places "import-replacing" economies at the center of why cities grow and develop, and in *The Nature of Economies* (2001) she makes an explicit analogy between the way cities operate economically and the way natural systems operate ecologically, considering for example how both cities and forests take in energy from outside and retain it within internal cycles. In a 2001 interview, Jacobs suggests her economic insights were the most important of her storied career, eclipsing her contributions in the planning field for which she is more often remembered; see Steigerwald (2001). Florida (2002) says cities succeed when they attract and retain creative people—scientists, architects, and the like; Florida (2008) provides a method to choose a city for you. Rappaport (2008a) concurs, finding that even small differences in amenities can lead to large differences in population density, which is good for the smaller cities in the country. Fainstein (2005) reminds us that social justice must accompany social diversity for cities to succeed, the first in a list of criteria for controlling the density-dependent negativities.

Other suggestions for managing unwanted aspects of urban density? Infectious diseases can be controlled through vaccination, regular monitoring, and rapid response (Morens et al. 2004). Jones et al. (2008) suggest giving animals wild places to live separate from cities will also help with zoonotic disease. Crime has been dropping for two decades

nationwide, but the spectacular safety gains (for example) in New York City appear to have been the simple result of adding cops and focusing them on criminal hotspots—see Zimring (2011). Garbage and sewage can largely be handled by modern sanitation techniques; the current move toward a service- and information-based economy decreases our dependence on polluting industrial processes (gate duties will provide a further push), and a renewable energy infrastructure will dramatically reduce greenhouse gas emissions. Further gains might be made by renewing the housing stock in cities and decreasing inequity.

Owen (2009) takes New York as his example of how urban density can lead to lower per capita resource use, backed up by statistics in the city's sustainability plan about energy use and carbon dioxide production (Office of Long-term Planning and Sustainability 2007). What's important to remember is that while per capita resource is lower in cities than in rural areas, total resource use depends on the total population, making it possible for cities both to be resource hogs, in terms of the total amount of resources, while at the same time being more efficient on a per-individual basis. See for example studies by Dodman (2009) and Norman et al. (2006). Newman and Kenworthy (1999) make similar points, from the transportation perspective.

Density 5K

The National Household Travel Survey or NHTS holds many secrets, including the distribution of US population by residential density in 2009 (US Federal Highway Administration 2011b). You can download it for yourself at nhts.ornl.gov.

City Island data quoted in the text are from the 2010 Census for Bronx County, New York, from American Factfinder at factfinder2.census.gov. To find the density of the place you live, type in "density" in the search box, then choose your "geography"

by name of a town, city, county, or zip code in the geography box. Focus on data from Population, Housing Units, Area, and Density: 2010 derived from Summary File 1 (ID: *GCT-PH1*).

How much population density is the question everyone should be asking, not just ecologists. Economists tend to see density in terms of balancing the good aspects of cities (economy, culture, healthcare, etc.) with the bad (price). See McDonald (1989) and Rappaport (2008a,b). There is a substantive economic literature on the advantages to firms which can lower their transportation costs: See Paul Krugman (1991)'s seminal article, and a review of Krugman's thinking twenty years on (Fujita and Thisse 2009). Dewey and Montes-Rojas (2009) produce a model of between city vs. within-city wage differentials. Etc.

Most studies find that high-density land use is correlated with decreased VMT and increased use of non-automobile modes (cf. Barnes 2001; Holtzclaw et al. 2002; Brownstone and Golob 2009). Newman and Kenworthy (1999) published results of a worldwide study between population density and automobile use; they observed that at higher densities, automobile use declined exponentially. These effects seem to be strongest above densities of fifty persons/ha (or approximately 13,000 people per square mile), at least as I read their data. Newman and Kenworthy (1999) are a bit more conservative in the text, writing: "In all cases [of the cities they examined] there appears to be a critical point (about 20 to 30 persons per ha [or 5,180–7,770 persons/sq mile]) below which automobile-dependent land use patterns appear to be an inherent characteristic of the city." Barnes (2001), in a review of travel patterns in cities across the United States, summarizes similarly: "A common result [in the travel minimization literature] is that significant reductions in auto travel only occur when densities exceed 10,000 people per square mile, which is nearly the upper bound of density in most [US] cities." Of course all of these depend on how you define population density. My recommendation of five thousand people per square mile includes some room for parks and open space, so the built part of the area might be about ten thousand people per square mile.

Economic productivity of American cities is from US Bureau of Economic Analysis (2012b) and can be compared to national productivities from the Central Intelligence Agency (2012).

Minimum Population, 10K

George K. Zipf was a linguist interested in the frequency of words in natural language (Zipf 1935). It turns out that city sizes follow a frequency distribution similar to Shakespeare's verse: Zipf's Law—see Gabaix (1999) and Reed (2001). Decker et al. (2007) wrote that urban size relationships may be a result of ecological limits, which suggests that taking better care of our resources may enable new forms of city size distribution to be expressed, but let's leave that as a hypothesis. For other power laws, see Newman (2005) on the earthquake and moon craters, and Humphries et al. (2010) on sharks.

Discussions of economic specialization and diversity among cities and the reasons they exist can be found in Richardson (1972), Rigby and Essletzbichler (2002), and Dobbs et al. (2011), among many others.
In a fascinating study, Bettencourt et al. (2007) showed that many of the economic phenomena desired, including wages, income, growth, domestic product, bank deposits, and rates of invention, scale superlinearly with city size, meaning as the city gets bigger, these other attributes increase faster than city size itself; many of these phenomena scale according to a power law distribution with a similar coefficient of 1.15–1.30. Other phenomena, like per capita resource consumption and infrastructure required, scale linearly with city size or decrease, meaning that as cities grow larger (and presumably

denser), we can, mathematically speaking, have our cake and eat it too.

Minimum viable populations for other species are discussed in Reed (2003) and Beissinger and McCullough (2002); for human communities in Indiana, see Tanaka et al. (2012). More research along these lines would be welcome.

Moving to Town

New Town districts are roughly modeled on business improvement districts and other geographically specific tax incentives. See Hanson (2009), Hanson and Rohlin (2011), and Mitchell (2008). For some advice on better, simpler, saner zoning, see Elliott (2008). Shoup (2005) can help on parking.

To measure how much motivation might be thrown this direction, consider that total federal government outlays in 2009 were $3,517,677,000,000 (US Office of Management and Budget 2012), while state and local government spent $2,966,613,957,000 (Barnett 2011). Directable to new towns are: federal expenditures including those for general science and basic research ($10 billion); ground transportation ($54 billion); community development ($8 billion); education, training, employment, and social services ($74 billion); veteran housing and education ($44 billion); government administration ($23 billion); and health research and training ($31 billion). State and local expenditures for highways ($152 billion), housing and community development ($47 billion), university education ($234 billion), and government administration ($123 billion) could in part be redirected to encourage and reward density.

Location-efficient mortgages and other such geographic tactics are discussed in National Resource Defense Council (2009), Giery et al. (2003), and Blackman and Krupnick (2001).

The home-to-work adjustment is a policy nudge aimed directly

at the journey to work, which is a minutely studied aspect of transportation behavior. See for example American Society of Planning Officials (1951), Clark and Burt (1980), Levinson and Kumar (1994), McGuckin and Srinivasan (2005), and Tilahun and Levinson (2008). Pisarki (2006) presents data showing that average home-to-work trip lengths increased almost 14 percent in 1990–2000 as the Clintonian incentives worked their magic. Fortunately many of us will be immediate beneficiaries or neutral to a home-to-work adjustment: According to analysis of the National Household Travel Survey (Federal Highway Administration 2011b) 142.5 million Americans (46 percent) already live within five miles of the place they work or don't work at all. Another 64.7 million Americans (21 percent) live five to ten miles from work—an easy streetcar ride or bike ride on a nice day. The population whose journey to work is costing us the greatest vehicle miles traveled are those living ten or more miles from work; these 99.6 million people (32 percent of the population) account for 45 percent of total national distance traveled, over 2.1 trillion miles in 2009.

The journey to work is an important, but not the dominant, fraction of travel for Americans. Examining the data for all households (working or not), the journey to work accounts for only 19 percent of total travel distance. The importance of the journey to work lies not so much in the total distance, but in its importance in structuring work and residence collocation decisions (Tilahun and Levinson 2008) and choice of personal travel mode (Maat and Timmermans 2009; Salon 2009; McDonald 2007). For why people do sometimes choose transit, see Taylor et al. (2009). Wardman et al. (2007) and Hunt and Abraham (2006) focus on why people decide to bicycle; Rodriguez et al. (2009) look at why people choose to walk. Some people *like* to commute—for positive utilities, see work by Redmond and Mokhtarian (2001) and Mokhtarian and Salomon (2001). Handy et al. (2005) try to distinguish "necessary" travel from "excess travel," travel we enjoy.

Finally, where we live and work sets up trip chains (McGuckin et al. 2005). A trip chain is a series of trips made in sequence, like leaving work, stopping at the store, and picking up the kid before arriving at home. What's significant about trip chains is that once someone starts driving, he or she is more likely to make all these trips by driving, rather than changing modes. See recent studies by Noland and Thomas (2007) and Lee et al. (2007).

One option to cope with the effects of the proposed home-to-work adjustment is for businesses to officially condone telecommuting. Although telecommuting hasn't been the boon for transportation it was hoped to be—e.g. Ory and Mokhtarian (2005) and Tang et al. (2008)—the real gains may be in terms of community and job performance—see Robert and Borjesson (2006), Fonner and Roloff (2010), and Dahlin et al. (2008).

Statistics on geographic mobility of Americans come from the US Census Bureau (2011b) at www.census.gov/hhes/migration. Although one might suspect that such high levels of mobility are a result of unsettled economic times, data from a decade before show similar patterns. Berkner and Faber (2003) report that 120 million Americans, or 46 percent of the population, moved house between 1995 and 2000 (including me!). Strauss-Kahn and Vives (2009) study the movements of corporate headquarters, which tend to relocate to metropolitan areas with good transportation links (especially airports) and near business partners.

Where We Come From

I get a lump in my throat every time I think about how much evolutionary anthropology has to teach us about making great cities. Zimmer (2007) provides a readable summary of human evolution; Norton and Jin (2009) provide a useful recent review of theories of how behavioral modernity developed and what it includes. Robin Dunbar's academic writings are unusually clear and accessible—his study of neocortex and group size is from 1992; his examination of human group sizes was published in 1993. He brings together two decades of work in an excellent account of the quirks of human evolution in 2010. Goncalves et al. (2011) show evidence for Dunbar-size groups in Twitter conversations, and the Facebook Data Team (2011) shows similar findings among "maintained relationships" on facebook.com.

Fletcher (1995) provides a thorough survey of settlement density across time and cultures. He presents data in Figure 4.3 of his book indicating hunter-gatherers living in camps/settlements with population densities equivalent to 500,000 people per square mile or more (though obviously over areas much less than one square mile). Draper (1973) studied camp sizes of !Kung bushmen, where each person had on average 181 square feet of space, an equivalent population density of 154,000 people per square mile. In the 1970s studies like Patricia Draper's led sociologists and anthropologists to question the previous assumption, derived from studies of animals, that high densities "naturally" led to aggressiveness, violence and other social pathologies. To their surprise, they found no or inconsistent evidence of density causing problems—see the review by Gillis (1974). Baldassare and Feller (1975) and Levinson (1979) provide reflections on anthropological aspects of density. In turn these studies led Fletcher (1995) to formulate his theory of the limits on settlement growth, which he sees as a function of limits of interactions and communication, nicely tying into Dunbar's studies cited above. For a mind-expanding series of essays on hunter-gatherer economics, see Gowdy (1998). Kimmelman (2011) reminds us that Aristotle's original definition of *polis* was circumscribed by the limits of a herald's cry.

Cities for People

The literature on livable, walkable, better cities is deep, active, and growing. Some of my favorites are Alexander et al. (1977), Hiss (1990), Crawford (2002), Register (2002), Gissen (2003), and of course, Gehl (2010). Standard texts of the New Urbanism movement are Duany et al. (2000), Calthorpe and Fulton (2001), and Duany et al. (2009); see also the Congress for New Urbanism's website at www.cnu.org. A larger scale view, more in keeping with natural systems and patterns, is landscape urbanism–see for example, Waldheim (2006) and Hung et al. (2010). Dunham-Jones and Williamson (2011) provide a useful guide to ripping up old suburbia and replacing it with something much better. For advice on designing with nature, the classic text is McHarg (1969). An updated, enjoyable, more anecdotal treatise in a similar vein is Courtenay (2011). Elliott (2008) provides some excellent advice on reforming zoning codes to create livable communities. For what's happening in Europe and Australia, see Beatley (2000 and 2008, respectively).

The quotes on walking in the city are mostly from Rebecca Solnit's 2001 book on wanderlust. Dickens's tendency to walk is documented by Tomalin (2011). Lincoln's penchant to walk is noted by all his biographers; I noticed it when reading Foote (1986). Gandhi extols the value of walking in his autobiography (1993 reprint and translation of original from 1927) Moody's quote is from *Harry Potter and the Goblet of Fire* (Rowling 2002). Hester's summary of what every great community needs is from his recent book on ecological democracy (2010). Many progressive municipalities do adopt urban growth limits—see discussion in Calthorpe and Fulton (2001) and legal examples in Oregon and Washington. Another example is "Nature-City," a design shown at the Museum of Modern Art as part of the "Foreclosed: Rehousing the American Dream" exhibition—see Pace (2012).

"Forget the Damned Motorcar"

Space usage varies by model of car. For example, my 1977 Volvo wagon was 191 inches long and 68 inches wide and occupied an area of approximately 90 square feet, while a neighborhood electric vehicle, like a GEM E4, might occupy only 60 square feet, an electric scooter about 18 square feet, a bicycle about 6 square feet, and a person about 2 square feet. The Manual on Uniform Traffic Control Devices (US Federal Highway Administration 2009), which is the national standard, shows streetside parking spaces dimensioned to be 8 feet x 20 feet, occupying 160 square feet, but actual spacing varies somewhat by municipality. Off-street parking requires parking aisles and access roads to navigate to spaces that add up to the 330 square feet. Dave King from Columbia University pointed out to me the ratio of office space to parking cited in the text. I had no idea.

For a surprisingly pleasant and remarkably lucid, while still quantitative, account of the true costs of "free" parking, see Shoup (2005). Vanderbilt (2008) deals with the psychology of driving. The Texas Transportation Institute at Texas A&M University is the leading source on the costs of traffic congestion. Each year they publish the annual urban mobility report (mobility.tamu.edu/ums); the 2011 study (Shrank et al.) illustrates congested conditions in 2010 on a number of levels.

Urban noise accompanies traffic congestion causing undesirable health effects and generally making cities much less pleasant. A World Health Organization 2011 report estimates that a million years of disability-adjusted life-years (or DALYs) are lost per year in Europe from noise aggravation, which seems extreme, but noise is annoying and unhealthy. Chepesiuk (2005) summarizes the US situation. Even urban robins have to sing at night to be heard, because it's too noisy during the day—see Fuller et al. (2007). Although the US Environmental Protection Agency (2010c) tries

to spin the current air pollution situation as positively as possible, the American Lung Association (2011) finds that 50.3 percent of Americans live in counties where they are exposed to unhealthy levels of ozone and particulate pollutants, mostly due to traffic.

I suppose starting this section with Michelangelo put me in mind of his statue of Moses in the San Pietro in Vincoli church on the Oppian Hill in Rome. We will need the strength of Moses *and* the wisdom of St. Peter to break our bondage to the car, I fear.

Chapter 11
Roads to Rails

The opening quote is from the famous play by Tennessee Williams in 1947.

A Brief Physics Lesson

Sir Isaac Newton haunts this chapter. A younger contemporary of John Locke, Newton made major contributions in mathematics, optics, astronomy, and mechanics, mostly from a brief, productive eighteen-month period. Later in life he was Master of the Mint, where he controlled the British money supply, and by fixing an exchange rate between silver and gold in 1717 put the United Kingdom effectively on the gold standard, which the Brits would adhere to until the horrors of World War I forced them off in 1914. Berlinski (2002) provides a readable biography; Dolnick (2011) describes his world.

The strange notion that an object in motion will stay in motion perpetually in a vacuum is a restatement of Newton's First Law of Motion. What a motor does is apply a force; Newton's Second Law of Motion says that the acceleration of an object is proportional to the force applied and inversely proportional to the object's mass, which follows from the conservation of momentum and applies to light as well as matter. Although energy was not understood while Newton was alive, the discovery of conservation of energy in the early

nineteenth century was entirely compatible with the foundations he had laid two centuries before.

My simplistic description of the physics of vehicles follows MacKay's (2008) lucid account, which you can read on-line at www.withouthotair.com or by purchasing his book; see in particular Chapter 3 and Technical Chapter A. You can also read more in Forinash (2010) or any standard undergraduate physics textbooks (e.g. Halliday et al. 2007).

The force of friction on a wheeled vehicle depends on the coefficient of rolling resistance. MacIsaac and Garrott (2002) provide details on how rolling resistance changes with changing tire pressure. Well-inflated car tires not only save gas, they are safer to drive on (Transportation Research Board 2006). Typical coefficients of rolling resistance for automobile tires vary from 0.0098 to 0.0138; steel wheels on steel rails have coefficients of 0.0015–0.0035 (Lindgreen and Sorenson 2005). Train cars on a level track have such small amounts of rolling resistance that they sometimes roll down the tracks on a windy day, even though they might weigh thirty tons or more. Did you know the study of friction, wear, and lubrication is called tribology? See tribologists Olofsson and Lewis (2006) for more.

The US Department of Energy and Environmental Protection Agency have collaborated on a useful website called fueleconomy.gov, where you can check out the fuel mileage for different car models in city, highway, and combined driving, back to the 1987 model year; they also have a nifty figure showing where the energy goes when you drive your car—see www.fueleconomy.gov/feg/atv.shtml, yet another reminder that it is not information that is wanting.

Forinash (2010) writes of electric motors: "The limits to efficiency of [electric] motors and generators due to the second law of thermodynamics are exceedingly small. An ideal motor with no friction or other loss can have a theoretical efficiency of more than 99% and real electric motors have been built with efficiencies close to this limit. . . . For real electric motors there are mechanical friction losses and resistance. . . . Well designed low-horsepower (<1,000 W) motors typically have efficiencies of about 80%, and larger motors (> 95 kW) have efficiencies as high as 95%" (p. 123). The Nissan Leaf's motor consumes 80 kW; and the Chevy Volt's consumes 110 kW, according to their respective websites: www.nissanusa.com/leaf-electric-car/key-features and www.chevrolet.com/volt-electric-car.html.

Smil (2008) makes a big deal over the amount of energy different fuels can contain, as do I. Unfortunately some misguided apologists for the fossil fuel industry use this data to argue that we can't replace the car ever (cf. Bryce 2010), but that's not to say we can't have something else (e.g. streetcars) instead. Smil provides some fun energetic comparisons: The energy of a flea hopping (1 x 10^{-7} J) to the annual global interception of solar energy (5.5 x 10^{24} J); the power of ephemeral phenomena, from a hummingbird's flight (7.0 x 10^{-1} W) to a magnitude 9 earthquake (1.6 x 10^{15} W); and the efficiency of common energy conversions, from some ecosystems that manage only a paltry 1–2 percent, to a large electric generator with efficiencies of 98–99 percent.

The search for more energy-dense batteries has been underway for a century now. See reviews of the twenty-first-century state of play by Karden et al. (2007) and Shukla et al. (2001). Don't hold your breath.

A Better Car

Vehicle occupancies can be found in Santos et al. (2011), based on calculations from the National Household Travel Survey. Occupancy varies by trip type: commuters average 1.13 people per trip, shoppers and errand-makers 1.78 and 1.84 people per trip, respectively, and socialities, 2.20 people per trip. Pisarski (2006) shows that commuting alone

varies dramatically between different American cities, from a low of 56.3 percent of trips in New York to a high of 84.2 percent in Detroit in 2000. With streetcars for the commute of the future, no one will have to travel alone.

I calculated energy consumptions per person per mile at usual and maximum occupancy for different modes of transportation (Figure No. 56). Walking, biking, and skating energy consumption were drawn from www.fitwatch.com. Automobile fuel consumptions were calculated for combined driving fuel efficiencies reported in fueleconomy.gov for the various models indicated; curb weights and maximum occupancies are from manufacturer websites. Usual vehicle occupancies for different automobile types were derived from averages calculated from the 2009 NHTS (US Federal Highway Administration 2009). Public transportation energy consumption and usual occupancy for the various transit lines indicated were calculated from the 2009 National Transit Database (NTD)—see US Federal Transit Administration (2011). Maximum occupancies for transit modes were estimated from transit authority websites or estimates from similar lines when I couldn't find the exact numbers. MTA subway occupancy is based on 200 passenger capacity for a ten-car train. The Staten Island Ferry maximum capacity is for the "Molinari" Class ferries. Vehicle weights for trains are car weights, not including the locomotive. Fuel consumption rates for the Boeing 737 and 747 aircraft were deduced from the graphs provided in Boeing (2010) and Boeing (2002), respectively, and are estimated for trip distances of 3,000 and 3,400 nautical miles, respectively. Aircraft usual occupancies are based on the maximum occupancies multiplied by the average passenger load factor for 2010 (US Bureau of Transportation Statistics 2012).

Sperling and Gordon (2008) provide an entertaining review of the recent developments of electric, hybrid, plug-in hybrid, and fuel cell cars. Sandalow (2009) and

colleagues make the case for plug-in hybrids; though wonkish, this book brings together some of the best thinking on how to generate an electric vehicle revolution; many of their recipes could be applied to streetcars and NEVs as well, where the physical challenges aren't so daunting. My issue with writers like Sandalow and Sperling is their fundamental, undeniable, unshakeable (it would seem) assumption that personal automobiles are the only way. It's a bit like the Catholic Church in 1517. Be careful who is knocking at your door!

A note on fuel cell vehicles: Rifkin (2003) gives an impassioned appeal for the hydrogen economy based on fuel cell technology for cars; however there are numerous debilitating technical problems, which, it seems, may keep chemical engineers busy for some decades (see Agrawal et al. 2005), the most important of which may be the small size of hydrogen gas molecules (literally just two protons), which means hydrogen is difficult to bottle up. Hydrogen fuel cells also are carriers of energy since hydrogen gas does not exist in any quantities in nature (it's too reactive to stay around long). So hydrogen gas as a fuel needs to be produced from another fuel, which might be renewable or might be a fossil fuel; either way each energy transition costs energy, which means, for now fuel cells are just another version of the Siren song, albeit a bubbly, explosive leitmotif.

Who Killed the Electric Car? was made by Chris Paine (2006). More details about the Nissan Leaf are available from the Nissan website, including costs of charging (www.nissanusa.com/leaf-electric-car); costs estimated for the battery follow comments Nissan executives made to the *Wall Street Journal* and other outlets—see Garthwaite (2010) and Loveday (2010).

Learn more about neighborhood electric vehicles (NEVs) in Abuelsamid (2009). Francfort and Carroll (2001) describe operational characteristics of NEV fleets, and Brayer et al. (2006) describe guidelines for deploying NEV fleets in the future, based on studies done by the Idaho National Laboratory. Brayer and colleagues write: "NEVS are designed to meet most light-duty applications, such as people movers and light utility use. NEVs are significantly faster than golf carts, which typically have top speeds of 12 to 15 mph. Typical NEV payload capabilities range from 600 pounds to 1,000 pounds (including passengers). When the batteries are functioning properly, a fully functional range is typically around 30 miles for each full charge in mild climates. In cold climates, the range can be reduced by as much as half. Options are available, such as fast charging, that allow the range to be extended to over 100 miles per day by opportunity charging in 20 to 30-minute increments throughout the day." Moawad et al. (2011) share a similar vision of smaller, lighter, more efficient vehicles in America through 2045 and back it up with simulation of over two thousand different vehicle types. For a beautiful vision of what is possible for these kinds of vehicles, see Mitchell et al. (2010).

A Better Streetcar

For the good news about streetcars, see Ohland and Poticha (2009). Tennyson (1998) makes the case for streetcars over buses; for more fun and less reverence, see The Infrastructurist (2010). A lot of writing about streetcars is nostalgic (e.g. Diers and Isaacs 2007) or dismissive (e.g. Jones 2010), but we have more than enough experience with streetcars to know what a lovely, efficient, cost-effective solution they are for urban transportation, which is why they have seen a renaissance, in spite of auto-dominated streets. In 2009 the United States had seventy-four urban/suburban railway systems in operation (commuter rail, heavy rail, light rail, including streetcars, cable car, and trolleybus). They collectively provided 4.5 billion rides covering 30.3 billion passenger-miles in 2009. What streetcars really need, though, is streetcar-only streets. One sign of the potential for streetcars is the success of bus rapid

transit (BRT), which is essentially running buses like trains, but without rails. I like streetcars better for reasons described in the text, but in a pinch will go with BRT, too. See Cervero (1994) and Weinstock et al. (2011) for more.

The streetcar counting game depends on the amount of energy required per vehicle-mile for cars vs. streetcars. For example, one Seattle Streetcar trundling down the street in 2009 used 7.98 kWh/vehicle-mile, which is equivalent to the energy used by 3.95 Ford F-150 pickups traveling the same mile, 5.99 Honda Accord LXs, or 10.93 Toyota Priuses. The actual streetcar-to-car count in your traffic depends on its vehicle composition; five is approximately what I see on City Island in the mornings, where there seems a proclivity toward pickup trucks and SUVs even though the Bronx is a long way from the countryside and rarely sees lasting snow any more.

Of course you could play the same game on a per-passenger basis, in which case at average occupancy, 1.29 streetcar passengers could go by for the same amount of energy as every Prius passenger, 2.19 streetcar passengers for every Accord passenger, and 3.74 streetcar-straphangers for every pickup truck rider. That is a potential 29 percent, 119 percent, and 274 percent improvement in energy efficiency of streetcars over those personal motor vehicle types, respectively.

According to the historical census from the US Bureau of the Census (1975; Series Q264-273), the apex of streetcar development in America was 1917, when the streetcar network extended over 44,835 miles of track servicing 32,548 miles of streetcar line (some lines had multiple tracks.) According to Steuart (1905), in 1902 there were 813 street railway companies serving 4,774,211,904 fare-paying passengers with 1,144,430,426 car-miles traveled (Steuart 1905; Table 7). Although one might suspect Steuart's precision, the numbers are impressive considering the national population in 1902 was

only 79,163,000, or just 26 percent of the 2010 American population, which means in 1902, the average person took sixty streetcar rides. Forty-three of forty-eight states plus the District of Columbia had streetcar service that year, not only in thirty-three large cities with population of 25,000–100,000 people, but also in forty-six towns with population less than 25,000).

Roads to Rails

Many works extol the advantages of walking, bicycling, and other forms of personal mobility: see Hurst (2004), Byrne (2009), Mapes (2009) on bicycling, the most energetically efficient form of personal transportation ever invented; Solnit (2001) on walking; and Alvord (2000) and Balish (2006) on getting out of your car. The number of short trips less than three miles is from analysis of the National Household Travel Survey (US Federal Highway Administration, 2011b). The current rail system, including freight trains, is described in the US Bureau of Transportation Statistics (2006), Association of American Railroads (2010), and the US Government Accounting Office (2011). Read Jarrett Walker's sage advice in Walker (2011).

Transportation planners use the concept of "level of service" (LOS) to determine transportation capacities. Streets can move more people but pay the price in delays, congestion, and pollution. To estimate maximum capacities, I used statistics on LOS D, which is not good, but not the worse it could be. Sidewalks with LOS D levels can accommodate 900 persons per hour per foot of width; cars move 11,000 vehicles per day per lane at the same LOS (US Federal Highway Administration 2009).

Gilbert and Perl (2010) provide a detailed analysis of the space and energy uses of freight and personal transportation compared to other modes. They conclude, as I do, that grid-connected electric rail is the most flexible and efficient way to move us and our stuff. Their perspective is more global than mine;

in particular, see their analysis for China. Highly recommended. Also see Crawford (2002).

To read the detailed difficulties of the California High-Speed Rail Plan see the newly released revised business plan (California High-Speed Rail Authority 2012). For some academic viewpoints on the current debate over high-speed rail, see Ryder (2012), Lane (2012), Kanafani et al. (2012), and Campos and de Rus (2009).

Transportation funding is summarized by the US Bureau of Transportation Statistics (2012) at www.bts.gov/publications/government_transportation_financial_statistics/2012. Construction costs for streetcars are from Ohland and Poticha (2009).

What Happened?

For more on empty forests, read Redford (1992).

Although I disagree with his interpretation that the streetcar's decline was inevitable or that they are forever gone, Jones (2010) nicely lays out the statistics, documenting the rise and fall of the street railways. See Bottles (1991) and Miller (1941) for additional details, and for individual lines in texts like Myers (2005) or Diers and Isaacs (2007). The list of streetcar magnates is from Phillips (2003). Senator Williams is perhaps more famous for his conviction for bribery and conspiracy in the "Abdul scam" or Abscam case of the late 1970s, wherein Federal Bureau of Investigations personnel disguised as a wealthy Middle Eastern sheik offered bribes to a number of US politicians, including gullible Pete.

In 1962 President Kennedy called on Congress to approve federal capital assistance for mass transportation, saying "To conserve and enhance values in existing urban areas is essential. But at least as important are steps to promote economic efficiency and livability in areas of future development. Our national welfare therefore requires the provision of good urban transportation, with the properly balanced use of private vehicles and modern mass transport to help shape as well as serve urban growth." In 1964 the Urban Mass Transportation Act passed and was signed by President Lyndon Johnson. This act required coordinated planning between mass transit and personal transport in all urban areas with more than fifty thousand people, and opened up the first federal funding sources for public transportation. For a detailed account of the "golden age" of urban transportation planning, see Danielson (1965).

What Business Does Best

The current financing model for transit is ripe with the problems of public managers, subject to the ballot box, trying to run a transportation company. Consider the case of New York City Transit (NYCT), managed by the Metropolitan Transportation Agency (MTA), by far the country's largest and best-used transit agency. Approximately one-third of all public transit trips made in the country each day are made on vehicles owned and operated by the MTA; a city the size of Seattle rides on the subway each night, and yet even with a massive customer base in the country's densest city, the New York subway and buses haven't been able to break even. The problem is not the energy costs, which in 2010 were less than 5 percent of the operating budget (a mere $131 million), or even depreciation of the rolling stock, switches and rails, estimated at $1.29 billion (or 15 percent of the budget); the problem is the labor costs, which are 70 percent of the budget ($5.76 billion including postemployment pensions and other costs paid to former transit workers). Similar high labor costs plague transit systems from Chicago to Denver to San Francisco. See Gwillim (2008) and the US Federal Transit Administration (2011). For more on the MTA, see Tri-State Transportation Campaign (2012), the *New York Times* MTA summary webpage, and then try to decipher the MTA's own budget numbers available on-line (www.mta.info/mta/budget).

To put these big numbers in perspective, consider them on a per-fare basis. To break even without government support, each of the 2.31 billion paying passengers on New York City Transit in 2009 would need to pay a full fare of $2.50 just to cover the costs of the people driving the trains, staffing the tollbooths, running the back office, and on retirement from the system. A fare of $3.60 would cover all operating costs. However current fares are $2.25 per ride, and after various discounts and reduced price schemes, the average fare paid plummets to only $1.50 per rider actually received by the system, which leaves a several billion dollar hole each year in the MTA budget—a gap currently plugged by dedicated taxes on property, mortgage recording, business licenses in a seven-county region around New York City, and the proceeds from the RFK Bridge connecting Manhattan, Queens, and the Bronx.

Meanwhile the longest commutes in the nation? Not stuck in traffic in Los Angeles. Not trapped on the highways around Atlanta. The longest commutes are for the poor straphangers in Queens County, New York, and Bronx County, New York, for whom public transportation is the right choice economically, patriotically, and environmentally, but which returns them long, slow, packed rides, based on schedules enforced by labor union rules and lack of investment in street-level infrastructure. See McKormick and Jones (2010) for overview, and McKenzie and Rapino (2011) for details. The entire story is remarkable, unsustainable, and in need of change.

Chapter 12
Invest in the Sun

Though the quote from Columbus should be easy to track down, I haven't been able to get to the bottom of it, but it's often cited: BrainyQuote.com.

Storage on the Mountain

Hansen and Lovins (2010) point out that it is not only the renewable sources that are intermittent: The big boys sometimes fail to show up, too. They analyzed data from the North American Electric Reliability Council for all US generators during 2003–07 to show that "coal-fired capacity was shut down an average of 12.3 percent of the time (4.2 percent without warning); nuclear, 10.6 percent (2.5 percent without warning); and gas-fired, 11.8 percent (2.8 percent without warning). Despite extraordinary efforts to keep nuclear plants running constantly, the average unit worldwide unexpectedly failed to produce 6.4 percent of its energy output through 2008 and 5.3 percent in 2008, not counting planned refueling or maintenance shutdowns." Unfortunately because these plants are so large, when one trips off, the blackouts can cascade across the system, as happened during the Northeast blackout of 2006.

Rastler (2010), Schoenung (2011), and Ribeiro et al. (2001) provide useful comparisons of various electrical storage options from industry, government, and academic perspectives, respectively; they basically agree that pumped storage is most cost effective over the lifetime of a project, given current technology. Divya and Ostergaard (2009) update the story for large-scale electrochemical batteries and *The Economist* (2012b) discusses new avenues in pumped hydrologic storage, including constructing a "green power island" in the ocean and a system of larger and smaller underground pipes that could be built anywhere. For a detailed description of a pumped hydrologic station, see Mock (1972)'s design specs for Raccoon Mountain, part of the Tennessee Valley Authority set-up; fifteen years later, Adkins (1987) provides a summary of operating performance at Raccoon Mountain for the TVA, with an average energy efficiency of 79 percent. The total operating capacity of pumped storage in the United States in 2009 was 20,538 MW,

with a summer capacity of 22,160 MW and winter capacity of 22,063 MW according to the US Energy Information Administration (2012a). The same administration (2010b) summarizes electric generators of all types through its Annual Electric Generator (EIA-860) data file, available online. Search in the "generator.xls" searching for "prime mover" = PS to find pumped storage facilities. There are currently 151 turbines in thirty-nine facilities scattered across eighteen states nationwide. Together pumped storage facilities had 58.9 percent as much nameplate capacity as listed wind and solar/photovoltaic producers did during 2010.

Pumped storage is not without environmental risks. The plants need to be well engineered to avoid collapse and they need to go easy on fish and other wildlife. The most infamous pumped storage plant was never built because of environmental concerns over fish in tidal portions of the Hudson River. The fight over the Storm King Mountain plant, proposed by Consolidated Edison on a mountain north of the West Point Military Academy in the early 1960s is retold in Robert Boyle's lively description from 1969. The engineers estimated how many fish would be chewed up for a Hudson River that flowed one way; an estuary biologist pointed out that because of the tides their estimates were off by a factor of four. I examined the potential distribution of additional pumped storage sites in the United States by analyzing the National Elevation Dataset and precipitation, major rivers, and aquifer data from the National Atlas, nationalatlas.gov.

Rules of the Grid

The "old grid" as described in the text was first engineered by Samuel Insull, a billionaire utility magnate of the early twentieth century, and former personal secretary to none other than Thomas Edison. British-born Insull gained Edison's trust and founded Edison General Electric, which is better known today as the publicly traded

company, General Electric. Passed over for the presidency of GE in 1892, Insull moved to Chicago, where he became president of the Chicago Edison Company (later renamed Commonwealth Edison) and became a major operator and consolidator of utility companies, including numerous electric streetcar lines. Insull believed in regulated utilities, economies of scale, and the one-way grid, where his companies generated the power and customers bought it, at rates set by regulators; regulations were required to protect the utility from charges of monopoly, which of course it was. Insull realized that without the regulators, the public would never allow a company like his to gather together so much power, especially in the wake of Standard Oil in 1911. How times have changed! See biography by Wasik (2006).

As with other aspects of the early-twentieth-century industrial infrastructure, Insull's worldview worked great then and makes problems for us now; it's not Insull's fault, but it will be ours if we don't act to change it. Fox-Penner (2010) gives a popular insider-outsider perspective on how changes like the smart grid might actually happen. Numerous government reports document the benefits and the challenges of the smart grid: See for example, Kannberg et al. (2003) and Myles et al. (2011). According to Myles et al. (2011), the current US electric power industry consists of 17,342 generating units, 164,000 miles of transmission lines, 3 million miles of distribution lines, and 12.3 million distributed generation units in homes and businesses; most of the latter are not directly connected to the grid because the grid cannot accommodate them. Currently power plants run at 47 percent of capacity across all fuel types, and transmission lines are loaded to 43 percent capacity on average, suggesting there is definitely room to grow. To see a fascinating interactive map view of the US electric grid, log on to National Public Radio's 2009 webpage on visualizing the US electric grid (www.npr.org/templates/story/story.php?storyId=110997398).

President Nixon and Congress adopted a number of measures to reduce dependency on oil in the 1970s, most of which have since been repealed or rendered toothless, including fixing the temperature for government buildings at 68 degrees, lowering the speed limit to 55 mph, moving Daylight Savings Time to February, even closing gas stations on weekends. In the winter of 1974, fuel shortages led to rules where only drivers with even-numbered license plates could buy gas on even-numbered days; odd-numbered license plate drivers could buy gas on the odd-numbered days of the month (a practice remembered in the aftermath of Hurricane Sandy in New York and New Jersey). In November 1973 Nixon gave a speech in which he exhorted Americans to undertake "Project Independence," saying "Let us set as our national goal, in the spirit of Apollo, with the determination of the Manhattan Project, that by the end of the decade we will have developed the potential to meet our energy needs without depending on any foreign energy source." Although failing at the larger goal, the other measures helped somewhat in reducing oil dependency in the short run. See description in Yergin (1991). For those of you who prefer your history with a comedic twist, check out this episode of *The Daily Show with Jon Stewart* (www.thedailyshow.com/watch/wed-june-16-2010/an-energy-independent-future)—highly recommended.
California's unpleasant experience of deregulation is explained in some detail in Breder (2003) and Blumstein et al. (2002). The list of gambits implemented by power brokers is from McCullough (2002). The details of Enron's scandalous rise and fall are told in McLean and Elkind (2004) and the subsequent documentary (Gibney 2006) about the "smartest guys in the room." Yep. Fox-Penner (2010) has an interesting discussion of how deregulation might continue to proceed under the smart grid.

The federal government, concerned about buying electricity in deregulated markets, provides advice for its own budget managers in Warwick (2002).

Charles River Associates (2005) summarize the benefits of customers knowing in real time how much they are paying for power. See further study of how household indications decrease use by Willis et al. (2010). Attari et al. (2010) provide some interesting data on how perception of energy use doesn't always line up with the reality. Some utilities are making real-time electricity prices available online: See www2.ameren.com/RetailEnergy/realtimeprices.aspx and edata2007.pjm.com/eData/index.html.

Where the Wind Blows

Mill technologies belong to an earlier industrial revolution, that of the Middle Ages; see Gimpel (1976). In the twelfth century, engineers in Europe began to explore the potential of "post-mills," unlike the horizontal windmills, mounted on a vertical axle, known in Iran and Afghanistan from the seventh century. Windmill technology flourished from 1180 onward wherever there was a lack of free-flowing streams and lots of wind, especially on the plains of northern Europe. Later they were adapted for tides, which allow the water to flow in on the flood then close dams; on the ebb, the tidal mills generate power. Gimpel cites a late-twelfth-century letter from Jocelin of Brakelond describing a dispute about a windmill in which the local abbot, with a local monopoly on grinding because of his control of water mills, objected to Herbert the Dean setting up a windmill nearby. The Dean responded "that he had the right to do this [build a windmill] on his free fief, and that free benefit of the wind ought not be denied to any man." I could not agree more, subject to gate duties, of course. An edited version of Jocelin's chronicle is available on the web; see Gasquet (1997).

The windmills in Dutch New Amsterdam, later New York

City, can be seen on many early maps—for example, see the inset on the Jansson-Visscher Map of 1651, reproduced in Cohen and Augustyn (1997). President Carter's support for wind energy came in the form of the Wind Energy Systems Act in 1980, which provided $60 million in research and development for wind. Goldburg (2000) estimates that wind power received approximately $1.12 billion in subsidies from the government over the course of the twentieth century, in comparison to the *$115.07 billion* for nuclear power (measured in constant 1999 dollars).

Cervantes took advantage of the windmills of La Mancha of the early seventeenth century to help us imagine the odd caricatures drawing by a nation at war. (Spain at the time of Cervantes writing was embroiled in the Eighty Years War with the Netherlands.) Cervantes (1605) writes:

"Just then they came in sight of thirty or forty windmills that rise from that plain. And no sooner did Don Quixote see them that he said to his squire, 'Fortune is guiding our affairs better than we ourselves could have wished. Do you see over yonder, friend Sancho, thirty or forty hulking giants? I intend to do battle with them and slay them. With their spoils we shall begin to be rich for this is a righteous war and the removal of so foul a brood from off the face of the earth is a service God will bless.'

"'What giants?' asked Sancho Panza.

"'Those you see over there,' replied his master, 'with their long arms. Some of them have arms well nigh two leagues in length.'

"'Take care, sir,' cried Sancho. 'Those over there are not giants but windmills. Those things that seem to be their arms are sails which, when they are whirled around by the wind, turn the millstone.'"

Judging by some of the hysterical accounts of industrial wind power on the Internet, modern people can still see giants that are not there.

Clever energy-geography analogies with Saudi Arabia come from T. Boone Pickens, before he abandoned his well-publicized Pickens Plan to save America with natural gas in the short term and wind power in the long, after reputedly buying up large tracts of land with natural gas underfoot and wind above. See Mooney (2008) for the quote and Montaigne (2011) for a more recent update on the new and improved, much less green, Pickens plan. The Woody Guthrie-esque geography is from Elliott et al. (1986), which summarizes the various regional atlases published in 1980 and 1981 thanks to federal support for wind-power research enacted during the Carter Administration. The most recent summary of the onshore wind resource potential is provided by a new analysis by the National Renewable Energy Laboratory and AWS Truepower (2011)—see their website "Wind Powering America" at www.windpoweringamerica.gov. They estimate for each of the fifty states and the total US windy land area using a gross capacity factor (without losses) of 30 percent and greater at 80 m height above ground; they estimate the amount of energy that could be generated from development of the "available" windy land area after exclusions.

The 2011 National Offshore Wind Strategy summarizes wind over US waters as follows: "Offshore wind resource data for the Great Lakes, US coastal waters, and the OCS [Outer Continental Shelf] indicate that for annual average wind speeds above 7 meters per second (m/s), the total gross resource of the United States is 4,150 GW, or approximately four times the generating capacity of the current U.S. electric power system" (Schwartz 2010). "Of this capacity, 1,070 GW are in water less than 30 meters (m) deep, 630 GW are in water between 30 m and 60 m deep, and 2,450 GW are in water deeper than 60 m" (US Department of Energy and Department of Interior 2011). For further details, see Schwartz et al. (2010) and Musial et al. (2010).

It's helpful when considering these numbers to remember that the US electricity generation in 2009 was 3.95 million GWh, based on a total electric power nameplate capacity of 1,028 GW according to the US Energy Information Administration (2012a). Tremendous gains in installed capacity have occurred over the last thirty years. Summaries of the lift and drag physics of wind turbines are provided in Grogg (2005) and Forinash (2010).

All new wind installations would be subject to the ecological use fees, which would add to these costs, and help build sensitivity into our environmental use. For more advice on being sensitive, consider National Resource Council (2007) and Manville (2005).

Where the Sun Shines

The "accessible resource" figure for solar renewable power comes from US Energy Information Administration (1993). Government energy analysts define an accessible resource to be: "That part of resources able to be accessed with existing technologies regardless of cost. The term 'accessible resources' is definitionally similar to the term 'recoverable resources' for oil and gas resources." It is equivalent to the potential resource as used in this book. The economic costs of renewable energy are changing so rapidly, and will change even faster once all natural resources are appropriately costed by gate duties, that any estimate of the "economically recoverable" resource becomes dated quickly. Nevertheless 172 billion GWh is a very impressive number and a significant fraction of that resource is available today.

To figure out how much sun there is where you live, go outside during daylight hours or log on to use the National Renewable Energy Laboratory's PVWatts program: www.nrel.gov/rredc/pvwatts. A map of the solar resource is available from www.nrel.gov/gis/solar.html, with separate calculations for

concentrating solar and photovoltaic panels.

The Nevada Test Site and Nellis Air Force Base example comes from Schaeffer (2008), who recommends Sharp 208-watt, 27-volt modules, rated in lab tests at 188.3 watts, generating an estimated 1.118 MacKays (kWh/day) per panel. Losses from converting direct current to alternating current would knock down output to 0.996 MacKays per panel. Allowing 23 square feet per panel, approximately 1.2 million panels could be packed into a square mile. At this panel density, meeting the 2005 US electricity demand, 9,448 square miles of Nevada desert would need to be covered, or a square area approximately 97 miles on a side. That would cover about 60 percent of the Nevada Test Site and surrounding Nellis Air Force Range. That's a lot of panels, to be sure: some 11.5 billion. Schaeffer (2008) lists Sharp 208-watt panels at $1,149 retail price, which means those panels would have cost some $13.2 trillion in 2008, approximately the same as annual GDP. Four years later, the price of the same panels has dropped to $320 retail (as of March 4, 2012; e.g. www.ecodirect.com). If we imagine negotiating a price closer to wholesale (say $160 per panel) and stretching out the investment over a decade, then we are talking about a capital investment of approximately $184 billion per year, labor, inverters, and transmission lines not included. It is a lot of panels, but then we also import a lot of oil. According to the State Energy Data System (SEDS) maintained by the US Energy Information Administration, in 2010 Americans spent a total of $376.5 billion on motor gasoline, $365.9 billion on electricity, and $1.2 trillion on energy overall (www.eia.gov/beta/state/seds/data.cfm?incfile=/state/seds/sep_prices/total/pr_tot_US.html&sid=US). It is a lift, but not an impossible one.

Although the Air Force isn't quite up to covering all of the bombing range in silicon, they are proud of their efforts: the press release from Nellis AFB estimates that the US taxpayer will save $1 million per year as a result of the new installation of 72,000 panels (Whitney 2007); Kanigher (2010) documents plans to double up on another 160 acres of solar panels at Nellis. Perhaps Las Vegas should build solar plants rather than suburbs and use the Hoover Dam as part of a pumped storage scheme rather than straight hydroelectric? They've got a good start already: One of the largest concentrating solar facilities in the country is Nevada Solar One (75.7 MW); See US Energy Information Administration (2011a) for operating solar generation capacity in Nevada and elsewhere.

Another tack is to think of distributed solar power (Denholm and Margolis 2006). The same authors in 2008 estimate the amount of photovoltaic panel required per person in different parts of the country, taking into account differences in solar insolation; nationwide they figure it would take on average about 100 square meters per person to supply everybody with enough electricity at current usage. They cite a report by Chaudhari et al. (2004) that tries to estimate the amount of roof space available for solar panels nationwide. Obviously the roof has to be the right material, strong enough to hold the weight, oriented in a reasonable direction (generally south in the Northern Hemisphere), and not shaded by trees or other buildings; shading is the biggest issue, especially in leafy suburbs. That said, these folks figure there is potentially 67.4 billion square feet of roof space available nationwide for photovoltaics in 2010 taking into account other building-top uses like air conditioning and green roofs.

Fuller descriptions of the operating characteristics of photovoltaic panels and concentrating solar can be found in Forinash (2010) and Bloom (2009). IEEE (1989) provides a summary of electric power generation from the Solar Electric Generating Stations in California; Stoddard et al. (2006) update the story. Stromsta (2011) reports the news on melting salt in Spain.

Finkelstein and Organ (2001) describe all that you might like to know about Stirling engines or air engines, as they like to call them. Wikipedia had a nice animation of the unconventional operation of these engines; see en.wikipedia.org/wiki/Stirling_engine.

Biofuels are a fraught topic. Hill et al. (2006) and Groom et al. (2008) provide careful comparisons of energy inputs and outputs and land requirements of various biofuel crops. In terms of land use, algae seems the most promising alternative going forward, especially if coupled with excess nutrients recycled from flue gases and wastewater—see Clarens et al. (2010). Otherwise biofuels are a bad idea economically and ecologically. The number of motor vehicles that would need to be replaced is from the US Federal Highway Administration (2011c) for 2009. Of course moving to electrified transportation also means swapping out the vehicle fleet for a fleet of streetcars and NEVs instead.

In a study of how ethanol subsidies would affect markets and land use, Westcott (2007) suggests that grain prices would double, while taking land from other crops, particularly soybeans; indeed we have seen increases in grain prices worldwide, which may have helped trigger the Arab Spring of 2011 (Walsh 2011). The Energy Policy Act of 2005, the Food, Conservation, and Energy Act of 2008, and the Energy Improvement and Extension Act of 2008 all include tax credits and other incentives for biofuel production; the Congressional Budget Office (2010) estimated that these credits amount to a subsidy of $6 billion in 2009. I borrow the phrase "political fuels" from Montgomery (2010).

The Warm, Generous Earth

Estimates of the geothermal potential resource are provided in Green and Nix (2006) and Massachusetts Institute of Technology (2006). One potential issue with shallow geothermal is it is possible to destroy the resource in the short term; the rate of heat flowing through rocks from the center of the earth is fixed and slow, and therefore must be respected. The Geysers geothermal facility in California began experiencing declines in natural steam in 1987 after twenty-seven years of production, probably because of over-depletion. The solution was to inject treated wastewaters into the wells to generate further steams; see Khan (2010) for the interesting details. Geothermal power generation is also more controllable than solar or wind power because the plant operator can decide how much carrying liquid or gas to inject into the well at a time.

Although the oil and gas industry have detailed data on the costs to drill deep wells, they mostly keep it to themselves; what is known about costs is summarized by Bloomfield and Laney (2006). Figure on at least $1 million per well.

Economies of Scale

It's not straightforward to calculate exactly how many windmills, solar panels, or geothermal wells are required to meet the goal of a hundred-fold increase in renewable generation because how much energy can be generated varies place to place and by generation type, as described in the reference materials. However if we assume ten windmills per farm, one hundred solar panels per installation, and three wells per geothermal facility, we may be talking about somewhere on the order of 300,000 windmills, 1.5 million solar panels (or equivalent concentrating solar set-ups or Stirling engines, if we can get them to work), and 60,000 geothermal wells. There is nothing magical about this mix; we could have more solar and less geothermal, or more wind and less solar. Geothermal and offshore wind seem to be the lowest hanging fruit; the first because it is relatively cheap given current technological know-how and the second because it is America's largest unexploited resource and is closest to coastal cities. Remember we have over 360,000 oil wells; maybe some of those can be converted to geothermal? Once the environmental concerns are taken into account, we can start on offshore wind. For a more conservative view, see National Academy of Sciences (2010) or the US Energy Information Administration's Annual Energy Outlook at www.eia.gov/forecasts/aeo.

Statistics on the development of railroads are from Series K1 of the Historical Statistics of the United States (US Census Bureau and Social Science Research Council 1949); statistics on street railways are from Series Q264-273 of the Bicentennial edition of the historical statistics (US Bureau of the Census 1975). Miles of surfaced road are collated from Table M-200 in US Federal Highway Administration (1987). Number of housing units built between 1950 and 2000 is from US Census Bureau (2012a), number of wars fought includes the Korean Conflict, the Vietnam War, the Gulf War, the Iraq War, and the War on Terrorism from Fischer et al. (2007), and number of baseball games played is an estimate. Assessments of American potential are up to you.

Finally for the economically minded, it's important to recognize that the costs of renewable energy have dropped considerably over the last decade, and costs are poised to continue changing. Europe and China have invested heavily in renewable energy generation because they already acknowledged that over the long run, these technologies are our best chance for lowering energy costs and generating sustainable societies. Meanwhile the costs of conventional fossil fuel sources are also changing, generally in the opposite direction from renewable energies, with ongoing risks to the global economy from each new oil shock. Gate duties will further enhance this disparity and get us moving in the right direction.

Part III: Ramifications

Chapter 13
A Future

All characters appearing in this chapter are fictitious. Any resemblance to real persons, living or dead, is purely coincidental.

For alternative future visions see Bostrom and Cirkovic (2011), Brand (2011), Diamond (2004), and the 2012 platforms of the Republican and Democratic Parties.

Chapter 14
Cost, Sacrifice, and Evolution

Chernow (1998) describes John D. Rockefeller's giving philosophy in some detail. It is both amazing how much money Rockefeller made and how much of it he gave away, in the process affecting large portions of the cultural and scientific landscape of America, from the University of Chicago to the Rockefeller Foundation to (via his daughter-in-law and son) the Museum of Modern Art. Indeed many of the major philanthropic foundations are based on proceeds of the Siren song: Ford Foundation (automobiles), J. Paul Getty Trust (oil), The Robert Wood Johnson Foundation (chemicals and pharmaceuticals), John D. and Catherine T. MacArthur Foundation (banking and real estate), the Andrew W. Mellon Foundation (oil, steel, construction), and The Pew Charitable Trusts (oil and shipbuilding, especially oil tankers).

Paying More for Gas

For some discussion of the inelasticity of demand to oil price changes, see Barsky and Kilian (2004). What is the magic demand destruction price? No one knows for sure save perhaps the oil companies, and they aren't telling. A Consumer Union survey in 2007 found that "on average, car owners said prices would have to rise to $3.90 per gallon before they would 'drastically' reduce their driving" and a survey by the same group in 2008 found that during times of higher gasoline prices, Americans

buy smaller, more fuel-efficient cars. I suspect the demand destruction price is around $4/gallon in 2012. Perhaps we can learn something from the public policy experiment of using taxation to discourage tobacco use, which has been broadly, though not universally, effective—see for example Tworek et al. (2010).

Giving Up the Car

Several pioneering spirits have written about a carless existence, typically in exultant, favorable terms (e.g. Alvord 2000; Balish 2006). Urban people often celebrate freedom from the car (e.g. Owen 2009). Young people already seem to think cars are "so twentieth century"!

Moving House

Some statistics on current patterns of American mobility are available from the US Census; see Berkner and Faber (2003) and US Census Bureau (2011b). Companies also move; see Strauss-Kahn and Vives (2009). For some of the benefits of telecommuting, see Robert and Borjesson (2006) and Fonner and Reloff (2010). There are a variety of good sources on downtown revitalization including Breen and Rigby (2005), Dunham-Jones and Williamson (2011), and Tachieva (2010). Also helpful might be some of the books on living creatively in smaller spaces, such as Bartolucci and Kurzaj (2003) and Salomon (2009). Florida (2008) describes the emerging competition between towns.

Living with People

What can I say? People are annoying. And wonderful too. Truth is nobody can be described by just one adjective all the time. There are many books on handling confrontational situations gracefully. I have found Fisher and Ury (1991) particularly helpful, and a lot of the good advice in Faber and Mazlish (2012) can help with adults as well as with children. Robert Frost (1914) gives good advice on boundaries in his poem "Mending Wall." This section is dedicated to my colleague Kim Fisher, who might be the least annoying person on the face of the earth, but who was the first to point out this potential cost of Terra Nova.

Small Moves

Fishman (2006) provides a fascinating look into how the big-box retail strategy works. Crawford (2002) describes a detailed model of how cities could operate without cars; his freight strategies are particularly interesting, using a combination of rail and small delivery stations, an idea I gladly endorse. Estill (2008) describes some of the pleasures of living in more local economies; see also Shuman (2007). The idea is to create neighborhoods that function like small towns, even in big, dense cities.

Industrial Changes

See the discussion of airplanes in MacKay (2009) for why transcontinental travel will probably remain fundamentally powered by petroleum (though Wassener 2011 writes about aviation biofuels). Though romantic and energy-efficient, sailing ships are slow, and none of the attempts to build solar aircraft get anywhere close to the kind of mass necessary for moving large numbers of people. Energy density is what is required to go long distances fast, and that requirement brings us right back to oil. Similarly it is hard to imagine international shipping without petroleum. But it's easy to imagine less international shipping if we weren't so dependent on oil.

Ohland and Poticha (2009) provide a starting place for transforming the automotive industry into the streetcar industry. Mitchell et al. (2010) provide some interesting ideas for neighborhood electric vehicles. You can read more about United Streetcar here: unitedstreetcar.com.

One justification for the size of the streets and various other apparatuses of the city is they are required for safety. Since everyone agrees safety is important, the conversation usually stops there. However if we are going to move into this new world, then we will have to ask: How can we make our emergency services work in tighter spaces? How do denser places (think old European cities) manage without large fire trucks or wide ambulances? I don't know the specifics, but from what I know of the FDNY and the NYPD in New York, there is abundant capacity for creativity when it comes to make our cities safe and secure.

Restoration ecology is a growing science and art designed to avoid emergencies in the future. Learn more through Falk et al. (2006) and Howell et al. (2011).

Intrusive Government

For nudges in public policy, see the brilliant book by Sunstein and Thaler (2008). A necessary complement is Kahneman (2011) on how easily we can be nudged. That we need government does not seem to be in dispute; the question is what we want our government to aim for and how we want to get there.

Shrinking Government

As the land is finite, so is the revenue that can be drawn from it for government expenditure under a system of gate duties. Though I appreciate many of the things that elected government does, I also see that government, like any bureaucracy, will grow to fill the available space. Government is formed of people, and some of those people enter government as a means to aggrandize power, which is why the writers of the Constitution insisted on a system of checks and balances. To see both sides of the argument, compare Madrick (2010) and Miller (2006).

The most interesting part of the natural world are constraints—the constraints on growth, power, and influence. Constraints are what make a place special, what determine what lives there, and how well those ecosystems fare (Odum and Barrett 2004; Aber and Melillo 2002). Same in politics— the conditions need not be the same everywhere, but everywhere we need a political system limited by and responsive to the people and the world.

Chapter 15
Collateral Benefits

The opening quote is modified from Ralph Waldo Emerson, grandfather of the American environmental movement. Read his essays assembled in an edited volume by Crase (2010).

A Safer Country

The cost of keeping the country safe by deploying the military in the Middle East is studied by Stern (2010), focusing on aircraft carrier deployments, and Delucchi and Murphy (2008), more generally and conservatively. Delucchi and Murphy (2008) also cite estimates of Persian Gulf precautionary defense costs by Copulos (2003) at \$52–62 billion per year, Ravenal (1991) at \$50 billion per year, Kaufmann and Steinbrunner (1991) at \$64.5 billion per year, and Moreland (1985) at \$54 billion per year. Foreign aid expenditures in the Middle East and other parts of the world are given in US Census Bureau (2011d).

PS 175 budget numbers come from schools.nyc.gov for FY2012. For comparison purposes with my local school, I estimated total Persian Gulf expenditures outside of war time at approximately \$50 billion per year.

Just before the first bombs fell in Iraq on March 19, 2003, President George W. Bush spoke to the nation by television, saying: "The people of the United States and our friends and allies will not live at the mercy of an outlaw regime that threatens the peace with weapons of mass murder. We will meet that threat now, with our Army, Air Force, Navy, Coast Guard, and Marines, so that we do not have to meet it later with armies of fire fighters and police and doctors on the streets of our cities." The full speech is available at georgewbushwhitehouse.archives.gov/news/releases/2003/03/20030319-17.html.

Eight years later, former Vice President Dick Cheney continued to defend the decisions made in the run-up to the Iraq War, saying to NBC News: "If you look back at the proposition that we faced after 9/11 with respect to Saddam Hussein, we were concerned with the prospects of terrorists like the 9/11 crowd acquiring weapons of mass destruction. I think that's still the biggest threat we face. At the time, to go after Saddam Hussein and take him down, we eliminated a major source of proliferation" (Jackson 2011). Is that the biggest threat we face? I don't think so, not by a long shot.

Costs of the Iraq War are from Bilmes and Stigliz (2008). Brown University's "Costs of War" Project, available at costsofwar.org, provides another accounting. Department of Homeland Security budget figures are available on-line at www.dhs.gov/xabout/budget/dhs-budget.shtm. The FY2011 budget summary, which has a magnificent bald eagle on the front, listed requested discretionary spending for Department of Homeland Security at $47 billion, and mandated another $9 billion in fee and trust spending, for a total budgetary authority of $56 billion, up a billion from FY2010 and about four billion from FY2009. In case you were wondering, the entire FY2011 Department of Defense budget was $687 billion. You can read all about it at comptroller.defense.gov/budget.html. Measures of the military budget as a function of GDP are available in Table 11 of Daggett and Belasco (2009).

The EIA Annual Energy Outlook for 2011 projects domestic oil production at twelve to thirteen quadrillion BTU each year through 2030. Twelve quadrillion BTU is known to ordinary folks as about two billion barrels of crude oil per year.

Kane (2004) provides a summary of US troop levels and foreign deployments from 1950 to 2005. Troop levels are currently down from approximately 3.5 million during the Korean Conflict and again during the Vietnam War. However they are up from the

pre–World War II levels considerably. President Eisenhower recognized the change, noting in a speech in 1961:

"Our military organization today bears little relation to that known by any of my predecessors in peacetime, or indeed by the fighting men of World War II or Korea. Until the latest of our world conflicts, the United States had no armaments industry. American makers of plowshares could, with time and as required, make swords as well. But now we can no longer risk emergency improvisation of national defense; we have been compelled to create a permanent armaments industry of vast proportions. Added to this, three and a half million men and women are directly engaged in the defense establishment. We annually spend on military security more than the net income of all United States corporations. This conjunction of an immense military establishment and a large arms industry is new in the American experience.

"The total influence—economic, political, even spiritual—is felt in every city, every State house, every office of the federal government. We recognize the imperative need for this development. Yet we must not fail to comprehend its grave implications. Our toil, resources and livelihood are all involved; so is the very structure of our society. In the councils of government, we must guard against the acquisition of unwarranted influence, whether sought or unsought, by the military industrial complex. The potential for the disastrous rise of misplaced power exists and will persist. We must never let the weight of this combination endanger our liberties or democratic processes. We should take nothing for granted. Only an alert and knowledgeable citizenry can compel the proper meshing of the huge industrial and military machinery of defense with our peaceful methods and goals, so that security and liberty may prosper together."

Eisenhower knew of what he spoke; read the entire speech at www.h-net.org/~hst306/documents/

indust.html. I will also point out that the conditions that Eisenhower was responding to—the Cold War with the Soviet Union—have long since changed.

A Healthier Populace

An excellent treatise on the negative health effects of the current land use and transportation system is Frumkin et al. (2004), or in shorter form, Frumkin (2002). The consequences of obesity are cribbed from Stanford Hospital & Clinics (2011).

How much of a difference would less sprawl make in decreasing obesity? Samimi et al. (2009) constructed a computer model connecting land use, transportation, and exercise and predicted a 0.4 percent decline in obesity for every one percent decline in automobile use. Eid et al. (2008) dispute the connection between obesity and urban sprawl; instead they believe obese people just self-select to live in sprawling communities. Everyone agrees however that regardless of weight, age, or sex, walking and bicycling actively will improve your health. Frumkin et al. (2004) devote a chapter to the relationship between sprawl and exercise and note that having nearby sidewalks and footpaths, enjoyable scenery, safe areas to walk, and the example of other people out walking and exercising all induce folks to get out and move around. Sallis et al. (2004) update the story.

Not only will your physical health improve, so will your mental health: See Teychenne et al.'s (2010) review for adults, and Biddle and Asare (2011) on kids. For a perspective on intellectual life and walking, see Solnit (2001; Chapter 2 "The Mind at Three Miles Per Hour").

Motor vehicle accidents killed 36,284 Americans in 2009 (Kochanek et al. 2011, Table 2), a down year because the recession meant people were driving less; in 2008, 39,831 Americans perished in the cars, in 2007, 43,945 (Minino et al. 2011, Table 2). Rates of mortality from motor vehicle accidents were similar to suicides in 2009 (36,547) and outpaced septicemia (35,587), chronic liver disease and cirrhosis (30,444), and homicides (16,591) on the top causes of death list. The National Center for Health Statistics, one of the Centers for Disease Control and Prevention, reports National Vital Statistics back to 1900 (www.cdc.gov/nchs/deaths.htm). Summarizing their data on motor vehicle related deaths from 1906 through 2009 indicates that approximately 3.626 million Americans have died behind the wheel or automotive related accidents; mortality peaked in 1972 when 56,333 Americans died. By way of grim comparison, Fischer et al. (2007) summarize US military casualties from the Revolutionary War through Operation Iraqi Freedom (up to June 2, 2007). I added in estimated Confederate casualties during the Civil War (258,000 deaths) and from Operation Enduring Freedom in Afghanistan (1,718 deaths as of fall 2011); all together I estimate that approximately 1.268 million souls have given everything for their nation on the battlefield or as a result of injuries or health risks from war. That is, about 2.8 Americans have died in motor vehicles for every one of the war dead. The odds of dying while transporting via various modes are from Injury Facts published by the National Safety Council, cited in Frumkin et al. (2004).

While you're out walking in the green dense towns of the future, I'm happy to report you'll have better air to breathe. Cars produce 80 percent of the carbon monoxide pollution in the United States and are significant contributors to various volatile organic compounds (notably benzene, a carcinogen), nitrogen oxides (commonly abbreviated NOx, and pronounced to rhyme with "fox"), ammonia, and particulate matter. The EPA classifies particulates as "inhalable coarse particulates" between 2.5 and 10 micrometers in diameter, and "fine particulates" as less than 2.5 micrometers. The recent good news is that air quality has improved noticeably in America since the late twentieth

century; the bad news is there is still a lot of air pollution. In 2008 more than 128 million Americans were exposed to air pollution levels considered unhealthy (US Environmental Protection Agency 2010c).

Brunekreef and Holgate (2002) provide a synthetic review of the health effects of air pollution, as does Chapter 4 in Frumkin et al. (2004). The factoid about reduced life expectancy is from Halperin (1993). Further details can be found from the websites of the Environmental Protection Agency (www.epa.gov/oar/airpollutants.html) and the Centers for Disease Control and Prevention (www.cdc.gov/nceh/airpollution).

You can learn about the health effects of water pollution in the indefatigable Frumkin (2004), Chapter 7, and from Schwarzenbach et al. (2010). Teal and Howarth (1984) summarize some of the ecological consequences. Toxicity of some of the chemicals used during the Deepwater Horizon clean-up in 2010 and a lot more are summarized by Restore the Gulf Task Force (2011) website.

A Greener Continent

There are several excellent summaries of the ecological consequences of roads. See: Benitez-Lopez et al. (2010), Jones et al. (2000), Spellerberg (1998), and Trombulak and Frissell (2000). Forman (2000) estimates that the direct ecological effect of roads falls on approximately one-fifth of the land area of the lower forty-eight states. The Globio Project estimates the biodiversity consequences of roads globally (Pereira et al. 2010). Fortunately because the effects of roads are fairly well understood there are a number of conservation strategies that can limit the negative effects, most of which can be applied with equal aplomb to rail projects. To read about what can be done, pick up tomes by my colleagues at WCS: Beckmann et al. (2010) and Hilty et al. (2006).

The Great American expansion (or "sprawl") has also seen its share of scholarly treatment by ecologists; see for example Johnson and Klemens (2005). Growing literatures are addressing urban ecology with forethought and data—see reviews by Pickett et al. (2011) and Grimm (2008). On the other end of the human footprint, an emerging science of exurban development is showing the manifold consequences of that second home in the woods—see Radeloff et al. (2010), Hansen et al. (2005), and Theobold et al. (2005).

The importance of disturbance to ecosystems is a deep topic and a life changing one—it helps us see "disasters" of all kinds in a new light. I wrote about this topic in *Mannahatta* and for a brief op-ed published in the *New York Times* on September 11, 2009; a more scholarly treatment of the same issues can be found in Pickett et al. (1997), Resh et al. (1988), and Gunderson (2000). Hurricanes Sandy and Katrina have made disturbance a nationwide topic. The flip side of disturbance is resilience: See Hey and Phillips (1995) for calculation of potential reduced flooding risk due to beaver activity. Such studies fall under the general title of ecosystem services that nature provides humanity; see Costanza et al. (1997, 1998) and Loomis et al. (2000).

A Cooler Climate

Scientists the world over have been meeting for two decades now to summarize the data about climate change every five years; the latest work of the Intergovernmental Panel on Climate Change is a series of reports published in 2007. Some highlights to convey the magnitude of the problem:
• Eleven of the last twelve years (1995–2006) rank among the twelve warmest years since 1850, when regular measurements of temperature began;
• The 100-year linear trend suggests an average global warming of +0.74 degrees C;
• Sea levels have been rising since 1961 at an average rate of 1.8 mm per year and the rate of sea level

rise has escalated since 1993 to 3.1 mm per year because of thermal expansion, melting glaciers, and disappearing ice caps;
• Arctic sea ice has decreased 2.7 percent per decade since 1978 according to satellite measurements, and overall snow cover and size of mountain glaciers have declined in both the Northern and Southern hemispheres;
• Tropical cyclone activity has increased in the North Atlantic since 1970. But you didn't need me to tell you that.

Why are these things happening? The climate is a physical system, and you know what energy is. Changing the amount of greenhouse gases and aerosols in the atmosphere means that the air holds in more energy than it otherwise would; that energy has to go somewhere and it does: into the land and the ocean in the form of warming. It's not the first time the climate has changed. What's different now is that human activity has added greenhouse gases to the atmosphere at a pace faster than nature ever could; though two hundred years seems like a long time to us, it's just a brief moment in the way the earth system works, which normally metes out such dramatic changes on the scale of millennia.

Although one might have hoped that having the cheap oil window closed, greenhouse gas emissions would have gone down, but instead they increased by 70 percent between 1970 and 2004. We know this fact and many others about greenhouse gas levels in the atmosphere because of direct measurements (for example, at the Mauna Loa observatory on Hawaii—see Tans and Keeling 2012) and because we can compare concentrations in the air today to bubbles caught in the ice on the poles and in glaciers (e.g., Lüthi et al. 2008). Cool, no? Ironically if it weren't for increased greenhouse gas emissions, the computer models suggest that the combination of slight decreases in solar radiation and dust from volcanic eruptions since 1950 would have resulted in cooling over the last fifty years.

And where do the greenhouse gas emissions come from? In the United States, the EPA inventories greenhouse gas emissions annually, with the 2011 report covering emissions from 1990 to 2009. In 2009 the transportation sector added 1,724.1 million metric tons of greenhouse gases to the atmosphere (measured in CO_2 equivalents), or 33 percent of all US greenhouse gases. The remaining 66 percent come from fossil fuels combusted for industrial processes and heating and for electricity generation (largely coal). Remember we are talking about tons of *gases*, not solids; 1.7 billion tons is a lot of gas! See where it all goes in Esser et al. (2011).

If you like your news bad rather than good about climate change and civilization, take your pick from Cullen et al. (2000), DeMenocal (2001), Diamond (2004), and Dalfes et al. (1996). Orr et al. (2005) provide a review of the nasty consequences of ocean acidification.

Happier Communities

For an account of the historical dwindling of American social capital over the twentieth century, absorb Robert Putnam's patient, detailed analysis in *Bowling Alone* (2001). If you prefer to be yelled at, try Kunstler (1994).

Frumkin et al. (2004) suggest three reasons why sprawling development undermines social capital: (1) people have less time to engage in civic activities outside the home because they are spending so much time on the commute; (2) because people are spending more time in their car, there are fewer opportunities for "spontaneous, informal social interaction," also known as bumping into each other; (3) sprawl privatizes the public realm. Why care about your local park or the plaza downtown if you never spend any time there? Putnam (2001) estimates that "each ten additional minutes in daily commuting time cuts involvement in community affairs by 10 percent." Thus commuting inflicts a "civic penalty" not only on commuters, who are too exhausted from the commute to

attend, but also on non-commuters like retirees and others outside the workforce who are left holding the bag at community events.

If you are looking for empirical evidence, Glynn (1981) compared two suburbs in Maryland: Greenbelt, a designed community with well-defined boundaries and a central mall area, and Hyattsville, a more sprawling, unbounded, less differentiated community. He found that people had a stronger sense of community in Greenbelt and connected it to walkability and the opportunity to get to know the neighbors. Related recent corroborating studies include Freeman (2001), Lund (2002), Leyden (2003), and Hester (2010).

Greater Freedom

If you don't like regulations, the best way to get rid of them is to obviate the need. Though laws are not inconsistent with freedom, and in many instances freedom is protected by the law, having fewer rules will make the rules that remain easier to enforce and simpler to follow (cf. Leoni 1991). One consequence of our many rules and punishments is a burgeoning criminal justice system, which has particularly sore effects on African-American and Latino men. See Stuntz (2010) for detailed treatment; Gopnik (2012) has a disturbing review.

Fresher Food

The argument for fresher foods has been made more eloquently elsewhere; favorites of mine include Nabhan (2002), Kains (1940), and everyone's favorite, Pollan (2006). To actually know how to garden and manage the land in and about the city, read with pleasure Donahue (2001) or Haeg et al. (2010). A delightful account of the relationship of city and field is Steel's *Hungry City* (2009).

A Better World

The International Energy Agency (IEA) (2010) estimates that total world energy demand in 2008 was equal to 12,271 million tonnes of oil equivalent (Mtoe), where one "toe" is equal to 11.63 MWh or 11,630 hours of microwaving energy. Thus the world used an equivalent of 142.7 trillion hours of microwaving energy in 2008, or 391 billion MacKays (kWh per day)—details in IEA (2010)'s Table 2.2. Depending on various scenarios, global energy demand could rise to 14,920 Mtoe by 2035 (climate responsive scenario—good) or 18,048 Mtoe (current policies scenario—bad). Global electricity consumption is given in IEA (2010)'s Table 7.1 as 16,819 TWh or 16.8 trillion hours of microwaving or 46 billion MacKays.

The estimate of global wind power potential is from Archer and Jacobson (2005), based on actual wind measurements at 8,199 locations around the world; it includes both onshore and offshore localities. Their calculation is considered conservative because it is based only on observed data (when some areas are not well sampled) and only uses areas with wind power estimates of class 3 or better. They use a capacity factor of 48 percent from their studies of actual turbine performance in these wind classes.

The estimate of concentrating solar power potential is from Breyer and Knies (2009), Table 1. They take into account the distribution of population and the potential range of high voltage direct current transmission lines and subtract transmission losses from their estimates; they assume 10 percent land use efficiency. The geothermal resource base potential estimate is from Mock et al. (1997) and includes four components on a global scale: hydrothermal (130,000 quads), geopressured (540,000 quads), magma (5 million quads), and two grades of "hot dry rock"—the low-grade resource could provide 78.5 million quads and the moderate grade resource 26.5 million quads. "Hot dry rock" is essentially found everywhere at various depths and depends on pumping water or gas into the rock to extract the heat to turn a turbine. Mock et al.'s geopressured estimate includes the energy available from burning the methane mixed in.

As a comparison to what is suggested in this book, MacKay (2009) summarizes his preferred energy path for the UK this way:

"First, we electrify transport. Electrification both gets transport off fossil fuels, and makes transport more energy-efficient. (Of course, electrification increases our demand for green electricity.) Second, to supplement solar-thermal heating, we electrify most heating of air and water in buildings using heat pumps, which are four times more efficient than ordinary electrical heaters. This electrification of heating further increases the amount of green electricity required. Third, we get all the green electricity from a mix of four sources: from our own renewables; perhaps from "clean coal"; perhaps from nuclear; and finally, and with great politeness, from other countries' renewables. Among other countries' renewables, solar power in deserts is the most plentiful option. As long as we can build peaceful international collaborations, solar power in other people's deserts certainly has the technical potential to provide us, them, and everyone with 125 kWh per day per person." For other visions, see Gilbert and Perl (2010) and Crawford (2002).

The European Community (EC) is applying strong language and policies that have effectively moved the energy landscape of Europe toward renewables, even though as a continent it is rather small, rather cloudy, and not very windy, except offshore. See EC (2005). Africa has huge untapped renewable energy potential that the Europeans are already greedily eyeing. See German Aerospace Center (2005). Kamkwamba (2009) tells a heartwarming tale of a windmill on-line. My wish is that developing countries, in Africa and elsewhere, skip immediately to a future based on renewables, especially as the populations continue to urbanize (Hanson 2007). For the renewable energy potential of other parts of the world, start your reading with Arango and Larsen (2010), Geoff (2011), Hoogwijk et al. (2004), Parikka (2004),

Ramachandra et al. (2011), and Yusaf et al. (2011). It's more than I can do to say how each country could change; each will have to figure it out on its own.

The concluding story of Turtle Island is adapted from Hitakonanu'laxk (2005). As Joseph Campbell (1949) reminds us: "The unconscious sends all sorts of vapors, odd beings, terrors, and deluding images into the mind—whether in dream, broad daylight, or insanity. . . destruction of the world that we have built and in which we live, and of ourselves within it; but then a wonderful reconstruction, of the bolder, cleaner, more spacious, and fully human life—that is the lure, the promise and terror, of these disturbing. . . visitants from the mythological realm that we carry within."

Bibliography

Aber, J. D. and Melillo, J. M., 2002. *Terrestrial Ecosystems,* Florence, KY: Brooks Cole.

Abuelsamid, S., 2009. What Is a Neighborhood Electric Vehicle (NEV)? green.autoblog.com/2009/02/06/greenlings-what-is-a-neighborhood-electric-vehicle-nev.

Acevedo, W., 2006. Rates, Trends, Causes, and Consequences of Urban Land-Use Change in the United States, Reston, VA: US Department of the Interior, US Geological Survey.

Adams, H., 1907. *The Education of Henry Adams: An Autobiography,* Boston: Privately published.

Adams, J. T., 1931. *The Epic of America,* Boston: Little, Brown.

Adkins, F., 1987. Raccoon Mountain Pumped-Storage Plant–Ten Years Operating Experience. *IEEE Transactions on Energy Conversion,* (3):361–368.

Agrawal, R., Offutt, M., and Ramage, M. P., 2005. Hydrogen Economy—An Opportunity for Chemical Engineers? *AIChE Journal,* 51(6):1582–1589.

Ahlfeldt, G. M. and Wendland, N., 2009. Looming Stations: Valuing Transport Innovations in Historical Context. *Economics Letters,* 105(1):97–99.

Alexander, C., Ishikawa, S., and Silverstein, M., 1977. *A Pattern Language: Towns, Buildings, Construction,* Oxford, UK: Oxford Univ. Press.

Altshuler, A. et al. eds., 1999. Governance and Opportunity in Metropolitan America: Committee on Improving the Future of US Cities through Improved Metropolitan Area Governance, Washington D.C.: National Academy Press.

Alvaredo, F. et al., 2012. The World Top Incomes Database, Paris: Paris School of Economics.

Alvord, K., 2000. *Divorce Your Car!: Ending the Love Affair with the Automobile,* Gabriola Island, BC: New Society.

Ambrose, S. E., 2001. *Nothing Like It In the World: The Men Who Built the Transcontinental Railroad 1863–1869,* New York: Simon & Schuster.

Amerada Hess Corporation, 2004. Material Safety Data Sheet. Gasoline, All Grades, Woodbridge, NJ: Amerada Hess Corporation.

American Automotive Association, 2012. Your Driving Costs: 2012 Edition, Heathrow, FL: AAA Association Communication.

American Lung Association, 2011. State of the Air 2011, Washington D.C.: American Lung Association.

American Society of Civil Engineers, 2009 Report Card for America's Infrastructure, Reston, VA: American Society of Civil Engineers.

American Society of Planning Officials, 1951. The Journey to Work: Relation Between Employment and Residence, Chicago, IL: American Society of Planning Officials.

American Transit Association, 1971. '70–'71 Transit Factbook, Washington, D.C.: American Transit Association.

Anderson, G. M. and Martin, D. M., 1987. The Public Domain and Nineteenth Century Transfer Policy. *Cato Journal,* 6(3):905–923.

Andrewartha, H. G. and Birch, L. C., 1961. *The Distribution and Abundance of Animals,* Chicago, IL: Univ. of Chicago Press.

Andrews, A. et al., 2009. Unconventional Gas Shales: Development, Technology, and Policy Issues, Washington D.C.: Congressional Research Service.

Appadurai, A., ed., 1988. *The Social Life of Things: Commodities in Cultural Perspective,* Cambridge UK: Cambridge Univ. Press.

Arango, S. and Larsen, E. R., 2010. The Environmental Paradox in Generation: How South America Is Gradually Becoming More Dependent on Thermal Generation. *Renewable and Sustainable Energy Reviews,* 14(9):2956–2965.

Archer, C. L. and Jacobson, M. Z., 2005. Evaluation of Global Wind Power. *Journal of Geophysical Research,* 110(D12110):1–20.

Argonne National Laboratory, 2012. Argonne GREET Model: The Greenhouse Gases, Regulated Emissions, and Energy Use in Transportation Model, Argonne, IL: Argonne National Laboratory.

Arthur, J. D., Bohm, B., and Layne, M., 2008. Hydraulic Fracturing Considerations for Natural Gas Wells of the Marcellus Shale. Ground Water Protection Council 2008 Annual Forum, Cincinnati, OH.

Associated Press, 2008. Tons of Drugs Dumped into Wastewater. Sept. 14.

Association of American Railroads, 2010. *Railroad Facts 2010.* Association of American Railroads, Washington D.C.

Atkinson, R., 1994. *Crusade: The Untold Story of the Persian Gulf War,* New York: Mariner.

Attari, S. Z. et al., 2010. Public Perceptions of Energy Consumption and Savings. Proceedings of the National Academy of Sciences, 107(37):16054–16059.

Aune, F. R. et al., 2010. Financial Market Pressure, Tacit Collusion, and Oil Price Formation. *Energy Economics,* 32(2):389–398.

Baigrie, B., 2006. *Electricity and Magnetism: A Historical Perspective,* Westport, CT: Greenwood.

Bailey, B. and Farber, D. eds., 2004. *America in the Seventies,* Lawrence, KS: Univ. Press of Kansas.

Baldassare, M. and Feller, S., 1975. Cultural Variations in Personal Space: Theory, Methods, and Evidence. *Ethos,* 3(4):481–503.

Baldwin, N., 2001. *Edison: Inventing the Century,* Chicago, IL: Univ. of Chicago Press.

Balish, C., 2006. *How to Live Well Without Owning a Car: Save Money, Breathe Easier, and Get More Mileage Out of Life,* Berkeley, CA: Ten Speed Press.

Balmford, A. et al., 2001. Economic Reasons for Conserving Wild Nature. *Science,* 297(5583):950–953.

Banner, S., 2011. *American Property: A History of How, Why, and What We Own,* Cambridge, MA: Harvard Univ. Press.

Barabasi, A.-L., 2003. *Linked:*

How Everything Is Connected to Everything Else and What It Means, New York: Plume.

Barber, L. B. et al., 2005. Chemical Loading into Surface Water along a Hydrological, Biogeochemical, and Land Use Gradient: A Holistic Watershed Approach. *Environmental Science & Technology*, 40(2):475–486.

Barnes, G., 2001. Population and Employment Density and Travel Behavior in Large US Cities, St. Paul, MN: Minnesota Department of Transportation.

Barnett, J. L., 2011. State and Local Government Finances Summary: 2009, Washington D.C.: US Census Bureau.

Barsky, R. S. and Kilian, L., 2004. Oil and the Macroeconomy Since the 1970s. *Journal of Economic Perspectives*, 18(4):115–134.

Bartolucci, M. and Kurzaj, R., 2003. *Living Large in Small Spaces: Expressing Personal Style in 100 to 1,000 Square Feet*, New York: Harry N. Abrams.

Bayar, T., 2011. World Wind Market: Record Installations, But Growth Rates Still Falling. Renewable Energy World. www.renewableenergyworld. com/rea/news/article/2011/08/ world-wind-market-record-installations-but-growth-rates-still-falling?cmpid=rss.

Beatley, T., 2000. *Green Urbanism: Learning from European Cities*, Washington, D.C.: Island Press.

———, 2008. *Green Urbanism Down Under: Learning from Sustainable Communities in Australia*, Washington, D.C.: Island Press.

Becker, J. et al., 2008. Bush Drive for Home Ownership Fueled Housing Bubble. *New York Times*, Dec. 21.

Beckmann, J. P. et al., eds. 2010. *Safe Passages: Highways, Wildlife, and Habitat Connectivity*, Washington, D.C.: Island Press.

Beder, S., 2003. *Power Play: The Fight to Control the World's Electricity*, New York: New Press; distrib. by W. W. Norton.

Beissinger, S. R. and McCullough,

D. R., eds., 2002. *Population Viability Analysis*, Chicago, IL: Univ. of Chicago Press.

Belasco, A., 2011. The Cost of Iraq, Afghanistan, and Other Global War on Terror Operations Since 9/11, Washington D.C.: Congressional Research Service.

Beller, E. E. et al., 2011. Historical Ecology of the Lower Santa Clara River, Ventura River, and Oxnard Plain: An Analysis of Terrestrial, Riverine, and Coastal Habitats.

Benitez-Lopez, A., Alkemade, R., and Verweij, P., 2010. The Impacts of Roads and Other Infrastructure on Mammal and Bird Populations: A Meta-analysis. *Biological Conservation*, 143(6):1307–1316.

Berkner, B. and Faber, C. S., 2003. Geographical Mobility: 1995 to 2000, Washington D.C.: US Census Bureau, Department of Commerce.

Berlinski, D., 2002. *Newton's Gift: How Sir Isaac Newton Unlocked the System of the World*, New York: Free Press.

Berman, B., 2011. *The Sun's Heartbeat: And Other Stories from the Life of the Star That Powers Our Planet*, New York: Little, Brown.

Bernanke, B. S., 1983. Irreversibility, Uncertainty, and Cyclical Investment. *Quarterly Journal of Economics*, 98(1):85–106.

———, 2006. Monetary Aggregates and Monetary Policy at the Federal Reserve: A Historical Perspective. Frankfurt, Germany: Fourth ECB Banking Conf.

Bernanke, B. S., Gertler, M., and Watson, M., 1997. Systematic Monetary Policy and the Effects of Oil Price Shocks. Brookings Papers on Economic Activity, 28(1):91–157.

Berry, W., 1996. *The Unsettling of America: Culture and Agriculture*, San Francisco, CA: Sierra Club.

Bettencourt, L. M. A. et al., 2007. Growth, Innovation, Scaling, and the Pace of Life in Cities. *Proceedings of the National Academy of Sciences*,

104(17):7301–7306.

Bezemer, D. J., 2009. "No One Saw This Coming": Understanding Financial Crisis Through Accounting Models. Munich Personal RePEc Archive, (15982).

Biddle, S. J. H. and Asare, M., 2011. Physical Activity and Mental Health in Children and Adolescents: A Review of Reviews. *British Journal of Sports Medicine*, 45(11):886–895.

Bierhorst, J. 1995. *Mythology of the Lenape: Guide and Texts*, Tucson, AZ: Univ. of Arizona Press.

Biewick, L. R. H., 2008. Areas of Historical Oil and Gas Exploration and Production in the United States, Reston, VA: US Geological Survey.

Billi, R. M. and Kahn, G. A., 2008. What Is the Optimal Inflation Rate? *Federal Reserve Bank of Kansas City Economic Review*, Second Quarter: 5–28.

Bilmes, L. J. and Stiglitz, J. E., 2008. *The Three Trillion Dollar War: The True Cost of the Iraq Conflict*, New York: W. W. Norton.

Bin Laden, O., 1996. "Declaration of War Against the Americans Occupying the Land of the Two Holy Places," Aug. 23, 1996. www. terrorismfiles.org/individuals/declaration_of_jihad1.html (English translation).

Black, B., 1998. Oil Creek as Industrial Apparatus: Re-creating the Industrial Process through the Landscape of Pennsylvania's Oil Boom. *Environmental History*, 3(2):210–229.

Black, E., 2007. *Internal Combustion: How Corporations and Governments Addicted the World to Oil and Derailed the Alternatives*, New York: St. Martin's Griffin.

Blackman, A. and Krupnick, A., 2001. Location Efficient Mortgages: Is the Rationale Sound? (Discussion Paper 99-49REV). Resources for the Future, Washington D.C.

Blaisdell, B. ed., 2000. *Great Speeches by Native Americans*, Mineola, NY: Dover.

Blank, D. M., 1954. *The Volume*

of Residential Construction, 1889–1950. Ann Arbor, MI: UMI Dissertation Publishing.

Blankenship, R. E., 2010. Early Evolution of Photosynthesis. *Plant Physiology,* 154:434–438.

Bloom, A. J., 2009. *Global Climate Change: Convergence of Disciplines,* Sunderland, MA: Sinauer.

Bloom, A. J., Chapin, F. S. I., and Mooney, H. A., 1985. Resource Limitation in Plants—An Economic Analogy. *Annual Review of Ecology and Systematics,* 16:363–392.

Bloomfield, K. K. and Laney, P. T., 2010. Estimating Well Costs for Enhanced Geothermal System Applications, Idaho Falls, ID: Idaho National Laboratory.

Blumstein, C., Friedman, L. S., and Green, R., 2002. The History of Electricity Restructuring in California. *Journal of Industry, Competition and Trade,* 2(1):9–38.

Boeing, 2002. 747-400, www.boeing.com/commercial/airports/plan_manuals.html.

———, 2010. 737, www.boeing.com/commercial/airports/plan_manuals.html.

Bohi, D. S., 1989. Energy Price Shocks and Macroeconomic Performance, Washington D.C.: Resources for the Future.

Bostrom, N. and Cirkovic, M. M., 2011. *Global Catastrophic Risks.,* New York: Oxford Univ. Press.

Bottles, S. L., 1991. *Los Angeles and the Automobile: The Making of the Modern City,* Berkeley, CA: Univ. of California Press.

Boyce, D. G., Lewis, M. R., and Worm, B., 2010. Global Phytoplankton Decline Over the Past Century. *Nature,* 466(7306):591–596.

Boyle, R. H., 1969. *The Hudson River: A Natural and Unnatural History,* New York: W. W. Norton.

Boyne, W. J., 2002. *The Two O'Clock War: The 1973 Yom Kippur Conflict and the Airlift That Saved Israel,* New York: Thomas Dunne.

BP, 2011. Statistical Review of World Energy 2011, London: BP.

Brand, S., 2011. *SALT Summaries,* Condensed Ideas About Long-term Thinking,* San Francisco, CA: Long Now.

Braungart, M., 2002. *Cradle to Cradle: Remaking the Way We Make Things,* New York: North Point.

Brayer, R. et al., 2006. Guidelines for the Establishment of a Model Neighborhood Electric Vehicle (NEV) Fleet, Idaho Falls, ID: Idaho National Laboratory.

Breen, A. and Rigby, D., 2005. *Intown Living: A Different American Dream,* Washington D.C.: Island Press.

Breyer, C. and Knies, G., 2009. Global Energy Supply Potential of Concentrating Solar Power. *Proc. SolarPACES,* 15–18.

Brinkley, D. G., 2004. *Wheels for the World: Henry Ford, His Company, and a Century of Progress,* New York: Penguin.

Brooks, B. W., Huggett, D. B., and Boxall, A. B. A., 2009. Pharmaceuticals and Personal Care Products: Research Needs for the Next Decade. *Environmental Toxicology and Chemistry,* 28(12):2469–2472.

Brown, A. C., 1999. *Oil, God, and Gold: The Story of Aramco and the Saudi Kings,* New York: Houghton Mifflin.

Brown, D. G. et al., 2005. Rural Land-use Trends in the Conterminous United States, 1950–2000. *Ecological Applications* 15(6): 1851–1863.

Brown, D. G. and Robinson, D. T., 2006. Effects of Heterogeneity in Residential Preferences on an Agent-Based Model of Urban Sprawl. *Ecology and Society,* 11(1):46.

Brown, D. H., Brennan, T. C., and Unmack, P. J., 2007. A Digitized Biotic Community Map for Plotting and Comparing North American Plant and Animal Distributions. *Canotia,* 3(1):1–12.

Brown, J. H., 1995. *Macroecology,* Chicago, IL: Univ. of Chicago Press.

Brownstone, D. and Golob, T. F., 2009. The Impact of Residential Density on Vehicle Usage and Energy Consumption. *Journal of Urban Economics,* 65(1):91–98.

Brunekreef, B. and Holgate, S. T.,

2002. Air Pollution and Health. *Lancet,* 360:1233–1242.

Bruno, M. and Sachs, J. D., 1985. *Economics of Worldwide Stagflation,* Cambridge, MA: Harvard Univ. Press.

Bryan, F. R. and Ford, B., 2002. *Friends, Families and Forays: Scenes from the Life and Times of Henry Ford,* Detroit, MI: Wayne State Univ. Press.

Bryce, R., 2004. *Cronies: Oil, The Bushes, and the Rise of Texas, America's Superstate,* New York: PublicAffairs.

———, 2005. Gas Pains. *The Atlantic.* May.

———, 2010. The Real Problem with Renewables. *Forbes.* May 11.

Buchanan, M., 2002. *Nexus: Small Worlds and the Groundbreaking Theory of Networks,* New York: W. W. Norton.

Buk-Swienty, T., 2008. *The Other Half: The Life of Jacob Riis and the World of Immigrant America,* New York: W. W. Norton.

Burrough, B., 2009. *The Big Rich: The Rise and Fall of the Greatest Texas Oil Fortunes,* New York: Penguin.

Byrne, D., 2009. *Bicycle Diaries,* New York: Viking Adult.

California Department of Conservation, 2009. Annual Report of the State Oil & Gas Supervisor, 2009, Sacramento, CA: Department of Conservation.

California High-Speed Rail Authority, 2012. Revised 2012 Business Plan. California High-Speed Rail Authority, Sacramento, CA.

Calthorpe, P. and Fulton, W., 2001. *The Regional City,* Washington, D.C.: Island Press.

Campbell, J., 1949. *The Hero with a Thousand Faces,* New York: Pantheon.

Campos, J. and de Rus, G., 2009. Some Stylized Facts about High-Speed Rail: A Review of HSR Experiences around the World, *Transport Policy* 16, 19–28.

Caplan, B., 2008. Externalities. *The Concise Encyclopedia of Economics.* www.econlib.org/library/Enc/Externalities.html.

Cappiello, N., 2011. "Drill, Baby, Drill" Remains Republican Solution Even After Oil Spill. *Huffington Post.* Oct. 27.

Carmalt, S. and St. John, B., 1986. Giant Oil and Gas Fields, in Halbouty, M. T., ed., *Future Petroleum Provinces of the World*, 11–53. Tulsa, OK: American Association of Petroleum Geologists.

Carnevale, M. L., 2008. Steele Gives GOP Delegates New Cheer: "Drill, Baby, Drill!" Washington Wire–*Wall Street Journal.* Sept. 3.

Carson, R., 1962. *Silent Spring*, Boston: Houghton Mifflin.

Case, K. E., Fair, R. C., and Oster, S. C., 2008. *Principles of Macroeconomics* 9th ed., Upper Saddle River, NJ: Prentice Hall.

Celiz, M. D., Tso, J., and Aga, D. S., 2009. Pharmaceutical Metabolites in the Environment: Analytical Challenges and Ecological Risks. *Environmental Toxicology and Chemistry*, 28(12):2473–2484.

Center for Housing Policy, Local Initiatives Support Corporation and Urban Institute, 2011. LISC Foreclosure Risk Scores. www.foreclosure-response.org/maps_and_data/lisc_data.html#sub3.

Central Intelligence Agency, 2012. The World Factbook, Washington, D.C.: Central Intelligence Agency Office of Public Affairs.

Cervantes Saavedra, M., 1605. *El Ingenioso Hidalgo Don Quixote De La Mancha* [Part I]. Madrid: Juan de la Cuesta.

Cervero, R., 1994. *The Transit Metropolis*, Washington, D.C.: Island Press.

Charles River Associates, 2005. Primer on Demand-Side Management, with an Emphasis on Price-Responsive Programs, Washington D.C.: World Bank.

Chaudhari, M., Frantzis, L. and Hoff, T. E., 2004. PV Grid Connected Market Potential under a Cost Breakthrough Scenario, Chicago, IL: Navigant Consulting for The Energy Foundation.

Chepesiuk, R., 2005. Decibel Hell: The Effects of Living in a Noisy World. *Environmental Health Perspectives*, 113, A34–A41.

Chernow, R., 1998. *Titan: The Life of John D. Rockefeller, Sr.*, New York: Random House.

Cheung, S., 1973. The Fable of the Bees: An Economic Investigation. *Journal of Law and Economics*, 16(1):11–33.

Chevron Corporation, 2006. Aviation Fuels Technical Review, San Ramon, CA: Chevron Corporation.

Childs, W. R., 2005. *The Texas Railroad Commission: Understanding Regulation in America to the Mid-Twentieth Century*, College Station, TX: Texas A&M Univ. Press.

Cho, E. J., Rodriguez, D., and Song, Y., 2008. The Role of Employment Subcenters in Residential Location Decisions. *Journal of Transport and Land Use*, 1(2):121–151.

City of New York Department of City Planning, 2011. MapPLUTO, New York: Department of City Planning.

Chura, P., 2011. *Thoreau the Land Surveyor*, Gainesville, FL: Univ. Press of Florida.

Clarens, A. F. et al., 2010. Environmental Life Cycle Comparison of Algae to Other Bioenergy Feedstocks. *Environmental Science & Technology*, 44:1813–1818.

Clark, W. A. V. and Burt, J. E., 1980. The Impact of Workplace on Residential Relocation. *Annals of the Association of American Geographers*, 70(1):59–67.

Coase, R., 1960. The Problem of Social Cost. *Journal of Law and Economics*, 3(1):1–44.

Coffman, C. and Gregson, M. E., 1998. Railroad Development and Land Value. *Journal of Real Estate Finance and Economics*, 16(2):191–204.

Coghlan, G. and Sowa, R., 2000. National Forest Road System and Use, Washington D.C.: US Forest Service.

Cohen, P. E. and Augustyn, R. T., 1997. *Manhattan in Maps 1527–1995*, New York: Rizzoli.

Coll, S., 2004. *Ghost Wars: The Secret History of the CIA, Afghanistan, and Bin Laden, from the Soviet Invasion to September 10, 2001*, New York: Penguin.

Committee on Foreign Affairs, Subcommittees on Foreign Economic Policy and Near East and South Asia, 1973. Oil Negotiations, OPEC, and the Stability of Supply. US Congress, Ninety-third session, first session, Washington D.C.

Common, M. and Stagl, S., 2005. *Ecological Economics: An Introduction*, Cambridge, UK: Cambridge Univ. Press.

Congressional Budget Office, 2000. Reforming the Federal Royalty Program for Oil and Gas, Washington D.C.: Congressional Budget Office.

———, 2010. Using Biofuel Fuel Tax Credits to Achieve Energy and Environmental Policy Goals, Washington, D.C.: Congressional Budget Office.

Conkin, P. K., 2008. *A Revolution Down on the Farm: The Transformation of American Agriculture Since 1929*, Lexington, KY: Univ. Press of Kentucky.

Conrad, J. M., 2010. *Resource Economics*, Cambridge, UK: Cambridge Univ. Press.

Conservation Biology Institute, 2010. Protected Areas Database US 1.1 (CBI Edition).

Consumer Union, 2007. Survey Finds Car Buyers Seek Fuel Efficiency. *Consumer Reports*, May.

———, 2008. Gas Prices Survey Shows Pain at the Pump Hurts at Home. *Consumer Reports*, June.

Converse, H. M., 1908. *Myths and Legends of the New York State Iroquois*. Albany, NY: Univ. of the State of New York.

Copulos, M. R., 2003. America's Achilles Heel: The Hidden Costs of Imported Oil, Washington, D.C.: The National Defense Council Foundation.

Cornes, R. and Sandler, T., 1996. *The Theory of Externalities, Public Goods, and Club Goods*, Cambridge, UK: Cambridge Univ. Press.

Costanza, R. et al., 1997. The Value of the World's Ecosystem

Services and Natural Capital. *Nature,* 387(6630):253–260.

———, 1998. The Value of Ecosystem Services: Putting the Issues in Perspective. *Ecological Economics,* 25(1):67–72.

Courtney, R. G., 2011. *My Kind of Countryside,* Chicago, IL: Center for American Places.

Crase, D., ed., 2010. *Ralph Waldo Emerson: Essays: The First and Second Series.* Des Moines, IA: Library of America.

Crawford, J. H., 2002. *Carfree Cities,* Utrecht, The Netherlands: International Books.

Cronon, W., 1983. *Changes in the Land: Indians, Colonists and the Ecology of New England,* New York: Hill & Wang.

———, 1995. The Trouble with Wilderness; or, Getting Back to the Wrong Nature, in W. Cronon, ed., *Uncommon Ground: Rethinking the Human Place in Nature.* New York: W. W. Norton.

Crosby, A. W., 1973. *The Columbian Exchange: Biological and Cultural Consequences of 1492,* Westport, CT: Greenwood.

Cullen, H. M. et al., 2000. Climate Change and the Collapse of the Akkadian Empire: Evidence from the Deep Sea. *Geology,* 28(4):379.

Daggett, S. and Belasco, A., 2009. Defense Budget for FY2003: Data Summary, Washington D.C.: Congressional Research Service.

Dahlin, E., Kelly, E. and Moen, P., 2008. Is Work the New Neighborhood? Social Ties in the Workplace, Family, and Neighborhood. *Sociological Quarterly,* 49:719–736.

Dalfes, H. N., Kukla, G., and Weiss, H., 1996. *Third Millennium BC Climate Change and Old World Collapse,* New York: Springer.

Daly, H. E. and Farley, J., 2010. *Ecological Economics: Principles and Applications,* 2nd ed., Washington D.C.: Island Press.

Danielson, M. N., 1965. *Federal-Metropolitan Politics and the Commuter Crisis,* New York: Columbia Univ. Press.

Dasgupta, P. and Heal, G., 1974. The Optimal Depletion of Exhaustible Resources. *Review of Economic Studies,* 41:3–28.

Davidson, A. and Blumberg, A., 2010. "Quantitative Easing" By The Fed, Explained. National Public Radio. Oct. 7.

Davis, M., 2010. Reflections on the Foreclosure Crisis. *Land Lines,* 2–8.

Davis, M. A., 2009. The Price and Quantity of Land by Legal Form of Organization in the United States. *Regional Science and Urban Economics,* 39(3):350–359.

Davis, M. A. and Palumbo, M. G., 2008. The Price of Residential Land in Large US Cities. *Journal of Urban Economics,* 63(1):352–384.

Davis, S. C., Diegel, S. W., and Boundy, R. G., 2011. Transportation Energy Databook: Edition 30, Oak Ridge, TN: Oak Ridge National Laboratory.

de Groot, R. S., Wilson, M. A., and Boumans, R. M., 2002. A Typology for the Classification, Description and Valuation of Ecosystem Functions, Goods and Services. *Ecological Economics,* 41(3):393–408.

Decker, E. H., Kerkhoff, A. J., and Moses, M. E., 2007. Global Patterns of City Size Distributions and Their Fundamental Drivers. *PLoS ONE,* 2(9), p.e934.

Deffeyes, K. S., 2006. *Beyond Oil: The View from Hubbert's Peak,* New York: Hill & Wang.

———, 2008. *Hubbert's Peak: The Impending World Oil Shortage,* Princeton, NJ: Princeton Univ. Press.

Dellinger, M., 2010. *Interstate 69: The Unfinished History of the Last Great American Highway,* New York: Scribner.

Delucchi, M. A. and Murphy, J. J., 2008. US Military Expenditures to Protect the Use of Persian Gulf Oil for Motor Vehicles. *Energy Policy,* 38(6):2553–2264.

DeMenocal, P. B., 2001. Cultural Responses to Climate Change during the late Holocene. *Science,* 292(5517):667.

Denholm, P. and Margolis, R. M., 2006. Very Large-Scale Deployment of Grid-Connected Solar Photovoltaics in the United States: Challenges and Opportunities, Golden, CO: National Renewable Energy Laboratory.

———, 2008. Impacts of Array Configuration on Land-Use Requirements for Large-Scale Photovoltaic Deployment in the United States, Golden, CO: National Renewable Energy Laboratory.

Dennen, R. T., 1977. Some Efficiency Effects of Nineteenth-Century Federal Land Policy: A Dynamic Analysis. *Agricultural History,* 51(4):718–736.

Dennis, S. M., 2007. A Decade of Growth in Domestic Freight: Rail and Truck Ton-Miles Continue to Rise, Washington D.C.: US Bureau of Transportation Statistics.

Dernbach, J. C., 1990. Industrial Waste: Saving the Worst for Last? *Environmental Law Reporter,* 20:10238.

Dewey, D. R., 1918. *Financial History of the United States,* New York: Longmans, Green.

Dewey, J. and Montes-Rojas, G., 2009. Inter-City Wage Differentials and Intra-City Workplace Centralization. *Regional Science and Urban Economics,* 39(5):602–609.

Diamond, J., 2004. *Collapse: How Societies Choose to Fail or Succeed,* 1st ed. New York: Viking Adult.

Diers, J. W. and Isaacs, A., 2007. *Twin Cities by Trolley: The Streetcar Era in Minneapolis and St. Paul,* Minneapolis, MN: Univ. of Minnesota Press.

Dietz, R., 1998. A Joint Model of Residential and Employment Location in Urban Areas. *Journal of Urban Economics,* 44:197–215.

District Department of Transportation, 2010. DC Transit Future System Plan—Final Report, Washington, D.C.: District Department of Transportation and Washington Metropolitan Area Transit Authority.

Divya, K. and Ostergaard, J.,

2009. Battery Energy Storage Technology for Power Systems—An Overview. *Electric Power Systems Research*, 79(4):511–520.

Dobbs, R. et al. 2011. Urban World: Mapping the Economic Power of Cities, New York: McKinsey Global Institute.

Dodman, D., 2009. Blaming Cities for Climate Change? An Analysis of Urban Greenhouse Gas Emissions Inventories. *Environment and Urbanization*, 21:185–201.

Dokko, J. et al., 2011. Monetary Policy and the Global Housing Bubble. *Economic Policy*, 26(66):237–287.

Dolnick, E., 2011. *The Clockwork Universe: Isaac Newton, the Royal Society, and the Birth of the Modern World*, New York: Harper.

Donahue, B., 2001. *Reclaiming the Commons: Community Farms and Forests in a New England Town*, New Haven: Yale Univ. Press.

Downey, M., 2009. *Oil 101*, New York: Wooden Table.

Draffan, G., 1998. Taking Back Our Land: A History of Railroad Land Grant Reform, www.land-grant.org.

Drake, J. D., 2011. *The Nation's Nature: How Continental Presumptions Gave Rise to the United States of America*, Charlottesville, VA: Univ. of Virginia Press.

Draper, P., 1973. Crowding among Hunter-Gatherers: The !Kung Bushmen. *Science*, 182(4109):301–303.

Droz, R. V., 2004. A History of the Standard Oil Company and Its Successors. www.us-highways.com/sohist.htm

Duany, A., Plater-Zyberk, E., and Speck, J., 2000. *Suburban Nation: The Rise of Sprawl and the Decline of the American Dream*, New York: North Point.

Duany, A., Speck, J., and Lydon, M., 2009. *The Smart Growth Manual*, New York: McGraw-Hill Professional.

Dunbar, R., 1992. Neocortex Size as a Constraint on Group Size in Primates. *Journal of Human*

Evolution, 22(6):469–493.

——, 1993. Coevolution of Neocortical Size, Group Size and Language in Humans. *Behavioral and Brain Science*, 16:681–735.

Dunbar, R., 2010. *How Many Friends Does One Person Need?: Dunbar's Number and Other Evolutionary Quirks*, Cambridge, MA: Harvard Univ. Press.

Dunham-Jones, E. and Williamson, J., 2011. *Retrofitting Suburbia, Updated Ed.: Urban Design Solutions for Redesigning Suburbs*, New York: Wiley.

Dyson, F. J., 1971. Energy in the Universe. *Scientific American*, 225(3):50–59.

Editors of *Bicycling Magazine*, 2003. *The Noblest Invention: An Illustrated History of the Bicycle*, Emmaus, PA: Rodale.

Editors of TIME magazine. 1957. Taxes: The Case of Aramco. *TIME*, May 6, 1957

Edmonson, N., 1975. Real Price and the Consumption of Mineral Energy in the United States, 1901–1968. *Journal of Industrial Economics*, 23(3):161–174.

Eggers, C. W., 2008. The Fuel Gauge of National Security. *Armed Forces Journal* (May 1, 2008): 12.

Eichengreen, B., 2008. *Globalizing Capital: A History of the International Monetary System*, 2nd ed., Princeton, NJ: Princeton Univ. Press.

Eid, J. et al., 2008. Fat city: Questioning the Relationship between Urban Sprawl and Obesity. *Journal of Urban Economics*, 63(2):385–404.

Eisenhower, D. D., 1957. Special Message to the Congress on the Situation in the Middle East. www.presidency.ucsb.edu.

Elliott, D. L., 2008. *A Better Way to Zone: Ten Principles to Create More Livable Cities*, Washington D.C.: Island Press.

Elliott, D. L. et al., 1986. Wind Energy Resource Atlas of the United States, Richland, WA: Pacific Northwest National Laboratory.

Esri, 2010. U.S. National Transportation Atlas Railroads,

Redlands, CA: Esri.

——. United States Populated Place Points, Redlands, CA: Esri.

Esser, G., Kattage, J., and Sakalli, A., 2011. Feedback of Carbon and Nitrogen Cycles Enhance Carbon Sequestration in the Terrestrial Biosphere. *Global Change Biology*, 17:819–842.

Estill, L., 2008. *Small Is Possible: Life in a Local Economy*, Gabriola Island, BC: New Society.

Etemand, B. and Luciani, J., 1991. *World Energy Production, 1800–1985*, Geneva, Switzerland: Librairie Droz.

European Commission, 2005. Communication from the Commission—The Support of Electricity from Renewable Energy Sources, Brussels: European Commission. europa.eu/legislation_summaries/energy/renewable_energy/l24452_en.htm

Faber, A. and Mazlish, E., 2012. *How to Talk So Kids Will Listen and Listen So Kids Will Talk Updated*, New York: Scribner.

Facebook Data Team, 2009. Maintained Relationships on Facebook. www.facebook.com/note.php?note_id=55257228858.

Fainstein, S. S., 2005. Cities and Diversity. *Urban Affairs Review*, 41(1):3–19.

Faiola, A., Nakashima, E., and Drew, J., 2008. What Went Wrong. *Washington Post*. Oct. 14.

Falk, D. A., Palmer, M., and Zedler, J., eds., 2006. *Foundations of Restoration Ecology*, Washington D.C.: Island Press.

Federal Reserve Bank of New York, 2007. Repurchase and Reverse Repurchase Transactions—Fedpoints—Federal Reserve Bank of New York, New York: Federal Reserve Bank of New York.

——. 2012. Primary Dealers List. www.newyorkfed.org/markets/pridealers_current.html.

Federal Reserve Bank of St. Louis, 2012c. Adjusted Monetary Base (AMBNS), St. Louis, MO: Federal Reserve Bank of St. Louis.

——, 2012a. Dow Jones Industrial Average (DJIA), St. Louis, MO:

Federal Reserve Bank of St. Louis.

———, 2012b. West Texas Intermediate (OILPRICE), St. Louis, MO: Federal Reserve Bank of St. Louis.

Federal Reserve Bank, 2012. H.3 Release—Aggregate Reserves of Depository Institutions, Washington, D.C.: Federal Reserve Bank.

Fein, G. S., 2011. News: ONR, Marine Corps Use Alternative Energy at Forward Operating Bases, Arlington, VA: Office of Naval Research. www. onr.navy.mil/Media-Center/ Press-Releases/2011/Forward-Operating-Base-Marine.aspx

Fernandez, L. E. et al., 2005. Characterizing Location Preferences in an Exurban Population: Implications for Agent-Based Modeling. *Environment and Planning B*, 32(6) 799.

Field, B. C., 2008. *Natural Resource Economics: An Introduction*, Long Grove, IL: Waveland.

Filkins, D., 2009. *The Forever War*, New York: Vintage.

Financial Crisis Inquiry Commission, 2011. *The Financial Crisis Inquiry Report, Authorized Edition: Final Report of the National Commission on the Causes of the Financial and Economic Crisis in the United States*, New York: PublicAffairs.

Finkelstein, T. and Organ, A. J., 2001. *Air Engines: The History, Science, and Reality of the Perfect Engine*, New York: ASME Press.

Fischer, H., Klarman, K,. and Oboroceanu, M.-J., 2007. American War and Military Operations Casualties: Lists and Statistics, Washington D.C.: Congressional Research Service.

Fisher, R. and Ury, W. L., 1991. *Getting to Yes: Negotiating Agreement Without Giving*, New York: Penguin.

Fishman, C., 2006. *The Wal-Mart Effect: How the World's Most Powerful Company Really Works—and How It's Transforming the American Economy*, New York: Penguin.

Fletcher, R., 1995. *The Limits of Settlement Growth: A Theoretical Outline*, Cambridge, UK: Cambridge Univ. Press.

Florida, R., 2002. *The Rise of the Creative Class: And How It's Transforming Work, Leisure, Community, and Everyday Life*, New York: Basic.

———. 2008. *Who's Your City?: How the Creative Economy Is Making Where to Live the Most Important Decision of Your Life*, New York: Basic.

Florida State Department of Transportation, 2012. *Generic Cost per Mile Models*, Tallahassee, FL: Department of Transportation.

Fonner, K. L. and Roloff, M. E., 2010. Why Teleworkers Are More Satisfied with Their Jobs than Are Office-Based Workers: When Less ContactIs Beneficial. *Journal of Applied Communication Research*, 38:336–361.

Foote, S., 1986. *The Civil War: A Narrative*, New York: Vintage.

Forinash, K., 2010. *Foundations of Environmental Physics: Understanding Energy Use and Human Impacts*, Washington D.C.: Island Press.

Forman, R. T. T., 2000. Estimate of the Area Affected Ecologically by the Road System in the United States. *Conservation Biology*, 14(1):31–35.

Fowler, C. W., 1981. Density Dependence as Related to Life History Strategy. *Ecology*, 62(3):602–610.

Fox, W., 2003. *History and Economic Impact*, in J. Janata, ed., *Sales Taxation*, Atlanta, GA: Institute for Professionals in Taxation.

Fox-Penner, P. S., 2010. *Smart Power: Climate Change, the Smart Grid, and the Future of Electric Utilities*, Washington, D.C.: Island Press.

Francfort, J. and Carroll, M., 2001. Field Operations Program: Neighborhood Electric Vehicle Fleet Use, Idaho Falls, ID: Idaho National Engineering and Environmental Laboratory.

Freeman, L., 2001. The Effects of Sprawl on Neighborhood Social Ties. *Journal of the American Planning Association*, 67:69–77.

Freyfogle, E. T., 2003. *The Land We Share: Private Property and the Common Good*, Washington D.C.: Island Press.

Friedman, M., 1970. *The Counter-Revolution in Monetary Theory*, Philadelphia, PA: Transatlantic Arts.

———, 2007. *Price Theory*, Piscataway, NJ: Transaction.

Friedman, M. and Schwartz, A. J., 1963. *A Monetary History of the United States, 1867–1960*, Princeton, NJ: Princeton Univ. Press.

Frost, R., 1914. *North of Boston*, London: D. Nutt.

Frumkin, H., 2002. Urban Sprawl and Public Health. *Public Health Reports*, 117:201–218.

Frumkin, H., Frank, L., and Jackson, D. R. J., 2004. *Urban Sprawl and Public Health: Designing, Planning, and Building for Healthy Communities*, Washington D.C.: Island Press.

Fry, B., 2008. All Streets. Cambridge, MA: Fathom Information Design.

Fry, J. et al., 2006. Completion of the 2006 National Land Cover Database for the Conterminous United States. *PE&RS*, 77 (9):858–864.

Fujita, M. and Thisse, J. F., 2009. New Economic Geography: An Appraisal on the Occasion of Paul Krugman's 2008 Nobel Prize in Economic Sciences. *Regional Science and Urban Economics*, 39(2):109–119.

Fuller, R. A., Warren, P. H., and Gaston, K. J., 2007. Daytime Noise Predicts Nocturnal Singing in Urban Robins. *Biology Letters*, 3(4):368–370.

Gabaix, X., 1999. Zipf's Law and the Growth of Cities. *The American Economic Review*, 89(2):129–132.

Gandhi, M. K., 1993. *An Autobiography, or My Experiments with Truth*, Boston, MA: Beacon Press.

Garrett, G., 1952. The World that Henry Ford Made. *Look*. Mar. 25.

Garthwaite, J., 2010. Nissan: LEAF, Like Other Electric Cars, Will Lose Money at First—Cleantech

News and Analysis. earth2tech. gigaom.com/cleantech/nissan-leaf-like-other-electric-cars-will-lose-money-at-first.

Gary, J. H., Handwerk, G. E., and Kaiser, M. J., 2007. *Petroleum Refining: Technology and Economics*, 5th ed., Boca Raton, FL: CRC.

Gasquet, F.A., ed., 1997. Jocelin of Brakelond: Chronicle of the Abbey of St Edmund's (1173–1202) (Cronica Joceline), Bronx, NY: Fordham Univ.

Gates, P. W., 1979. *Public Land Policies: Management and Disposal*, New York: Arno.

Gautier, D. L. et al., 2009. Assessment of Undiscovered Oil and Gas in the Arctic. *Science*, 324(5931):1175–1179.

Gehl, J., 2010. *Cities for People*, Washington D.C.: Island Press.

Geiser, K., 2001. *Materials Matter: Toward a Sustainable Materials Policy*, Cambridge, MA: MIT Press.

Geoff, K., 2011. History and Potential of Renewable Energy Development in New Zealand. *Renewable and Sustainable Energy Reviews*, 15(5):2501–2509.

George, H., 1879. *Progress and Poverty: An Inquiry into the Cause of Industrial Depressions and of Increase of Want with Increase of Wealth, The Remedy*, New York: D. Appleton.

German Aerospace Center, 2005. Concentrating Solar Power for the Mediterranean Region, Stuttgart, Germany: Deutsches Zentrum für Luft- und Raumfahrt e.V. (DLR).

Gesch, D. et al., 2009. The National Map-Elevation, Reston, VA: US Geological Survey.

Gesner, A., 1861. *A Practical Treatise on Coal, Petroleum, and Other Distilled Oils*, New York: Bailliere Brothers.

Gibney, A., 2006. *Enron: The Smartest Guys in the Room*, Pineville, LA: Magnolia.

Giery, M., Catala, M., and Winters, P. L., 2003. Proximate Commuting: Potential Benefits and Obstacles, Tampa, FL: Center for Urban Transportation Research.

Gilbert, R. and Perl, A., 2010.

Transport Revolutions: Moving People and Freight Without Oil, Gabriola Island, BC, Canada: New Society Publishers.

Gillham, O., 2002. *The Limitless City: A Primer on the Urban Sprawl Debate*, Washington D.C.: Island Press.

Gillis, A. R., 1974. Population Density and Social Pathology: The Case of Building Type, Social Allowance and Juvenile Delinquency. *Social Forces*, 53(2):306–314.

Gimpel, J., 1976. *The Medieval Machine: The Industrial Revolution of the Middle Ages*, London: Penguin.

Gissen, D., 2003. *Big and Green: Toward Sustainable Architecture in the 21st Century*, Princeton, NJ: Princeton Architectural Press.

Glaeser, E. L., 2011. *Triumph of the City: How Our Greatest Invention Makes Us Richer, Smarter, Greener, Healthier, and Happier*, New York: Penguin.

Glaeser, E .L. and Gyourko, J., 2005. Urban Decline and Durable Housing. *Journal of Political Economy*, 113(2):345–375.

Glaeser, E. L., Kahn, M. E., and Rappaport, J., 2008. Why Do the Poor Live in Cities: The Role of Public Transportation. *Journal of Urban Economics*, 63(1):1–24.

Glasby, G. P., 2006. Abiogenic Origin of Hydrocarbons: An Historical Overview. *Resource Geology*, 56(1):83–96.

Glynn, T., 1981. Psychological Sense of Community: Measurements and Application. *Human Relations*, 34(7):789–818.

Goldberg, M., 2000. Federal Energy Subsidies: Not all Technologies Are Created Equal, Washington D.C.: Renewable Energy Policy Project.

Goncalves, B., Perra, N., and Vespignani, A., 2011. Validation of Dunbar's number in Twitter Conversations. Preprint at *arXiv*:1105.5170.

Gonick, L., 1992. *The Cartoon Guide to Physics*, New York: Harper Perennial.

Gonick, L. and Criddle, C., 2005. *The Cartoon Guide to Chemistry*,

London: Collins Reference.

Goodwyn, L., 1976. *Democratic Promise: The Populist Movement in America*, New York: Oxford Univ. Press.

Gopnik, A., 2012. The Caging of America. *New Yorker*. Jan. 30

Goralski, R. and Freeburg, R. W., 1987. *Oil and War: How the Deadly Struggle for Fuel in WWII Meant Victory or Defeat*, New York: William Morrow.

Gordon, J. S., 2005. *Empire of Wealth: The Epic History of American Economic Power*, New York: Harper Perennial.

Goudie, A. and Viles, H., 2010. *Landscapes and Geomorphology: A Very Short Introduction*, Oxford, UK: Oxford Univ. Press.

Gowdy, J., 1998. *Limited Wants, Unlimited Means: A Reader on Hunter-Gatherer Economics and the Environment*, Washington D.C.: Island Press.

Graeber, D. 2011. *Debt: The First 5,000 Years*, Brooklyn, NY: Melville House.

Grayson, S., 2001. *Beautiful Engines: Treasures of the Internal Combustion Century*, Chicago, IL: Devereux.

Green, B. D. and Nix, R. G., 2006. Geothermal—The Energy Under Our Feet Geothermal Resource Estimates for the United States, Golden, CO: National Renewable Energy Laboratory.

Greenleaf, W., 1961. *Monopoly on Wheels: Henry Ford and the Selden Automobile Patent*, Detroit, MI: Wayne State Univ. Press.

Grimm, N. B. et al., 2008. Global Change and the Ecology of Cities. *Science*, 319(5864):756–760.

Grogg, K., 2005. *Harvesting the Wind: The Physics of Wind Turbines*, Northfield, MN: Carleton College.

Groom, M. J., Gray, E. M., and Townsend, P. A., 2008. Biofuels and Biodiversity: Principles for Creating Better Policies for Biofuel Production. *Conservation Biology*, 22(3):602–609.

Groom, M. J., Meffe, G. K., and Carroll, C. R., 2005. *Principles of Conservation Biology*, 3rd ed.,

Sunderland, MA: Sinauer.

Grossinger, R., 2012. *Napa Valley Historical Ecology Atlas: Exploring a Hidden Landscape of Transformation and Resilience,* Berkeley, CA: Univ. of California Press.

Grossinger, R., Beller, E. E. et al., 2007a. The Historical Ecology of Contra Costa County: An Illustrated Preview and Guide, San Francisco: CA: San Francisco Estuary Institute.

Grossinger, R., Striplen, C. J. et al., 2007b. Historical Landscape Ecology of an Urbanized California Valley: Wetlands and Woodlands in the Santa Clara Valley. *Landscape Ecology,* 103–120.

Grube, A. et al., 2011. Pesticides Industry Sales and Usage: 2006 and 2007 Market Estimates, Washington D.C.: US Environmental Protection Agency, Office of Prevention, Pesticides, and Toxic Substances.

Gunderson, L. H., 2000. Ecological Resilience—In Theory and Application. *Annual Review of Ecology and Systematics,* 31:425–439.

Gutfreund, O. D., 2005. *Twentieth-Century Sprawl: Highways and the Reshaping of the American Landscape,* New York: Oxford Univ. Press.

Gwilliam, K., 2008. A Review of Issues in Transit Economics. *Research in Transportation Economics,* 23(1):4–22.

Gyourko, J., Saiz, A., and Summers, A., 2008. A New Measure of the Local Regulatory Environment for Housing Markets: The Wharton Residential Land Use Regulatory Index. *Urban Studies,* 45:693–729.

Haeg, F., Allen, W., and Balmori. D., 2010. *Edible Estates: Attack on the Front Lawn,* rev. ed., New York: Metropolis.

Halbouty, M. T., 2001. Giant Oil and Gas Fields of the Decade 1990–2000: An Introduction, Denver, CO: American Association of Petroleum Geologists Annual Convention.

Hales, P. B., 2009. Levittown: Documents of an Ideal American Suburb. tigger.uic.edu/~pbhales/Levittown.html.

Halliday, D., Resnick, R., and Walker, J., 2007. *Fundamentals of Physics,* 8th ed., Hoboken, NJ: Wiley.

Halperin, K., 1993. Comparative Analysis of Six Methods for Calculating Travel Fatality Risk, A. Risk: *Issues in Health and Safety,* 4:15.

Hamilton, J. D., 1983. Oil and the Macroeconomy since World War II. *The Journal of Political Economy,* 228–248.

———, 1988. A Neoclassical Model of Unemployment and the Business Cycle. *Journal of Political Economy,* 96(3):593–617.

———, 2009. Causes and Consequences of the Oil Shock of 2007–08, Washington D.C.: Brookings Institution.

———, 2011. Historical Oil Shocks (NBER Working Paper No. 16790), National Bureau of Economic Research, Washington D.C.

Hammer, R. B. et al., 2004. Characterizing Dynamic Spatial and Temporal Residential Density Patterns from 1940–1990 across the North Central United States. *Landscape and Urban Planning,* 69(2–3):183–199.

Handy, S., Weston, L., and Mokhtarian, P. L., 2005. Driving by Choice or Necessity? *Transportation Research Part A: Policy and Practice,* 39(2–3):183–203.

Hansen, A. J. et al., 2005. Effects of Exurban Development on Biodiversity: Patterns, Mechanisms, and Research Needs. *Ecological Applications,* 15(6):1893–1905.

Hansen, L. and Lovins, A. B., 2010. Keeping the Lights On While Transforming Electric Utilities. *Solutions,* 3(1).

Hanson, A., 2009. Local Employment, Poverty, and Property Value Effects of Geographically-Targeted Tax Incentives: An Instrumental Variables Approach. *Regional Science and Urban Economics,* 39:721–731.

Hanson, A. and Rohlin, S., 2011. Do Location-Based Tax Incentives Attract New Business Establishments? *Journal of Regional Science,* 51:427–449.

Hanson, S., 2007. Urbanization in Sub-Saharan Africa–Council on Foreign Relations, Washington D.C.: Council on Foreign Relations.

Hardin, G., 1968. The Tragedy of the Commons. *Science,* 162(3859):1243–1248.

Hardwick, E., ed., 1993. *The Selected Letters of William James,* New York: Anchor/Doubleday.

Harper, R. M., 1911. The Hempstead Plains: A Natural Prairie on Long Island. *Bulletin of American Geographical Society,* 43:351–360.

Harvard University Graduate School of Design, [1920]. U.S. Western Railway Land Grants, GSD lantern slide 36471, Washington, D.C.: Library of Congress.

Hawken, P., Lovins, A., and Lovins, L. H., 2000. *Natural Capitalism: Creating the Next Industrial Revolution,* New York: Back Bay.

Hayden, D., 2004. *Building Suburbia: Green Fields and Urban Growth, 1820–2000,* New York: Vintage.

Head, I. M. et al., 2003. Biological Activity in the Deep Subsurface and the Origin of Heavy Oil. *Nature,* 426(6964):344–352.

Heilbroner, R. L., 1999. *The Worldly Philosophers: The Lives, Times and Ideas of the Great Economic Thinkers,* New York: Touchstone.

Heilbroner, R. L. and Singer, A., 1998. *The Economic Transformation of America: 1600 to the Present,* 4th ed., Florence, KY: Wadsworth.

Henderson, D. R., 2008. Arthur Cecil Pigou. *The Concise Encyclopedia of Economics.* www.econlib.org/library/Enc/bios/Pigou.html.

Herlihy, D. V., 2006. *Bicycle: The History,* New Haven, CT: Yale Univ. Press.

Hester, R. T., 2010. *Design for Ecological Democracy,* Boston, MA: MIT Press.

Hey, D. L. and Philippi, N. S., 1995. Flood Reduction through Wetland Restoration: The Upper Mississippi River Basin as a Case History. *Restoration Ecology*, 3:4–17.

Hill, J. et al., 2006. Environmental, Economic, and Energetic Costs and Benefits of Biodiesel and Ethanol Biofuels. *Proceedings of the National Academy of Sciences*, 103(30):11206–11210.

Hilty, J. A., Lidicker Jr., W. Z, and Merenlender, A. M., 2006. *Corridor Ecology: the Science and Practice of Linking Landscapes for Biodiversity Conservation*, 323, Washington, D.C.: Island Press.

Hinckley, J., 2005. *The Big Book of Car Culture: The Armchair Guide to Automotive Americana*, Minneapolis, MN: Motorbooks.

Hiss, T., 1990. *The Experience of Place*, New York: Vintage.

Hitakonanu'laxk, 2005. *The Grandfathers Speak: Native American Folk Tales of the Lenapé People*, New York: Interlink.

Holtzclaw, J. et al., 2002. Location Efficiency: Neighborhood and Socio-Economic Characteristics Determine Auto Ownership and Use—Studies in Chicago, Los Angeles and San Francisco. *Transportation Planning and Technology*, 25(1):1–27.

Homer, C. et al., 2004. Development of a 2001 National Landcover Database for the United States. *Photogrammetric Engineering and Remote Sensing*, 70(7):829–840.

Hoogwijk, M., de Vries, B., and Turkenburg, W., 2004. Assessment of the Global and Regional Geographical, Technical and Economic Potential of Onshore Wind Energy. *Energy Economics*, 26(5):889–919.

Höök, M., Söderbergh, B., Jakobsson, K., and Aleklett, K., 2009. The Evolution of Giant Oil Field Production Behavior. *Natural Resources Research* 18, 39–56.

Horn, J., 2011. Arnold Schwarzenegger's "Governator" Officially a No-Go. *Los Angeles Times*, May 20.

Horsman, P. V., 1985. *Seawatch: The Seafarer's Guide to Marine Life*. New York: Facts on File.

Hotelling, H., 1931. The Economics of Exhaustible Resources. *Journal of Political Economy*, 39(2):137–175.

Houser, T. et al., 2008. Leveling the Carbon Playing Field: International Competition and US Climate Policy Design, Washington D.C.: Peterson Institute/Worldwatch Institute.

Howell, E. A., Harrington, J. A., and Glass, S. B., 2011. *Introduction to Restoration Ecology*, Washington D.C.: Island Press.

Hoy, P., 2008. The World's Biggest Fuel Consumer. *Forbes*, June 5.

Hubbert, M. K., 1956. Nuclear Energy and the Fossil Fuels. In American Petroleum Institute Drilling and Production Practice Proceedings (spring 1956). American Petroleum Institute, 5–75.

———, 1982. Techniques of Prediction as Applied to the Production of Oil and Gas, Special Publication 631. Washington D.C.: National Bureau of Standards.

Huber, P., and Mills, M. P., 2006. *The Bottomless Well: The Twilight of Fuel, the Virtue of Waste, and Why We Will Never Run Out of Energy*, New York: Basic.

Hudson, H. K., 1963. Is the "Song of Plenty" a Siren Song? *Oil and Gas Journal* 61:131–136.

Humphries, M., 2002. Mining on Federal Lands, Washington D.C.: Congressional Research Service.

———, 2004. Oil and Gas Exploration and Development on Public Lands, Washington D.C.: Congressional Research Service.

———, 2006. Outer Continental Shelf: Debate Over Oil and Gas Leasing and Revenue Sharing, Washington D.C.: Congressional Research Service.

———, 2007. Royalty Relief for US Deepwater Oil and Gas Leases, Washington D.C.: Congressional Research Service.

Humphries, N. E. et al. 2010.

Environmental Context Explains Lévy and Brownian Movement Patterns of Marine Predators. *Nature*, 465(7301, June 24):1066–1069.

Hung, Y.-Y. et al., 2010. Landscape Infrastructure: Case Studies by SWA, Basel, Switzerland: Birkhäuser Architecture.

Hunt, J. D. and Abraham, J. E., 2006. Influences on Bicycle Use. *Transportation*, 34:453–470.

Hurst, R., 2004. *The Art of Urban Cycling: Lessons from the Street*, Guilford, CT: Globe Pequot.

Ickes, H., 1943. *Fightin' Oil*, New York: Knopf.

IEEE, 1989. Solar Electric Generating Stations (SEGS). IEEE Power Engineering Review, 9(8):4–8.

IHS Cambridge Energy Research Associates, 2009. Growth in the Canadian Oil Sands: Finding a New Balance, Cambridge, MA: IHS Cambridge Energy Research Associates.

Ingraham, M. W. and Foster, S. G., 2008. The Value of Ecosystem Services Provided by the US National Wildlife Refuge System in the Contiguous US. *Ecological Economics* 67:608–618.

Intergovernmental Panel on Climate Change, 2007. Climate Change 2007: Synthesis Report. Contribution of Working Groups I, II and III to the Fourth Assessment Report of the Intergovernmental Panel on Climate Change, Geneva, Switzerland: IPCC. www.ipcc.ch/publications_and_data/publications_and_data_reports.shtml#1.

International Energy Agency, 2010. World Energy Outlook 2010, Paris: International Energy Agency.

Ioannides, Y. M. and Zabel, J. E., 2008. Interactions, Neighborhood Selection and Housing Demand. *Journal of Urban Economics*, 63(1):229–252.

Israel, P., 2000. *Edison: A Life of Invention*, New York: Wiley.

Isser, S., 1996. *The Economics and Politics of the United States Oil Industry 1920–1990*, New York: Garland.

Jackson, D., 2011. Cheney Defends Iraq War. *USA Today*. Aug. 30.

Jackson, K. T., 1987. *Crabgrass Frontier: The Suburbanization of the United States,* New York: Oxford Univ. Press.

———, 1995. *The Encyclopedia of New York City,* New Haven, CT: Yale Univ. Press.

Jacobs, J., 1961. *The Death and Life of Great American Cities,* New York: Random House.

———, 1985. *Cities and the Wealth of Nations,* New York: Vintage.

———, 2001. *The Nature of Economies,* New York: Vintage.

Jakabovics, A., 2010. FDR Solves The Mortgage Crisis. CBS News. Aug. 19.

Johnson, E. A. and Klemens, M. W., eds., 2005. *Nature in Fragments: The Legacy of Sprawl,* New York: Columbia Univ. Press.

Johnson, S. L., 1951. The Fight for the Pre-emption Law of 1841. *Proceedings of the Arkansas Academy of Science,* 165–172.

Johnston, D. C., 2003. *Perfectly Legal: The Covert Campaign to Rig Our Tax System to Benefit the Super Rich—and Cheat Everybody Else,* New York: Portfolio.

Johnston, I. 2000. Homer. *The Iliad: a New Translation,* Arlington, VA: Richer Resources Publications. records.viu.ca/~johnstoi/homer.

Jones, D. W., 2010. *Mass Motorization and Mass Transit: An American History and Policy Analysis,* Bloomington, IN: Indiana Univ. Press.

Jones, J. A., Swanson, F. J., Wemple, B. C., and Snyder, K., 2000. Effects of Roads on Hydrology, Geomorphology, and Disturbance Patches in Stream Networks. *Conservation Biology,* 14:76–85.

Jones, K. E. et al., 2008. Global Trends in Emerging Infectious Diseases. *Nature,* 451(7181):990.

Jonnes, J., 2002. *South Bronx Rising: The Rise, Fall, and Resurrection of an American City,* Bronx, NY: Fordham Univ. Press.

Jortner, J., 2006. Conditions for the Emergence of Life on the Early Earth: Summary and Reflections. *Philosophical Transactions of the Royal Society B: Biological Sciences,* 361(1474):1877–1891.

Joule, J. P., 1845. The Mechanical Equivalent of Heat, Cambridge, UK: British Association for the Advancement of Science meeting.

Juhasz, A., 2011. *Black Tide: The Devastating Impact of the Gulf Oil Spill,* Indianapolis, IN: Wiley.

Kahneman, D., 2011. *Thinking, Fast and Slow,* New York: Farrar, Straus & Giroux.

Kains, M., 1940. *Five Acres and Independence,* Miami, FL: BN.

Kamkwamba, W., 2009. How I Harnessed the Wind. www.ted.com/talks/william_kamkwamba_how_i_harnessed_the_wind.html.

Kanafani, A., Wang, R., and Griffin, A., 2012. The Economics of Speed—Assessing the Performance of High Speed Rail in Intermodal Transportation. *Procedia—Social and Behavioral Sciences,* 43:692–708.

Kane, T., 2004. Global US Troop Deployment, 1950–2003, Washington D.C.: Heritage Foundation.

Kanigher, S., 2010. Nellis Wants to Double the Base's Solar Energy Output. *Las Vegas Sun.* Nov. 23.

Kannberg, L. D. et al., 2003. GridWiseTM: The Benefits of a Transformed Energy System, Richland, WA: Pacific Northwest National Laboratory.

Karbuz, S., 2006. The US Military Energy Consumption. karbuz.blogspot.com.

Karden, E. et al., 2007. Energy Storage Devices for Future Hybrid Electric Vehicles. *Journal of Power Sources,* 168(1):2–11.

Katz, A., 2010. *Our Lot: How Real Estate Came to Own Us,* New York: Bloomsbury USA.

Kaufman, B. I., 1978. *The Oil Cartel Case: A Documentary Study of Antitrust Activity in the Cold War Era,* Westport, CT: Greenwood.

Kaufmann, W. W. and Steinbruner, J. D., 1991. *Decisions for Defense: Prospects for a New Order,* Washington, D.C.: Brookings Institution Press.

Kay, J. H., 1998. *Asphalt Nation: How the Automobile Took Over America and How We Can Take It Back,* Berkeley, CA: Univ. of California Press.

Kazis, N., 2011. New York's Car Ownership Rate Is on The Rise. Streetsblog New York City. www.streetsblog.org/2011/04/06/new-yorks-car-ownership-rate-is-on-the-rise.

Kelly, T. D. and Matos, G. R., 2010. Historical Statistics for Mineral Commodities in the United States, Data Series 2005-140, Reston, VA: US Geological Survey.

Kenny, J. F. et al., 2009. Estimated Water Use in the United States in 2005, Reston, VA: US Geological Survey.

Kestenbaum, D. and Joffe-Walt, C., 2010. How to Spend $1.25 Trillion. National Public Radio. Aug. 26.

Khan, A., 2010. The Geysers Geothermal Field, an Injection Success Story. *Proceedings World Geothermal Congress 2010.* Bali, Indonesia.

Kimmelman, M., 2011. Wall Street Protest Shows Power of Place. *New York Times.* Oct. 16.

Kirk, M., *Frontline.* Feb. 20, 2003. Boston: WGBH Educational Foundation.

Kirsch, D. A., 2000. *The Electric Vehicle and the Burden of History,* New Brunswick, N.J.: Rutgers Univ. Press.

Klare, M. T., 2005. *Blood and Oil: The Dangers and Consequences of America's Growing Dependence on Imported Oil,* New York: Holt.

Klein, J., 1996. *Taken for a Ride,* Harriman, NY: New Day Films.

Kochanek, K. D. et al., 2011. Deaths: Preliminary Data for 2009. National Vital Statistics Reports, 59(4):1–51.

Kocherlakota, N. R., 2010. Two Models of Land Overvaluation and Their Implications, Minneapolis, MN: Federal Reserve Bank of Minneapolis.

Kostof, S., 1993. *The City Shaped: Urban Patterns and Meanings Through History,* New York: Bulfinch.

Krakowski, J. and Payne, N. F., 1986. Population Ecology of

Rock Doves in a Small City. *Transactions of the Wisconsin Academy of Sciences*, Arts, and Letters, 74:50–57.

Krugman, P., 1991. Increasing Returns and Economic Geography. *Journal of Political Economy*, 99, 483–499.

——, 2010. Why Is Deflation Bad? *New York Times*. Aug. 2, 2010.

Kümmerer, K. ed., 2008. *Pharmaceuticals in the Environment: Sources, Fate, Effects and Risks*, New York: Springer.

Kunstler, J. H., 1994. *The Geography of Nowhere: The Rise and Decline of America's Man-Made Landscape*, New York: Free Press.

——, 2006. *The Long Emergency: Surviving the End of Oil, Climate Change, and Other Converging Catastrophes of the Twenty-First Century*, New York: Grove.

Kushner, D., 2009. *Levittown: Two Families, One Tycoon, and the Fight for Civil Rights in America's Legendary Suburb*, New York: Bloomsbury USA.

Lacey, R., 1988. *Ford: The Men and the Machine*, New York: Little, Brown.

Lakoff, G., 2006. *Whose Freedom?: The Battle Over America's Most Important Idea*, New York: Picador.

Lakoff, G. and Johnson, M., 2003. *Metaphors We Live By*, Chicago, IL: Univ. of Chicago Press.

Lane, B. W., 2012. On the Utility and Challenges of High-Speed Rail in the United States. *Journal of Transport Geography*, 22:282–284.

Lawrence Livermore National Laboratory, 2012. Estimated Energy Use in 2011, Livermore, CA: Department of Energy.

Lee, Y., Hickman, M., and Washington, S., 2007. Household Type and Structure, Time-Use Pattern, and Trip-Chaining Behavior. *Transportation Research Part A: Policy and Practice*, 41(10):1004–1020.

Leoni, B., 1991. *Freedom and the Law*. Indianapolis, IN: Liberty Fund.

Leopold, A., 1949. *A Sand County Almanac, and Sketches Here and There*, New York: Oxford Univ. Press.

——, 1953. *Round River; from the Journals of Aldo Leopold*, New York: Oxford Univ. Press.

——, 1999. *For the Health of the Land: Previously Unpublished Essays and Other Writings*, J. B. Callicott and E. T. Freyfogle, eds., Washington D.C.: Island Press.

Leu, M., Hanser, S. and Knick, S., 2008. The Human Footprint in the West: A Large-Scale Analysis of Anthropogenic impacts. *Ecological Applications*, 18(5):1119–1139.

Levine, J., Inam, A. and Torng, G.-W., 2005. A Choice-Based Rationale for Land Use and Transportation Alternatives. *Journal of Planning Education and Research*, 24(3):317–330.

Levinson, D., 1979. Population Density in Cross-Cultural Perspective. *American Ethnologist*, 6(4):742–751.

Levinson, D. and Kumar, A., 1994. The Rational Locator: Why Travel Times Have Remained Stable. *Journal of the American Planning Association*, 60(3):319–332.

Lewin, P., 1982. Pollution Externalities: Social Cost and Strict Liability. *Cato Journal*, 2(1):205–229.

Lewis, M., 2011. California and Bust. *Vanity Fair*. Nov.

Leyden, K., 2003. Social Capital and the Built Environment: The Importance of Walkable Neighborhoods. *American Journal of Public Health*, 93:1546–1551.

Lindeman, R. L., 1942. The Trophic-Dynamic Aspect of Ecology. *Ecology*, 23:399–418.

Linder, M. and Zacharias, L. S., 1999. *Of Cabbages and Kings County: Agriculture and the Formation of Modern Brooklyn*. Iowa City, IA: Univ. of Iowa Press.

Lindgreen, E. and Sorenson, S. C., 2005. Driving Resistance from Railroad Trains, Lyngby, Denmark: Technical Univ. of Denmark.

Linklater, A., 2003. *Measuring America: How the United States Was Shaped By the Greatest Land Sale in History*, New York: Plume.

Livi-Bacci, M., 2006. *A Concise History of World Population*, Hoboken, NJ: Wiley-Blackwell.

Lloyd, H. D., 1894. *Wealth Against Commonwealth*, New York: Harper and Brothers.

Locke, J., 1960. *Locke: Two Treatises of Government*. P. Laslett, ed., Cambridge, UK: Cambridge Univ. Press.

Lomborg, B., 2001. *The Skeptical Environmentalist: Measuring the Real State of the World* 2nd ed., Cambridge, UK: Cambridge Univ. Press.

Loomis, J. et al., 2000. Measuring the Total Economic Value of Restoring Ecosystem Services in an Impaired River Basin: Results from a Contingent Valuation Survey. *Ecological Economics*, 33(1):103–117.

Loveday, E., 2010. WSJ: Nissan Leaf Profitable by Year Three; Battery Cost Closer to $18,000. autobloggreen. green.autoblog. com/2010/05/15/nissan-leaf-profitable-by-year-three-battery-cost-closer-to-18.

Lowe, J. S., 2009. *Oil and Gas Law in a Nutshell*, 5th ed., Eagan, MN: WestLaw.

Lowenstein, R., 2010. *The End of Wall Street*, New York: Penguin.

Lund, H., 2002. Pedestrian Environments and Sense of Community. *Journal of Planning, Education and Research*, 21:301–312.

Lüthi, D., et al., 2008. High-resolution Carbon Dioxide Concentration Record 650,000–800,000 Years Before Present. *Nature* 453 (7193): 379–382.

Lynch, K., 1960. *The Image of the City*, Cambridge, MA: MIT Press.

Maag, C. 2009. From the Ashes of '69, Cleveland's Cuyahoga River Is Reborn. *New York Times*, June 20.

Maantay, J. and Ziegler, J., 2006. *GIS for the Urban Environment*, Redlands, CA: Esri.

Maat, K. and Timmermans, H. J. P., 2009. Influence of the Residential and Work Environment on Car Use in Dual-Earner Households.

Transportation Research Part A: Policy and Practice, 43(7):654–664.

MacIsaac, J. D. and Garrott, W. R., 2002. Preliminary Findings of the Effect of Tire Inflation Pressure on the Peak and Slide Coefficients of Friction, Washington D.C.: National Highway Traffic Safety Administration.

MacKay, D. J., 2009. *Sustainable Energy—Without the Hot Air*, Cambridge, UK: UIT Cambridge.

Mackenzie, F. T., 2010. *Our Changing Planet: An Introduction to Earth System Science and Global Environmental Change*, 4th ed., New York: Prentice Hall.

MacLean, H. L. and Lave, L. B., 2003. Evaluating Automobile Fuel/Propulsion System Technologies. *Progress in Energy and Combustion Science*, 29(1):1–69.

MacLulich, D. A., 1937. *Fluctuations in the Numbers of the Varying Hare (Lepus americanus)*, Toronto: Univ. of Toronto Press.

Madrick, J., 2010. *The Case for Big Government*, Princeton, NJ: Princeton Univ. Press.

Mahler, J., 2006. *Ladies and Gentlemen, the Bronx Is Burning: 1977, Baseball, Politics, and the Battle for the Soul of a City*, New York: Picador.

Maniaque, C. and Russell, H., 2008. *Sorry, Out of Gas*, Montreal, Canada: Canadian Centre for Architecture.

Mankiw, N. G., 2008. *Principles of Economics*, 5th ed., Mason, OH: South-Western Cengate Learning.

——, 2009. Smart Taxes: An Open Invitation to Join the Pigou Club. *Eastern Economic Journal*, 35:14–23.

Mann, C. C., 2005. *1491: New Revelations of the Americas before Columbus*, New York: Knopf.

——, 2001. *1493: Uncovering the New World Columbus Created*, New York: Viking.

Mann, K. and Lazier, J., 2005. *Dynamics of Marine Ecosystems: Biological-Physical Interactions in the Oceans* 3rd ed., Hoboken, NJ: Wiley-Blackwell.

Mann, P., Horn, M., and Cross, I., 2007. Emerging Trends from 69 Giant Oil and Gas Fields Discovered from 2000–2006, Long Beach, CA: American Association of Petroleum Geologists Annual Convention.

Manville, A. M., 2005. Bird Strikes and Electrocutions at Power Lines, Communication Towers, and Wind Turbines: State of the Art and State of the Science—Next Steps Toward Mitigation, Arlington, VA: US Fish and Wildlife Service.

Mapes, J., 2009. *Pedaling Revolution: How Cyclists Are Changing American Cities*, Corvallis, OR: Oregon State Univ. Press.

Marco, G. J. et al., eds., 1987. *Silent Spring Revisited*, Washington, D.C.: American Chemical Society.

Margonelli, L., 2008. *Oil on the Brain: Petroleum's Long, Strange Trip to Your Tank*, New York: Broadway.

Marshall, A., 1890. *Principles of Economics*, London: Macmillan.

Martin, I. W., 2009. After the Tax Revolt: California's Proposition 13 Turns 30, Berkeley, CA: Institute of Governmental Studies Press.

Massachusetts Institute of Technology, 2006. The Future of Geothermal Energy: Impact of Enhanced Geothermal Systems (EGS) on the United States in the 21st Century: An Assessment by an MIT-Led Interdisciplinary Panel, Idaho Falls, ID: Idaho National Laboratory.

Masters, M. W., 2008. Testimony before Committee on Homeland Security and Governmental Affairs, Washington D.C.: US Senate.

McCartney, L., 2009. *The Teapot Dome Scandal: How Big Oil Bought the Harding White House and Tried to Steal the Country*, New York: Random House.

McCollough, J. and Check, H. F., 2010. The Baleen Whales' Saving Grace: The Introduction of Petroleum Based Products in the Market and Its Impact on the Whaling Industry. *Sustainability*, 2(10), 15:3142–3157.

McCormick, J. and Jones, T., 2010. New York City Area Has Among Longest US Commutes, Census Estimates Show. Bloomberg News, Dec. 14.

McCullough, R., 2002. Congestion Manipulation in ISO California, Portland, OR: McCullough Research.

McDonald, J. F., 1989. Econometric Studies of Urban Population Density: A Survey. *Journal of Urban Economics* 26(3, Nov.):361–385.

McDonald, N. C., 2007. Children's Mode Choice for the School Trip: The Role of Distance and School Location in Walking to School. *Transportation*, 35(1):23–35.

McDonald, R. I., Forman, R. T. T., and Kareiva, P., 2010. Open Space Loss and Land Inequality in United States' Cities, 1990–2000. *PLoS One*, 5(3), p.e9509.

McDonalds.com, 2012. McDonald's USA Nutrition Facts for Popular Menu Items, Oak Brook, IL: McDonald's Corporation.

McGuckin, N. and Srinivasan, N., 2005. The Journey-to-Work in the Context of Daily Travel. In Census Data for Transportation Planning Conference. Washington D.C.: Transportation Research Board.

McGuckin, N., Zmud, J., and Nakamoto, Y., 2005. Trip-Chaining Trends in the United States: Understanding Travel Behavior for Policy Making. *Transportation Research Record: Journal of the Transportation Research Board*, 1917(-1):199–204.

McHarg, I., 1969. *Design with Nature*, Garden City, NJ: Natural History Press.

McKenzie, B. and Rapino, M., 2011. Commuting in the United States: 2009, Washington D.C.: US Census Bureau, Department of Commerce.

McKibben, B., 2008. *Deep Economy: The Wealth of Communities and the Durable Future*, New York: St. Martin's Griffin.

McLean, B. and Elkind, P., 2004.

The Smartest Guys in the Room: The Amazing Rise and Scandalous Fall of Enron, New York: Portfolio Trade.

McNab, B. K., 2002. *The Physiological Ecology of Vertebrates: A View from Energetics,* Ithaca, NY: Comstock.

McNeill, J. R., 2001. *Something New Under the Sun: An Environmental History of the Twentieth-Century World,* New York: W. W. Norton.

Meadows, D. H. et al., 1972. *The Limits to Growth,* New York: Signet.

Meine, C. D., 2010. *Aldo Leopold: His Life and Work,* Madison, WI: Univ. of Wisconsin Press.

Meyers, S. L., 2005. *Manhattan's Lost Streetcars,* Charleston, SC: Arcadia.

Millennium Ecosystem Assessment, 2005. Ecosystems and Human Well-Being: Biodiversity Synthesis, Washington, D.C.: World Resources Institute.

Miller, J. A., 1941. *Fares, Please! A Popular History of Trolleys, Horse-Cars, Street-Cars, Buses, Elevateds, and Subways,* New York: D. Appleton Century.

Miller, J. J., 2006. *Size Matters: How Big Government Puts the Squeeze on America's Families, Finances, and Freedom,* Nashville, TN: Thomas Nelson.

Miller, K., 2002. How Important Was Oil in World War II? Fairfax, VA: George Mason Univ. History News Network. hnn.us/articles/339.html.

Miller, S. L. and Urey, H. C., 1959. Organic Compound Synthesis on the Primitive Earth. *Science,* 130(3370):245–251.

Minino, A. M., Xu, J. Q., and Kochanek, K. D., 2011. Deaths: Preliminary Data for 2008. National Vital Statistics Reports, 59(2):1–52.

Mitchell, J., 2008. *Business Improvement Districts and the Shape of American Cities,* Albany, NY: SUNY Press.

Mitchell, W. J., Borroni-Bird, C. E., and Burns, L. D., 2010. *Reinventing the Automobile: Personal Urban Mobility for the 21st Century,* Cambridge, MA: MIT Press.

Moawad, A., Sharer, P., and Rousseau, A., 2011. *Light-Duty Vehicle Fuel Consumption Displacement Potential up to 2045,* Argonne, IL: Argonne National Laboratory.

Mock, C. H., 1972. Electrical Features of Raccoon Mountain Pumped-Storage Plant. *IEEE Transactions on Power Apparatus and Systems,* 5:1875–1880.

Mock, J. E., Tester, J. W., and Wright, P. M., 1997. Geothermal Energy from the Earth: Its Potential Impact as an Environmentally Sustainable Resource. *Annual Review of Energy and the Environment,* 22:305–356.

Mokhtarian, P. L. and Salomon, I., 2001. How Derived Is the Demand for Travel? Some Conceptual and Measurement Considerations. *Transportation Research Part A: Policy and Practice,* 35(8):695–719.

Montaigne, F., 2011. A New Pickens Plan: Good for the US or Just for T. Boone? *Energy Bulletin.* April 15.

Montgomery, S. L., 2010. *The Powers That Be: Global Energy for the Twenty-first Century and Beyond,* Chicago: Univ. of Chicago Press.

Mooney, M., 2008. Pickens Building World's Largest Wind Farm. *Good Morning America.* July 8.

Moreland, H., 1985. A Few Billion for Defense: Plus 250 Billion for Overseas Military Intervention, Washington, D.C.: Coalition for a New Military Policy.

Morens, D. M., Folkers, G. K., and Fauci, A. S., 2004. The Challenge of Emerging and Re-emerging Infectious Diseases. *Nature,* 430(6996):242.

Morison, S. E., 1971. *The European Discovery of America–the Northern Voyages A. D. 500–1600,* Oxford, UK: Oxford Univ. Press.

Mumford, L., 1961. *The City in History: Its Origins, Its Transformations, and Its Prospects,* New York: Harcourt.

——, 1968. *The Urban Prospect,* New York: Harcourt, Brace and World.

——, 1979. *My Works and Days: A Personal Chronicle,* 1st ed., New York: Houghton Mifflin Harcourt.

Musial, W. and Ram, B., 2010. Large-Scale Offshore Wind Power in the United States: Assessment of Opportunities and Barriers, Golden, CO: National Renewable Energy Laboratory.

Myers, A. C., 1970. *William Penn's Own Account of the Lenni Lenape or Delaware Indians,* Moorestown, NJ: Middle Atlantic.

Myles, P. et al., 2011. 430.01.03 Electric Power System Asset Optimization, Pittsburgh, PA: National Energy Technology Laboratory.

Nabhan, G. P., 2002. *Coming Home to Eat: The Pleasures and Politics of Local Foods,* New York: W. W. Norton.

Nabokov, P., ed., 1999. *Native American Testimony: A Chronicle of Indian-White Relations from Prophecy to the Present, 1492–2000,* New York: Penguin.

National Academy of Sciences, 2010. Electricity from Renewable Resources: Status, Prospects, and Impediments, Washington D.C.: National Academies Press.

National Bureau of Economic Research, 2012. US Business Cycles Expansions and Contractions, Washington, D.C.: National Bureau of Economic Research.

National Center for Chronic Disease Prevention and Health Promotion, 1999. Physical Activity and Health: A Report of the Surgeon General, US Department of Health and Human Services. www.cdc.gov/nccdphp/sgr/index.htm.

National Commission on Terrorist Attacks Upon the United States, 2004. 9-11 Commission Report. Government Printing Office, Washington D.C.

National Commission on the BP

Deepwater Horizon Oil Spill and Offshore Drilling. 2011. Deep Water: The Gulf Oil Disaster and the Future of Offshore Drilling, National Commission on the BP Deepwater Horizon Oil Spill and Offshore Drilling.

National Institute of Standards and Technology, 2000. International System of Units from NIST. Gaithersburg, MD: NIST.

National Institute of Standards and Technology, 2012. Pentane, 2, 2, 4—trimethyl—. NIST Chemistry Webbook, Gaithersburg, MD: National Institute of Standards and Technology.

National Renewable Energy Laboratory, 2008. United States Photovoltaic Resource: Flat Panel Tilted at Latitude, Golden, CO: US Department of Energy.

National Research Council, 2007. Environmental Impacts of Wind-Energy Projects, Washington, D.C.: National Academies.

National Resource Defense Council, 2009. Location Efficient Mortgages. www.nrdc.org/cities/smartgrowth/qlem.asp.

National Security Council. 2002. The National Security Strategy 2002. Washington D.C.: White House.

Nelson, E., 2007. Milton Friedman and US Monetary History: 1961–2006. *Federal Reserve Bank of St. Louis Review*, 89(3): 153–182.

Neunzert, G. M., 2010. *Subdividing the Land: Metes and Bounds and Rectangular Survey Systems*, 1st ed., Boca Raton, FL: CRC.

Newman, M. E. J., 2005. Power Laws, Pareto Distributions and Zipf's Law. *Contemporary Physics* 46 (5): 323–351.

Newman, P. and Kenworthy, J., 1999. *Sustainability and Cities: Overcoming Automobile Dependence*, Washington D.C.: Island Press.

Newton, J. L., 2008. *Aldo Leopold's Odyssey: Rediscovering the Author of A Sand County Almanac*, Washington D.C.: Shearwater.

Nichols, J. B., Oliner, S. D., and Mulhall, M. R., 2010. Commercial and Residential Land Prices Across the United States, Washington D.C.: Federal Reserve Board.

Noland, R. B. and Thomas, J. V., 2007. Multivariate Analysis of Trip-Chaining Behavior. *Environment and Planning B: Planning and Design*, 34(6):953–970.

Norman, J., MacLean, H. L., and Kennedy, C. A., 2006. Comparing High and Low Residential Density: Life-Cycle Analysis of Energy Use and Greenhouse Gas Emissions. *Journal of Urban Planning and Development*, 132:10.

Norton, C. J. and Jin, J. J. H., 2009. The Evolution of Modern Human Behavior in East Asia: Current Perspectives. *Evolutionary Anthropology: Issues, News, and Reviews*, 18(6):247–260.

O'Flaherty, B., 2005. *City Economics,* Cambridge, MA: Harvard Univ. Press.

Odum, E., and Barrett, G. W, 2004. *Fundamentals of Ecology,* 5th ed, Independence, KY: Brooks Cole.

Office of Long-term Planning and Sustainability, 2007. plaNYC: A greener, greater New York, New York: City of New York.

Ohland, G. and Poticha, S., eds., 2009. *Street Smart: Streetcars and Cities in the Twenty-first Century,* 2nd ed, Oakland, CA: Reconnecting America.

Olofsson, U. and Lewis, R., 2006. Tribology of the Wheel–Rail Contact, in Iwnicki, S., ed., *Handbook of Railway Vehicle Dynamics*. Milton Park, UK: Taylor and Francis, 121–141.

Olson, D. M. et al., 2001. Terrestrial Ecoregions of the World: A New Map of Life on Earth. *BioScience,* 51(11):933–938.

Olson, M., 1988. The Productivity Slowdown, the Oil Shocks, and the Real Cycle. *Journal of Economic Perspectives,* 2(4):43–69.

Opie, J., 1998. *Nature's Nation: An Environmental History of the United States*, Florence, KY: Wadsworth Publishing.

Ormsby, T. et al., 2010. *Getting to Know ArcGIS Desktop,* Redlands, CA: ESRI.

Orr, J. C. et al., 2005. Anthropogenic Ocean Acidification over the Twenty-first Century and Its Impact on Calcifying Organisms. *Nature,* 437:681–686.

Ory, D. T. and Mokhtarian, P. L., 2005. An Empirical Analysis of Causality in the Relationship Between Telecommuting and Residential and Job Relocation, Davis, CA: Institute of Transportation Studies, Univ. of California, Davis.

Ostrom, E., Gardner, R., and Walker, J., 1994. *Rules, Games, and Common-Pool Resources.* Ann Arbor, MI: Univ. of Michigan Press.

Owen, D., 2009. *Green Metropolis: Why Living Smaller, Living Closer, and Driving Less Are the Keys to Sustainability,* New York: Riverhead.

Pace, G., 2012. Does Your Suburb Look Like THIS? *New York Daily News.* March 2.

Paine, C., 2006. *Who Killed the Electric Car?,* Culver City, CA: Sony Pictures Home Entertainment.

Parikka, M., 2004. Global Biomass Fuel Resources. *Biomass and Bioenergy,* 27(6):613–620.

Paulson, H. M., 2010. *On the Brink: Inside the Race to Stop the Collapse of the Global Financial System,* New York: Business Plus.

Pereira, H. M. et al., 2010. Scenarios for Global Biodiversity in the 21st Century. *Science,* 330(6010):1496–1501.

Perera, F. and Herbstman, J., 2011. Prenatal Environmental Exposures, Epigenetics, and Disease. *Reproductive Toxicology,* 31(3):363–373.

Phillips, K., 2003. *Wealth and Democracy: A Political History of the American Rich,* New York: Broadway.

Phillipson, N., 2010. *Adam Smith: An Enlightened Life,* New Haven, CT: Yale Univ. Press.

Pickett, S. et al., 2011. Urban Ecological Systems: Scientific Foundations and a Decade of Progress. *Journal of Environmental Management,* 92:331–362.

Pickett, S. T. A. et al., 1997. *The Ecological Basis of Conservation: Heterogeneity, Ecosystems and Biodiversity,* New York: International Thomson.

———, 2011. Urban Ecological Systems: Scientific Foundations and a Decade of Progress. *Journal of Environmental Management,* 92:331–362.

Pigou, A. C., 1932. *The Economics of Welfare,* London: Macmillan.

Pinjari, A. R. et al., 2007. Modeling Residential Sorting Effects to Understand the Impact of the Built Environment on Commute Mode Choice. *Transportation,* 34(5):557–573.

Pisarski, A., 2006. Commuting in America III, Washington D.C.: Transportation Research Board.

Platt, R. H., 2004. *Land Use and Society: Geography, Law, and Public Policy,* Washington D.C.: Island Press.

Pogue, J. E., 1921. *The Economics of Petroleum,* New York: Wiley.

Pollakowski, H. O. and Wachter, S. M., 1990. The Effects of Land-Use Constraints on Housing Prices. *Land Economics,* 66(3):315–324.

Pollan, M., 2006. *The Omnivore's Dilemma: A Natural History of Four Meals,* New York: Penguin.

Pozzi, F. and Small, C., 2001. Explanatory Analysis of Suburban Land Cover and Population Density in the U.S.A. IEEE/ISPRS Joint Workshop on Remote Sensing and Data Fusion, Rome, Italy.

Puentes, R. and Tomer, A., 2008. The Road...Less Traveled: An Analysis of Vehicle Miles Traveled Trends in the US, Washington D.C.: Brookings Institution.

Putnam, R. D., 2001. *Bowling Alone: The Collapse and Revival of American Community,* New York: Touchstone.

Radchenko, S., 2005. Oil Price Volatility and the Asymmetric Response of Gasoline Prices to Oil Price Increases and Decreases. *Energy Economics,* 27(5):708–730.

Radeloff, V. C. et al., 2010. Housing Growth in and near United States Protected Areas Limits Their Conservation Value. *Proceedings of the National Academy of Sciences,* 107(2):940.

Ramachandra, T. V., Jain, R., and Krishnadas, G., 2011. Hotspots of Solar Potential in India. *Renewable and Sustainable Energy Reviews,* 15(6):3178–3186.

Rappaport, J., 2008a. A Productivity Model of City Crowdedness. *Journal of Urban Economics,* 63 (2, Mar.):715–722.

———, 2008b. Consumption Amenities and City Population Density. *Regional Science and Urban Economics,* 38 (6): 533–552.

Rastler, D., 2010. Electricity Energy Storage Technology Options: A White Paper Primer on Applications, Costs, and Benefits, Palo Alto, CA: Electric Power Research Institute.

Ravanel, E. C., 1991. Designing Defense for a New World Order: The Military Budget in 1992 and Beyond, Washington, D.C.: Cato Institute.

Redford, K. H., 1992. The Empty Forest. BioScience 42(6): 412–422.

Redford, K. H. et al., 2008. What Is the Role for Conservation Organizations in Poverty Alleviation in the World's Wild Places? *Oryx,* 42(04):516–528.

Redmond, L. S. and Mokhtarian, P. L., 2001. The Positive Utility of the Commute: Modeling Ideal Commute Time and Relative Desired Commute Amount. *Transportation,* 28(2):179–205.

Reed, D. H., O'Grady, J. J., Brook, B. W., Ballou, J. D., and Frankham, R., 2003. Estimates of Minimum Viable Population Sizes for Vertebrates and Factors Influencing Those Estimates. *Biological Conservation* 113: 23–34.

Reed, W. J., 2001. The Pareto, Zipf and Other Power Laws. *Economics Letters,* 74(1):15–19.

Register, R., 2002. *EcoCities: Building Cities in Balance with Nature,* Berkeley, CA: Berkeley Hills.

Resh, V. H. et al., 1988. The Role of Disturbance in Stream Ecology. *Journal of the North American Benthological Society,* 7(4):433–455.

Restore the Gulf Task Force, 2011. RestoreTheGulf.gov. www.restorethegulf.gov.

Rhem, K. T., 2003. US Not Interested in Iraqi Oil, Rumsfeld Tells Arab World, Washington D.C.: American Forces Press Service. www.defense.gov/news/newsarticle.aspx?id=29374.

Ribeiro, P. F. et al., 2001. Energy Storage Systems for Advanced Power Applications. *Proceedings of the IEEE,* 89(12):1744–1756.

Richardson, H. C., 2008. *West from Appomattox: The Reconstruction of America after the Civil War,* New Haven, CT: Yale Univ. Press.

Richardson, H. W., 1972. Optimality in City Size, Systems of Cities and Urban Policy: A Sceptic's View. *Urban Studies* 9(1): 29–48.

Ricketts, T. H. et al., 1999. *Terrestrial Ecoregions of North America: A Conservation Assessment,* Washington, D.C.: Island Press.

Rifkin, J., 2003. *The Hydrogen Economy,* New York: Tarcher/Penguin.

Rigby, D. L. and Essletzbichler, J., 2002. Agglomeration Economies and Productivity Differences in US Cities. *Journal of Economic Geography,* 2(4):407–432.

Riis, J. A. and Museum of the City of New York, 1971. *How the Other Half Lives; Studies Among the Tenements of New York. With 100 Photos. from the Jacob A. Riis Collection, the Museum of the City of New York, and a New Pref. by Charles A. Madison,* New York: Dover.

Roach, J., 2003. Are Plastic Grocery Bags Sacking the Environment? *National Geographic News,* Sept. 2.

Robelius, F., 2007. Giant Oil Fields—The Highway to Oil. Giant Oil Fields and their Importance for Future Oil Production. Thesis. Uppsala, Sweden: Uppsala Univ.

Robèrt, M. and Börjesson, M., 2006. Company Incentives and Tools for Promoting Telecommuting. *Environment and Behavior,* 38(4):521–549.

Roberts, A., 2011. *The Storm of War: A New History of the Second World War,* New York: Harper.

Roberts, B. J., 2009. Geothermal Resource of the United States. Golden, CO: National Renewable Energy Laboratory.

Roberts, P., 2004. *The End of Oil: On the Edge of a Perilous New World,* New York: Houghton Mifflin Harcourt.

Roberts, T., *Frontline.* Feb. 22, 2005. Boston: WGBH Educational Foundation.

Rockefeller, J. D., 1909. *Random Reminiscences of Men and Events,* New York: Doubleday, Page.

Rodkin, D. 2005. "Can It Last?" *Chicago Magazine,* Oct.

Rodríguez, D. A. et al., 2009. Land Use, Residential Density, and Walking: The Multi-Ethnic Study of Atherosclerosis. *American Journal of Preventive Medicine,* 37(5):397–404.

Rosen, W., 2010. *The Most Powerful Idea in the World: A Story of Steam, Industry, and Invention,* New York: Random House.

Rotemberg, J. J. and Woodford, M., 1996. Imperfect Competition and the Effects of Energy Price Increases on Economic Activity. *Journal of Money, Credit, and Banking,* 28(4):550–577.

Rowling, J. K. 2002. *Harry Potter and the Goblet of Fire,* New York: Scholastic Paperbacks.

Russell, D., 2011. *Towards Ecological Taxation: The Efficacy of Emissions-Related Motor Taxation,* Burlington, VT: Gower.

Rybczynski, W., 2007. *Last Harvest: How a Cornfield Became New Daleville,* New York: Scribner.

Ryder, A., 2012. High Speed Rail. *Journal of Transport Geography* 22:303–305.

Sageman, B. B. et al., 2003. A Tale of Shales: The Relative Roles of Production, Decomposition, and Dilution in the Accumulation of Organic-Rich Strata, Middle-Upper Devonian, Appalachian Basin. *Chemical Geology,* 195(1–4):229–273.

Sahr, R., 2012. Consumer Price Index (CPI) Conversion Factors 1774 to Estimated 2019 to Convert to Dollars of 2007. Political Science Department, Oregon State Univ., Corvallis, OR. oregonstate.edu/cla/polisci/sahr/sahr.

Sallis, J. F. et al., 2004. Active Transportation and Physical Activity: Opportunities for Collaboration on Transportation and Public Health Research. *Transportation Research—Part A Policy and Practice,* 38(4):249–268.

Salomon, S., 2009. *Little House on a Small Planet: Simple Homes, Cozy Retreats, and Energy Efficient Possibilities,* Guilford, CT: Lyons.

Salon, D., 2009. Neighborhoods, Cars, and Commuting in New York City: A Discrete Choice Approach. *Transportation Research Part A: Policy and Practice,* 43(2):180–196.

Samimi, A., Mohammadian, A., and Madanizadeh, S., 2009. Effects of Transportation and Built Environment on General Health and Obesity. *Transportation Research Part D: Transport and Environment,* 14(1):67–71.

Sampson, A., 1975. *The Seven Sisters: The Great Oil Companies and the World They Shaped,* New York: Viking Adult.

Samuelson, P. and Nordhaus, W., 2009. *Economics,* 19th ed., Blacklick, OH: McGraw-Hill/Irwin.

Samuelson, P. A., 1954. The Pure Theory of Public Expenditure. *The Review of Economics and Statistics,* 36(4):387–389.

Sandalow, D. B. ed., 2009. Plug-In Electric Vehicles: What Role for Washington?, Washington D.C.: Brookings Institution Press.

Sanderson, E. W., 2009. After the Storms, an Island of Calm—and Resilience. *New York Times.* Sept. 11.

——. *Mannahatta: A Natural History of New York City,* New York: Abrams.

Sanderson, E. W. et al., 2002. The Human Footprint and the Last of the Wild. *Bioscience,* 52(10):891–904.

——, 2006. *The Human Footprint: Challenges for Wilderness and Biodiversity,* Mexico: CEMEX—Agrupacion Sierra Madre—Wildlife Conservation Society.

Sandia National Laboratory, 2008. Sandia, Stirling Energy Systems Set New World Record for Solar-to-Grid Conversion Efficiency, Albuquerque, NM: Sandia National Laboratories. share.sandia.gov/news/resources/releases/2008/solargrid.html.

Santos, A. et al., 2011. Summary of Travel Trends: 2009 National Household Travel Survey, Washington, D.C.: US Department of Transportation, Federal Highway Administration.

Sauer, C. O., 1971. *Sixteenth-Century North America: The Land and People as Seen by Europeans,* Berkeley, CA: Univ. of California Press.

——, 1980. *Seventeenth-Century North America,* Berkeley, CA: Turtle Island.

Schaeffer, J., 2008. *Real Goods Solar Living Source Book–Special 30th Anniversary Edition: Your Complete Guide to Renewable Energy Technologies and Sustainable Living* 30th ed., Gabriola Island, BC, Canada: New Society.

Schiffer, M. B., 1994. *Taking Charge: The Electric Automobile in America,* Washington D.C.: Smithsonian Institution Press.

——, 2006. *Draw the Lightning Down: Benjamin Franklin and Electrical Technology in the Age of Enlightenment,* Berkeley, CA: Univ. of California Press.

Schoenung, S., 2011. Energy Storage Systems Cost Update: A Study for the DOE Energy Storage Systems Program, Albuquerque, NM: Sandia National Laboratories.

Schopenhauer, A., 1901. *The Wisdom of Life, and Other Essays by Arthur Schopenhauer,* Washington D.C.: M.W. Dunne.

Schorn, D., 2006. The Oil Sands of Alberta. *60 Minutes.* Feb. 11.

Schurr, S. H., 1977. *Energy in the American Economy, 1850–1975: An Economic Study of Its History* and Prospects, Baltimore: Johns Hopkins Press.

Schwartz, M. et al., 2010. Assessment of Offshore Wind Energy Resources for the United States, Golden, CO: National Renewable Energy Laboratory.

Schwarzenbach, R. P. et al., 2010. Global Water Pollution and Human Health. *Annual Review of Environment and Resources*, 35:109–136.

Seewald, J. S., 2003. Organic-Inorganic Interactions in Petroleum-Producing Sedimentary Basins. *Nature*, 426(6964):327–333.

Seiler, C., 2008. *Republic of Drivers: A Cultural History of Automobility in America*, Chicago, IL: Univ. of Chicago Press.

Selby, D. and Creaser, R. A., 2005. Direct Radiometric Dating of Hydrocarbon Deposits Using Rhenium Osmium Isotopes. *Science*, 308(5726):1293–1295.

Sexton, S. E., Wu, J. J., and Zilberman, D., 2012. How High Gas Prices Triggered the Housing Crisis: Theory and Empirical Evidence, Berkeley, CA: Center for Energy and Environmental Economics, Univ. of California.

Sherlock, M. F., 2010. Energy Tax Policy: Historical Perspectives on the Current Status of Energy Tax Expenditures, Washington D.C.: Congressional Research Service.

Shiller, R. J., 2006. *Irrational Exuberance*, 2nd ed., New York: Crown Business.

Shirer, W. L., 1960. *The Rise and Fall of the Third Reich: A History of Nazi Germany*. New York: Simon & Schuster.

Shoumatoff, A., 2008. The Arctic Oil Rush. *Vanity Fair*. May.

Shoup, D. C. and American Planning Association, 2005. *The High Cost of Free Parking*, Chicago: Planners Press, American Planning Association.

Shrank, D., Lomax, T., and Eisele, B., 2011. 2011 Urban Mobility Report, College Station, TX: Texas Transportation Institute.

Shukla, A., Aricò, A., and Antonucci, V., 2001. An

Appraisal of Electric Automobile Power Sources. *Renewable and Sustainable Energy Reviews*, 5(2):137–155.

Shuman, M. H., 2007. *The Small-Mart Revolution: How Local Businesses Are Beating the Global Competition*, Williston, VT: Berrett-Koehler.

Silliman, B. (Jr.), 1855. *A Report on the Rock Oil, Or Petroleum from Venango County, Pennsylvania*, New Haven, CT: J. H. Benham's Steam Power Press.

Simberloff, D. S. and Abele, L.G., 1982. Refuge Design and Island Biogeographic Theory—Effects of Fragmentation. *American Naturalist*, 120:41–56.

Simmons, M. R., 2006. *The World's Giant Oil Fields*. Houston: Simmons.

———, 2005. *Twilight in the Desert: The Coming Saudi Oil Shock and the World Economy*, New York: Wiley.

Simon, J. 1996. The Ultimate Resource II: People, Materials, and Environment, Princeton, NJ: Princeton Univ. Press.

Skeet, I., 1991. *OPEC: Twenty-Five Years of Prices and Politics*, Cambridge, UK: Cambridge Univ. Press.

Skumatz, L. A. and Freeman, D. J., 2006. Pay As You Throw (PAYT) in the US: 2006 Update and Analyses, Superior, CO: Skumatz Economic Research Associates, Inc. for the US Environmental Protection Agency.

Slaats, J., 2000. TNC Ecoregions and Divisions of the Lower 48 United States. Washington, D.C.: The Nature Conservancy.

Slemrod, J. and Bakija, J., 2008. *Taxing Ourselves, 4th Edition: A Citizen's Guide to the Debate over Taxes*, Cambridge, MA: MIT Press.

Smil, V., 2007. *Energy in Nature and Society: General Energetics of Complex Systems*, Cambridge, MA: MIT Press.

Smith, A., 1776. *An Inquiry into the Nature and Causes of the Wealth of Nations*, London: Printed for W. Strahan; and T. Cadell.

———, 1987. *The Essential Adam Smith*. R. L. Heilbroner, ed., New York: W. W. Norton.

Smith, D., 1999. *Report from Engine Co. 82*, New York: Grand Central.

Smith, J. C., 2012. Entries for the "Spraberry-Dean Sandstone Fields" and "East Texas Oilfield" in *The Handbook of Texas Online*, Denton, TX: Texas State Historical Association.

Smith, M. and Gaviria, M. 2010. The Spill, *Frontline*. Oct. 26. Boston: WGBH Educational Foundation.

Smith, R. and Lourie, B. 2011. *Slow Death by Rubber Duck: The Secret Danger of Everyday Things*. Berkeley, CA: Counterpoint.

Smith, W. B. et al., 2009. Forest Resources of the United States, 2007, Washington D.C.: US Forest Service, US Department of Agriculture.

Snyder, G. 1974. *Turtle Island*. New York: New Directions.

Solnit, R., 2001. *Wanderlust: A History of Walking*, New York: Penguin.

Solow, R. M., 1956. A Contribution to the Theory of Economic Growth. *Quarterly Journal of Economics*, 70(1):65–94.

———, 1957. Technical Change and the Aggregate Production Function. *Review of Economics and Statistics*, 39(3):312–320.

———, 1974. The Economics of Resources or the Resources of Economics. *American Economic Review*, 64(2):1–14.

Sorkin, A. R., 2009. *Too Big to Fail: The Inside Story of How Wall Street and Washington Fought to Save the Financial System—and Themselves First*, New York: Viking Adult.

Soto, A. M. and Sonnenschein, C., 2010. Environmental Causes of Cancer: Endocrine Disruptors as Carcinogens. *Nature Reviews Endocrinology*, 6(7):364–371.

Spellerberg, I. F. 1998. Ecological Effects of Roads and Traffic: A Literature Review. *Global Ecology and Biogeography Letters*, 7:317–333.

Sperling, D. and Gordon, D., 2008. Advanced Passenger Transport Technologies. *Annual Review of Environment and Resources*, 33:63–84.

Standard and Poors, Inc., 2012. SandP/Case-Shiller Home

Price Indices. www.standard-andpoors.com/indices/sp-case-shiller-home-price-indices/en/us/?indexId=spusa-cashpidff--p-us.

Stanford Hospital & Clinics, 2011. Health Effects of Obesity. stanfordhospital.org/clinicsmed-Services/COE/surgicalServices/generalSurgery/bariatricsurgery/obesity/effects.html.

Steel, C., 2009. *Hungry City: How Food Shapes Our Lives*, London: Random House UK.

Steigerwald, B., 2001. City Views: Urban Studies Legend Jane Jacobs on Gentrification, the New Urbanism, and Her Legacy. Sept.

Stein, B. A., Kutner, L. S., and Adams, J. S., eds., 2000. *Precious Heritage: The Status of Biodiversity in the United States*, New York: Oxford Univ. Press.

Steingraber, S., 2010. *Living Downstream: An Ecologist's Personal Investigation of Cancer and the Environment*. Jackson, TN: Da Capo.

Stern, R. J., 2010. United States Cost of Military Force Projection in the Persian Gulf, 1976–2007. *Energy Policy*, 38(6):2816–2825.

Steuart, W. M., 1905. *Street and Electric Railways, 1902*, Washington, D.C.: US Bureau of the Census.

Stigler, G. J., 1987. *Theory of Price*, 4th ed., New York: Macmillan.

Stiglitz, J., 1974. Growth with Exhaustible Natural Resources: Efficient and Optimal Growth Paths. *Review of Economic Studies*, 41(5):123–137.

Stoddard, L., Abiecunas, J., and O'Connell, R., 2006. Economic, Energy, and Environmental Benefits of Concentrating Solar Power in California, Golden, CO: National Renewable Energy Laboratory.

Stoft, S., 2008. *Carbonomics: How to Fix the Climate and Charge It to OPEC*, Albuquerque, NM: Diamond.

Strauss-Kahn, V. and Vives, X., 2009. Why and Where Do Headquarters Move? *Regional Science and Urban Economics*, 39(2):168–186.

Strohl, D., 2010. Ford, Edison and the Cheap EV That Almost Was. Autopia: Road to the Future.

www.wired.com/autopia/2010/06/henry-ford-thomas-edison-ev.

Stromsta, K.-E., 2011. Big Day for CSP as Gemasolar Feeds the Grid for 24 Hours. *Recharge*. www.rechargenews.com/energy/solar/article265281.ece.

Stuntz, W. J., 2011. *The Collapse of American Criminal Justice*, Cambridge, MA: Belknap Press of Harvard Univ. Press.

Sunstein, C. R., and Thaler, R. H., 2008. *Nudge: Improving Decisions About Health, Wealth, and Happiness*, New Haven, CT: Yale Univ. Press.

Sutton, P. C. et al., 2012. The Real Wealth of Nations: Mapping and Monetizing the Human Ecological Footprint. *Ecological Indicators*, 16(0):11–22.

Swift, E., 2011. *The Big Roads: The Untold Story of the Engineers, Visionaries, and Trailblazers Who Created the American Superhighways*, New York: Houghton Mifflin Harcourt.

Tachieva, G., 2010. *Sprawl Repair Manual*, Washington D.C.: Island Press.

Taiz, L., and Zeiger, E., 2010. *Plant Physiology*, 5th ed., Sutherland, MA: Sinauer.

Tanaka, A. M., Fricker, J. D., and Haddock, J. E., 2012. Determining Viable Sizes for Indiana Communities Based on Essential Establishments and Services. *Journal of Urban Planning and Development* 86 (pre-publication notice).

Tang, W., Mokhtarian, P. L., and Handy, S. L., 2008. The Role of Neighborhood Characteristics in the Adoption and Frequency of Working at Home: Empirical Evidence from Northern California, Davis, CA: Institute of Transportation Studies, Univ. of California, Davis.

Tans, P. and Keeling, R., 2012. Trends in Carbon Dioxide: Mauna Loa, Hawaii, Boulder, CO: National Oceanic and Atmospheric Administration, US Department of Commerce.

Tarbell, I., 1904. *The History of the Standard Oil Company*, New York: McClure, Phillips.

Tarr, J. and McShane, C., 1997. The Centrality of the Horse to the Nineteenth-Century American

City, in R. Mohl, ed., *The Making of Urban America*, New York: SR, 105–130.

Taylor, B. D. et al., 2009. Nature and/or Nurture? Analyzing the Determinants of Transit Ridership across US Urbanized Areas. Transportation Research Part A: Policy and Practice, 43:60–77.

Taylor, P. et al., 2009. For Nearly Half of America, Grass Is Greener Somewhere Else—Pew Research Center, Washington D.C.: Pew Research Center.

Teal, J. M. and Howarth, R. W., 1984. Oil Spill Studies: A Review of Ecological Effects. *Environmental Management*, 8:27-43.

Teller, E., 1979. *Energy from Heaven and Earth*. New York: W.H. Freeman and Co.

Tennyson, E. L., 1998. Impact on Transit Patronage of Cessation or Inauguration of Rail Service, Washington D.C.: Transportation Research Board.

Teychenne, M., Ball, K., and Salmon, J., 2010. Sedentary Behavior and Depression Among Adults: A Review. *International Journal of Behavioral Medicine*, 17:246–254.

The Economist, 1999. Drowning in Oil. *The Economist*. March 4.

———, 2011. Rising from the Ruins: Housing and the Economy. *The Economist*, Nov. 5.

———, 2012a. The Future of Driving: Seeing the Back of the Car. *The Economist*, Sept. 12.

———, 2012b. Energy Storage: Packing Some Power. *The Economist*, March 3–9.

The Infrastructurist, 2010. 36 Reasons Streetcars Are Better Than Buses. *The Infrastructurist*. www.infrastructurist.com/2009/06/03/36-reasons-that-streetcars-are-better-than-buses.

Theobald, D. L., 2010. A Formal Test of the Theory of Universal Common Ancestry. *Nature*, 465(7295):219–222.

Theobald, D. M. et al., 2005. Ecological Support for Rural Land-Use Planning. *Ecological Applications*, 15(6):1906–1914.

Thoisy, B. et al., 2010. Rapid Evaluation of Threats to Biodiversity: Human Footprint Score and Large Vertebrate Species Responses in French Guiana. *Biodiversity and Conservation*, 19(6):1567–1584.

Thornton, D., 2010. Downsides of Quantitative Easing. *Economic Synopses (Federal Reserve Bank of St. Louis)*, 34:1–2.

Tietenberg, T. and Lewis, L., 2008. *Environmental and Natural Resource Economics*, 8th ed., Boston, MA: Addison Wesley.

Tilahun, N. and Levinson, D., 2008. Home Relocation and the Journey to Work, Social Science Research Network. ssrn.com/abstract=1736153.

——, 2011. Work and Home Location: Possible Role of Social Networks. Transportation Research Part A: Policy and Practice, 45(4):323–331.

Tomalin, C., 2011. *Charles Dickens: A Life*, New York: Penguin.

Transportation Research Board, 2006. Tires and Passenger Vehicle Fuel Economy: Informing Consumers, Improving Performance, Washington D.C.: National Research Council.

Tri-State Transportation Campaign, 2012. Tri-State Transportation Campaign—Transportation 101. www.tstc.org/101/mta.php.

Trombulak, S. C., and Frissell C. A. 2000. Review of Ecological Effects of Roads on Terrestrial and Aquatic Communities. *Conservation Biology*, 14:18–30.

Turchin et al., 2001. Are Lemmings Predator or Prey? *Nature*, 405: 562–565.

Tworek, C. et al., 2010. State-Level Tobacco Control Policies and Youth Smoking Cessation Measures. *Health Policy*, 97(2–3):136–144.

United Nations Population Division, 2011. World Population Prospects, the 2010 Revision, New York: United Nations Department of Economic and Social Affairs.

US Bureau of Economic Analysis, 2011a. National Economic Accounts, Corporate Profits After Tax by Industry, Washington D.C.: US Department of Commerce.

——, 2011b. National Economic Accounts, Fixed Asset Table. Current-Cost Net Stock of Residential Fixed Assets by Type of Owner, Legal Form of Organization, and Tenure Group, Washington D.C.: US Department of Commerce.

——, 2012b. National Economic Accounts, Government Current Receipts and Expenditures, Washington, D.C.: US Department of Commerce.

——, 2012a. National Economic Accounts, Gross Domestic Product, Washington, D.C.: US Department of Commerce.

——, 2012c. National Economic Accounts, Personal Consumption Expenditures by Type of Product, Washington, D.C.: US Department of Commerce.

US Bureau of the Census, 1975. Historical Statistics of the United States: Colonial Times to 1970. Bicentennial Edition., Washington D.C.: US Government Printing Office.

US Bureau of Labor Statistics, 2012b. Consumer Expenditure Survey, Washington D.C.: US Department of Labor.

——, 2012a. Consumer Price Index, All Urban Consumers—CPI-U, All Items, Washington D.C.: US Department of Labor.

——, 2012c. Current Employment Statistics—CES (National). Washington D.C.: US Department of Labor.

——, 2012d. Time Use Survey, Comparison of All Employees, Seasonally Adjusted, Before and After the March 2011 Benchmark. Washington D.C.: US Department of Labor.

US Bureau of Transportation Statistics, 2006. Freight in America: A New National Picture, Washington D.C.: US Department of Transportation.

——, 2012. National Transportation Statistics 2012, Washington, D.C.: US Department of Transportation. www.rita.dot.gov/bts/sites/rita.dot.gov.bts/files/publications/national_transportation_statistics/index.html

US Census Bureau, 1995. Urban and Rural Definitions, Washington, D.C.: US Department of Commerce.

——, 2002. Vehicle Inventory and Use Survey—2002 VIUS Data Releases, Washington, D.C.: US Department of Commerce.

——, 2010. US Census Bureau TIGER/Line Shapefiles, Washington, D.C.: US Department of Commerce. www.census.gov/geo/maps-data/data/tiger.html

——, 2011b. Geographical Mobility/Migration: 2009–2010 (CPS 2010), Washington, D.C.: US Department of Commerce.

——, 2011c. Historical Census of Housing Tables—Home Values, Washington, D.C.: US Department of Commerce.

——, 2011d. Statistical Abstract of the United States, Washington, D.C.: US Department of Commerce.

——, 2011a. Urban Area Criteria for the 2010 Census. *US Federal Register*, 76(164):53030–53043, Washington, D.C.: US Department of Commerce.

——, 2012c. Housing Vacancies and Homeownership (CPS/HVS), Washington D.C.: US Department of Commerce.

——, 2012b. New Residential Construction, Washington D.C.: US Department of Commerce.

——, 2012a. Population and Housing Unit Estimates, Washington D.C.: US Department of Commerce.

——, 2012d. Poverty Data—Historical Poverty Tables: Families, Washington, D.C.: US Department of Commerce.

US Census Bureau and Social Science Research Council, 1949. Historical Statistics of the United States, 1789–1945: A Supplement to the Statistical Abstract, Washington, D.C.: US Department of Commerce.

US Department of Defense, 2009. Defense Energy Support Center Fact Book: FY 2008: 31st edition, Washington D.C.: US Department of Defense.

US Department of Energy,

1997. Executive Summary: Plowshare Program. Office of Nuclear and National Security Information, Washington D.C.: US Department of Energy.

US Department of Energy and US Department of the Interior, 2011. A National Offshore Wind Strategy: Creating an Offshore Wind Energy Industry in the United States, Washington, D.C.: Several agencies.

US Department of Energy and US Environmental Protection Agency, 2011. Fuel Economy: Where the Energy Goes, Washington, D.C.: US Department of Energy and US Environmental Protections Agency.

US Department of Housing and Urban Development, 1995. National Homeownership Strategy. Urban Policy Brief, Washington, D.C.: US Department of Housing and Urban Development.

US Department of Justice, 1952. Memorandum for the Attorney General Relative to a Request for Grand Jury Authorization to Investigate the International Oil Cartel—June 24, 1952, Washington D.C.: US Department of Justice.

US Energy Information Administration, 1993. Renewable Resources in the US Electricity Supply, Washington D.C.: US Department of Energy.

——, 2010a. EIA Energy Kids—Oil (petroleum), Washington D.C.: US Department of Energy. www.eia.gov/kids/

——, 2010b. Annual Electric Generator data – EIA-860 data file, Washington D.C.: US Department of Energy.

——, 2010c. Updated Capital Cost Estimates for Electricity Generation Plants, Washington D.C.: US Department of Energy.

——, 2011a. Petroleum and Other Liquids, Domestic Crude Oil First Purchase Price by Area (Dollars per Barrel), 1859–2011, Washington D.C.: US Department of Energy.

——, 2011b. Short-Term Energy Outlook, Annual Average Motor Gasoline Retail Price, 1919–2010, Washington D.C.: US Department of Energy. Also see explore.data.gov/Energy-and-Utilities/Short-Term-Energy-Outlook-Real-Petroleum-Prices/zinz-q5tn.

——, 2011c. Crude Oil and Natural Gas Exploratory and Development Wells, Washington D.C.: US Department of Energy.

——, 2012a. Annual Energy Review 2011, Washington, D.C.: US Department of Energy.

——, 2012b. Petroleum and Other Liquids, Washington, D.C.: US Department of Energy. www.eia.gov/petroleum.

——, 2012c. International Energy Statistics, Washington, D.C.: Department of Energy. www.eia.gov/countries/

US Environmental Protection Agency, 1987. Management of Wastes from the Exploration, Development, and Production of Crude Oil, Natural Gas, and Geothermal Energy, Washington D.C.: US Environmental Protection Agency.

——, 2010d. Cap and Trade, Washington D.C.: US Environmental Protection Agency.

——, 2010b. Guide For Industrial Waste Management, Washington D.C.: US Environmental Protection Agency.

——, 2010a. Municipal Solid Waste in the United States: 2009 Facts and Figures, Washington D.C.: US Environmental Protection Agency.

——, 2010c. Our Nation's Air—Status and Trends through 2008, Washington, D.C.: US Environmental Protection Agency.

——, 2011. US Greenhouse Gas Inventory Report, Washington D.C.: US Environmental Protection Agency.

US Environmental Protection Agency and US Department of Energy, 2012. How Can 6 Pounds of Gasoline Create 19 Pounds of Carbon Dioxide? Washington, D.C.: US Environmental Protection Agency and US Department of Energy.

US Federal Highway Administration, 1967. US Highway Statistics Summary to 1965, Washington D.C.: US Department of Transportation.

——, 1977. US Highway Statistics Summary to 1975, Washington D.C.: US Department of Transportation.

——, 1987. Highway Statistics: Summary to 1985, Washington D.C.: US Department of Transportation.

——, 1997. US Highway Statistics Summary to 1995, Washington D.C.: US Department of Transportation.

——, 2009. Manual on Uniform Traffic Control Devices for Streets and Highways: 2009 Edition, Washington D.C.: US Department of Transportation.

——, 2010. Public Road Mileage-VMT-Lane Miles. Highway Statistics Series 2010, Washington, D.C.: US Department of Transportation.

——, 2011b. 2009 National Household Travel Survey, Washington D.C.: US Department of Transportation. nhts.ornl.gov.

——, 2011a. Our Nation's Highways 2011, Washington D.C.: US Department of Transportation.

——, 2011c. State Motor-Vehicle Registrations-2009, Washington D.C.: US Department of Transportation.

——, 2012. Historical Monthly VMT Report, Washington D.C.: US Department of Transportation.

US Federal Housing Administration, 1938. Underwriting Manual: Underwriting and Valuation Procedure under Title II of the National Housing Act with Revisions to February, 1983, Washington, D.C.: US Federal Housing Administration.

US Federal Transit Administration, 2011. National Transit Database, Rockville, MD: Federal Transit Administration. www.ntdprogram.gov

US Government Accounting Office, 2011. Surface Freight Transportation: A Comparison of the Costs of Road, Rail, and

Waterways Freight Shipments That Are Not Passed on to Consumers. Report to the Subcommittee on Select Revenue Measures, Committee on Ways and Means, House of Representatives, Washington D.C.: US Government Accounting Office.

US Internal Revenue Service, 2011. Tax Code, Regulations and Official Guidance: Internal Revenue Code. Washington D.C.: US Department of the Treasury. www.irs.gov/Tax-Professionals/Tax-Code,-Regulations-and-Official-Guidance.

US National Renewable Energy Laboratory and AWS Truepower, 2011. 80-Meter Wind Maps and Wind Resource Potential, Washington D.C.: US Department of Energy.

US Office of Federal Housing Enterprise Oversight, 2005. 2005 Report to Congress, Washington D.C.: US Office of Federal Housing Enterprise Oversight.

US Office of Management and Budget, 2012. Historical Tables, Washington D.C.: The White House. www.whitehouse.gov/omb/budget/Historicals

US Office of Technology Assessment, 1992. Managing Industrial Solid Wastes from Manufacturing, Mining, Oil and Gas Production, and Utility Coal Combustion, Washington D.C.: Congress of the United States.

US Public Roads Administration, 1947. Highway Statistics Summary to 1945, Washington D.C.: US Federal Works Agency.

Vanderbilt, T., 2008. *Traffic: Why We Drive the Way We Do*, New York: Knopf.

Wagner, F. H. et al., 2007. *Adaptive Management: The US Department of Interior Technical Guide*, Washington D.C.: US Department of the Interior.

Waldheim, C., ed., 2006. *The Landscape Urbanism Reader*, Princeton, NJ: Princeton Architectural Press.

Wallace, T. A., Martin, D. N., and Ambs, S., 2011. Interactions among Genes, Tumor Biology and the Environment in Cancer Health Disparities: Examining the Evidence on a National and Global Scale. *Carcinogenesis*, 32(8):1107–1121.

Walker, J., 2011. *Human Transit: How Clearer Thinking about Public Transit Can Enrich Our Communities and Our Lives*, Washington, D.C.: Island Press.

Walsh, B., 2011. Why Biofuels Help Push Up World Food Prices. *Time*. Feb. 14.

Wang, M., Lee, H., and Molburg, J., 2004. Allocation of Energy Use in Petroleum Refineries to Petroleum Products. *The International Journal of Life Cycle Assessment*, 9(1):34–44.

Wardman, M., Tight, M., and Page, M., 2007. Factors Influencing the Propensity to Cycle to Work. *Transportation Research Part A: Policy and Practice*, 41(4):339–350.

Warner, S. B. and Joint Center for Urban Studies, 1962. *Streetcar Suburbs: The Process of Growth in Boston, 1870–1900*, Cambridge: Harvard Univ. Press.

Warwick, W. M., 2002. *A Primer on Electric Utilities, Deregulation, and Restructuring of US Electricity Markets*, Richland, WA: Pacific Northwest National Laboratory.

Wasik, J. F., 2006. *The Merchant of Power: Samuel Insull, Thomas Edison, and the Creation of the Modern Metropolis*, New York: Palgrave Macmillan.

Wassener, B., 2011. Airlines Weigh the Advantages of Using More Biofuel. *New York Times*, Oct. 9.

Weber, D. J., 1992. *The Spanish Frontier in North America*, New Haven, CT: Yale Univ. Press.

Weber, M., 1966. *The City*, New York: Free Press.

Weightman, G., 2010. *The Industrial Revolutionaries: The Making of the Modern World 1776–1914*, New York: Grove.

Weinstock, A. et al., 2011: *Recapturing Global Leadership in Bus Rapid Transit: A Survey of Select US Cities*, New York: Institute for Transportation and Development Policy.

Wells, D. A., 1889. *Recent Economic Changes, and Their Effect on the Production and Distribution of Wealth and the Well-Being of Society*, New York: D. Appleton.

Wescott, P. C., 2007. Ethanol Expansion in the United States: How Will the Agricultural Sector Adjust?, Washington D.C.: Economic Research Service, US Department of Agriculture.

West, S. E. and Williams, R. C., 2004. Empirical Estimates for Environmental Policy Making in a Second-Best Setting, Washington D.C.: National Bureau of Economic Research.

White, R., 2011. *Railroaded: The Transcontinentals and the Making of Modern America*, New York: W. W. Norton.

Whitney, R., 2007. Nellis Activates Nation's Largest PV Array. Nellis Air Force Base. www.nellis.af.mil/news/story.asp?id=123079933.

Wilcox, B. A. and Murphy, D. D., 1985. Conservation Strategy—Effects of Fragmentation on Extinction. *American Naturalist*, 125:41–56.

Wilgen, B. W. van, Cowling, R. M., and Burgers, C. J., 1996. Valuation of Ecosystem Services. *BioScience*, 46(3):184–189.

Williams, F. 2012. *Breasts: A Natural and Unnatural History*. New York: W. W. Norton.

Williams, T., 1947. *A Streetcar Named Desire*, New York: New Directions.

Willis, G., 2006. How to Buy and Build on Rural Land. *Money Magazine*. Dec. 29, 2005.

Willis, R. M. et al., 2010. Alarming Visual Display Monitors Affecting Shower End Use Water and Energy Conservation in Australian Residential Households. *Resources, Conservation and Recycling*, 54(12):1117–1127.

Wiser, R. and Bolinger, M., 2012. 2011 Wind Technologies Market Report, Oak Ridge, TN: US Department of Energy.

Wolf, W., 1996. *Car Mania: A Critical History of Transport*, London: Pluto Press. Translated from German by G. Fagan.

Wollen, P. and Kerr, J., 2004. *Autopia: Cars and Culture*, London: Reaktion.

Woolfson, M., 2000. The Origin

and Evolution of the Solar
System. *Astronomy and
Geophysics,* 41(1):1.12–1.19.

World Health Organization,
2011. Burden of Disease
from Environmental Noise.
Quantification of Healthy Life
Years Lost in Europe, Geneva,
Switzerland: World Health
Organization.

Yergin, D., 1991. *The Prize: The
Epic Quest for Oil, Money, and
Power Trade.* New York: Simon
& Schuster.

———, 2011. *The Quest: Energy,
Security, and the Remaking of
the Modern World,* New York:
Penguin.

Young, P., 2010. Upward Trend in
Vehicle-Miles Resumed During
2009: A Time Series Analysis,
Washington D.C.: US Bureau of
Transportation Statistics.

Yusaf, T., Goh, S., and Borserio, J.
A., 2011. Potential of Renewable
Energy Alternatives in Australia.
*Renewable and Sustainable
Energy Reviews,* 15(5):2214–2221.

Zeilik, M., 2002. *Astronomy: The
Evolving Universe,* 9th ed.,
Cambridge UK: Cambridge
Univ. Press.

Zimmer, C., 2007. *Smithsonian
Intimate Guide to Human
Origins,* New York: Harper
Paperbacks.

———, 2009. *The Tangled Bank:
An Introduction to Evolution.*
Greenwood Village, CO: Roberts.

Zimring, F. E., 2011. *The City that
Became Safe: New York's Lessons
for Urban Crime and Its Control,*
New York: Oxford Univ. Press.

Zipf, G. K., 1935. *The Psychobiology
of Language,* New York:
Houghton Mifflin.

Table of Unit Conversions

Studies of energy, transportation, and land use take in a bewildering variety of quantities: distances, areas, weights, volumes, energy, and dollars and cents. Half the battle to understanding what these studies say is mastering the units of measurement they involve. In writing *Terra Nova,* I used whatever units seemed easiest to understand in whatever context I needed them, rather than standardizing to a constant set of measures throughout, which is undoubtedly what my science teachers would have wanted me to do. Worse, I defined my own units where that seemed to be the shortest path to fun and understanding. So as a form of scientific penance, and so you may find the units you like best, I include this abbreviated table of conversions.

For additional conversions and sources of unit information, please consult Davis et al. (2011) or Clark and Denton (1994). Abbreviations are included in parentheses after each unit name.

Distance

1 mile (mi)
= 63,360 inches (in)
= 5,280 feet (ft)
= 1,760 yards (yd)
= 1,609 meters (m)
= 1.609 kilometers (km)

Area

1 square mile (sq mi or mi²)
= 640 acres (ac)
= 258.9 hectares (ha)
= 2.589 square kilometers (km²)
= 2.788 x 10⁷ square feet (ft²)

Volume

1 U.S. gallon (gal)
= 231 cubic inches (cu in or in³)
= 3.785 liters (l)
= 0.0238 barrel (bbl)
= 0.003785 cubic meters (m³)

Mass

1 pound (lb)
= 0.4536 kilograms (kg)
= 5.0 x 10-4 short ton (t)
= 4.5362 x 10-4 metric ton (t)

Note: One US gallon of gasoline weighs approximately 6.2 pounds.

Energy (quantities)

1 kilowatt-hour (kWh)
= 3,412 British Thermal units (BTU)
= 2.655 x 106 foot-pounds (ft-lb)
= 3.600 x 106 joules (J)
= 860.4207 kilocalories (kcal)
= 60 minutes of microwaving (min-mw)

Note: 1 US gallon of conventional gasoline contains approximately 125,000 BTU/gal (gross or higher heating value) or 115,400 BTU/gal (net or lower heating value.) I typically use the lower heating value, which does not include heat lost through the vaporization of water during combustion.

Energy (flows)

1 kilowatt (kW)
= 0.9478 British Thermal units
= per second (BTU/sec)
= 737.6 foot-pound per second (ft-lb/sec)
= 1000 joules per second (J/s)
= 1.341 horsepower (hp)
= 0.239 kilocalorie per second (kcal/sec)
= 24 MacKays (=1 kWh/day)

Temperature

32 degrees Fahrenheit (°F)
= 0 degrees Celsius (°C)
= 273.15 Kelvin (K)

212 degrees Fahrenheit (°F)
= 100 degrees Celsius (°C)
= 373.15 Kelvin (K)

Orders of Magnitude

Sometimes scientific units are not quite of the right magnitude, so scientists pile them up, using standard prefixes to indicate the size of the pile. As shown in the table below, kilo- indicates a thousand somethings, which means that a kilogram is a thousand grams and a kilometer is a thousand meters, and so forth. This table provides pre-names for thirty-six orders of magnitude (where one order is ten times something), from exa- to atto-, to wow your friends, impress your mom, and more importantly, to speak cogently about the huge numbers of joules, watts, MacKays, and dollars that drive the US economy.

Prefix	Definition	Magnitude	SI symbol
EXA-	One quintillion	10^{18}	E
PETA-	One quadrillion	10^{15}	P
TERA-	One trillion	10^{12}	T
GIGA-	One billion	10^{9}	G
MEGA-	One million	10^{6}	M
KILO-	One thousand	10^{3}	k
HECTO-	One hundred	10^{2}	—
DECA-	Ten	10^{1}	—
—	One	10^{0}	—
DECI-	Tenth	10^{-1}	—
CENTI-	One hundredth	10^{-2}	c
MILLI-	One thousandth	10^{-3}	m
MICRO-	One millionth	10^{-6}	
NANO-	One thousand millionth	10^{-9}	n
PICO-	One million millionth	10^{-12}	p
FEMTO-	One thousand million millionth	10^{-15}	f
ATTO-	One million million millionth	10^{-18}	a

Note: The US national debt on June 30, 2012 was \$15,856,367,214,324.44 or approximately 15.9 petadollars (PUSD) according to the US Treasury website (www.treasurydirect.gov/NP/NPGateway).

Acknowledgments

All books are conversations, and this one is no exception. Although I can't predict everyone I will be talking to in the coming years, here are some of the folks who have helped advance the conversation thus far.

First, I wish to thank my colleagues at the Wildlife Conservation Society, who created the space for me to express myself in this way: John Robinson, Joshua Ginsberg, Steven Sanderson, and Cristián Samper. Speaking with them and many other colleagues provided insights that extend far beyond wildlife conservation. I would like to especially mention Will Banham, Tim Bean, Jon Beckmann, Elizabeth Bennett, London Davies, James Deutsch, Mario Gampieri, Jodi Hilty, J. Carter Ingram, Jerry Jenkins, Liana Joseph, Heidi Kretser, Danielle LaBruna, Caleb McClennen, Fiona McKibben, Leticia Orti, Graeme Patterson, Justina Ray, Robert Rose, Todd Stevens, James Watson, Lily Wendle, David Wilkie, and Steve Zack. I thank God every day for Kim Fisher, who provides not only cheerful and excellent technical advice, but also counsel and friendship to outlast any problem. Thank you, Kim.

Second, I would like to thank the wider community of scholars in New York City and elsewhere, who intentionally or otherwise have contributed to my thinking: Rohit Aggarwala, Amale Andraos, Paul Beier, Andrew Bowman, David Bragdon, Hillary Brown, Bill Browning, Linda Cox, Skye Duncan, Nancy Faxla-Raymond, Eugene Fine, Lisa Fine, Erica Fleishman, Bob Fox, Adam Freed, Gerald Frug, Chris Garvin, Bram Gunther, Owen Gutfreund, Eric Himmel, Tony Hiss, Cas Holloway, Elizabeth Johnson, Don Kelly, David King, Carol Kuester, Trent Lethco, Maya Lin, Bruce Lourie, Jim Lyons, Reinhold Martin, Mary Miss, Joseph Mori, Heidi Neilson, Annie Novak, Kate Orff, Thaddeus Pawlowski, Phillip Pond, Ali Sant, Carl Skelton, William Solecki, Michael Sorkin, Erika Svendson, Jennifer Wade, Julianne Warren, Alexandros Washburn, June Williamson, Logan Winston, Dan Wood, Tyler Volk, Charles Yackulic, and Darryl Young. Also my mother, Diana Sanderson, and my father, Wayne Sanderson, who brought me up in the suburbs and set me on this path. We solicited external reviews from a still wider group of scholars of the Siren song: David Levinson, Nancy McGuckin, and Daniel Sperling sent in thoughts, which I appreciate. Many others conversed with me through their own books on a panoply of subjects; they are acknowledged mainly in the notes, but let me particularly thank Morris Davis, Howard Frumkin, Ed Glaeser, James Hamilton, Dolores Hayden, Robert Heilbroner, Kenneth Jackson, Alyssa Katz, David MacKay, Gregory Mankiw, Robert Putnam, and Daniel Yergin for their insightful and generous scholarship. Countless others speak to me through datasets, compiled laboriously in office cubicles in Washington D.C. and elsewhere; my gratitude goes to the anonymous bean counters and webmasters who make the data and then make it available. Thanks to all for talking with me, one way or the other. Any errors in fact or judgment that remain are stubbornly mine alone.

Third, I am indebted to the many people who actually helped create the book object you hold in your hands, in particular the folks at Abrams: Deborah Aaronson, who signed the book and conveyed it through its stages; Sharon AvRutick, whose sharp editing made it a hundred times better; Katrina Weidknecht, for help with publicity; Kat Kopik, who helped with the reviews; and Michael Jacobs, for his vision of publishing books that can not only turn a profit but can change the world. Eddie Opara, Brankica Harvey, Ken Deegan, and Yo-E Ryou at Pentagram designed the object and drew and redrew all the figures based on my pitiful sketches and enigmatic data summaries— thanks to them for their many talents and long hours. I am grateful also to my literary agents, Martha Kaplan and Renée Zuckerbrot, who never shy away from telling me what's what.

Fourth and finally, daily and ultimately, I owe this book and much more to my family on City Island, Han-Yu Hung and Everett Sanderson, for whom each conversation with the outside world is one less I have with them. They have endured the long hours of quiet concentration and sudden, loud expostulations with equanimity, patience, and their own brands of thoughtfulness; this dissertation on the ways of the nation would not have been possible without their loving ways at home. Thank you.

Index

PROJECT MANAGER: Deborah Aaronson
PROJECT EDITOR: Sharon AvRutick
DESIGN: Pentagram, New York
PRODUCTION MANAGER: True Sims

Library of Congress Control
Number: 2013007203

ISBN: 978-1-4197-0434-5

Printed and bound in the
United States

10 9 8 7 6 5 4 3 2 1

Abrams books are available at
special discounts when purchased
in quantity for premiums and
promotions as well as fundrais-
ing or educational use. Special
editions can also be created to
specification. For details, contact
specialsales@abramsbooks.com
or the address below.

THE ART OF BOOKS SINCE 1949
115 West 18th Street
New York, NY 10011
www.abramsbooks.com

This book was set in Tiempos
(2010) and Calibre (2011), both
created by type designer Kris
Sowersby. It was printed on 70
lb. Finch Opaque Vellum at RR
Donnelley, Ohio.